Principles of Molecular Virology

2nd Edition

For
Joseph & Julian

Principles of Molecular Virology

2nd Edition

Alan J. Cann
University of Leicester, UK

ACADEMIC PRESS
San Diego London Boston
New York Sydney Tokyo Toronto

This book is printed on acid-free paper.

First edition published 1993
Second edition 1997

Academic Press, Inc.
525 B Street, Suite 1900, San Diego, California 92101-4495, USA
http://www.apnet.com

Academic Press Limited
24–28 Oval Road, London NW1 7DX, UK
http:// www.hbuk.co.uk/ap/

ISBN 0–12–158532–8

Cann, Alan.
Principles of molecular virology / Alan Cann. — 2nd ed.
p. cm.
Includes bibliographical references and index.
ISBN 0–12–158532–8 (alk. paper)
1. Molecular virology. I. Title.
[DNLM: 1. Viruses—physiology. 2. Genome, Viral. 3. Virus Replication.
4. Gene Expression Regulation, Viral. 5. Virus Diseases. QW 160 C224p 1997]
QR389.C36 1997
579.2—dc21
DNLM/DLC
for Library of Congress 96–47478
CIP

A catalogue record for this book is available from the British Library

Cover photo courtesy of Eye of Science/Science Photo Library

Typeset by J&L Composition Ltd, Filey, North Yorkshire.
Printed in Great Britain by WBC Book Manufacturers Ltd, Bridgend, Mid Glamorgan

98 99 00 01 02 EB 9 8 7 6 5 4 3 2

Contents

Preface to the Second Edition

Much has changed since the first edition of *Principles of Molecular Virology* was published in 1993, but the need for this book is now stronger than ever. This edition is completely revised and updated and contains many new figures and tables. As a result, it is somewhat longer, but not, I hope, so long that it loses sight of the original objective – to present the concerns of contemporary virology to students in a concise and digestible manner.

Many people should be thanked for helping inspire this edition, but I will single out a few for particular mention. I am very grateful to all those who provided feedback on the first edition and I have tried to include as many of their comments/requests as possible; in particular, coverage of the rapidly developing area of plant virology has been greatly expanded. I am also grateful to my teaching colleagues in Leicester and elsewhere for their suggestions. Finally, the person who should perhaps receive most thanks is my long-suffering editor at Academic Press, Tessa Picknett.

Alan J. Cann
Department of Microbiology & Immunology
University of Leicester
July 1996

Preface to the First Edition

This book came about through my own need for a text to augment my undergraduate lectures on virology. Not that there is a particular shortage of books on virology, but even during my own relatively short career, the subject has expanded so rapidly that most professionals feel pressurized by the task of keeping up with current trends, let alone the responsibility of initiating a new generation in such a rapidly changing discipline. Many excellent and recent texts exist that deal with the subject in a detailed and somewhat traditional manner; the second edition of *Fields' Virology* (Raven Press, 1990), the third edition of Matthews' *Plant Virology* (Academic Press, 1991) and, in a more general sense, the second edition of *Molecular Biology of the Cell* (Garland Press, 1989) immediately spring to mind. Unfortunately, these books have two major disadvantages for their use in teaching. First, the volume of material they contain is overwhelming for a student discovering the subject for the first time and who may be unable to sort out the information they require from the mountain of detail they encounter. Second, in these times, few students can afford to purchase one of these tomes. Even persuading university libraries, whose budgets are under constant pressure, to purchase more than one or two copies is difficult and an inadequate solution in a time of expanding student numbers. Better then to reserve these texts for the reference purpose for which they are best suited and to introduce students to the subject in a gentler way.

In discussion with many of my colleagues at this and other universities, it was clear that most felt there was a place for a text which would cover the current emphasis and concerns of virology. In these conversations, there was no doubt as to what was required. My contemporaries have no difficulty with the label 'molecular virology,' but most when pressed would have difficulty in expressing a comprehensive definition. Perhaps the best way to describe how I have approached the subject here is to consider it as 'virology at a molecular level' or even better, 'molecules and viruses.' Having already damned the 2000-page reference source as unsuitable for my purpose and yet set such an all-embracing definition of 'molecular virology,' the problem was therefore how to resolve these two apparently conflicting issues. My chosen solution is to outline the *principles* of the subject with reference to specific *examples* chosen to illustrate the matter under discussion. The onus is therefore firmly on the reader to pursue particular matters on which he or she requires more information in more detailed 'reference texts' or in the immense volume of research publications appearing annually.

I would like to have spent much more time discussing the history of virology in Chapter 1, a subject I find to be a fascinating as well as a valuable insight as to how we got to where we are today. In the event, it was only possible to provide a brief overview and to refer the reader to one of the many other texts which have been published on this subject – perhaps the one area of virology where an author's work does not become outdated in a short period. It is only my intention to arm readers with the framework that makes it possible for them to achieve this task successfully. Anyone who complains that this book does not spend sufficient time dealing with (or even mention) their pet area of interest has therefore missed the point.

To disarm the jargon (which I have avoided wherever possible) and unavoidable technical terms, I have included a glossary as an appendix to the book. Terms shown in the text in **bold** print are defined in this glossary.

Cliché it may be, but there are genuinely too many people to acknowledge for the creation of this book to make it possible to do this individually. It will have to suffice for me to thank all my colleagues in Leicester and elsewhere for helpful discussions; all the people who have helped and influenced my career over the years, and the undergraduates on whom I have field-tested the material, and for whom this book is intended.

Alan J. Cann
Microbiology Department
University of Leicester
October 1992

Chapter 1

Introduction

'Those who cannot remember the past are condemned to repeat it'.
(George Santayana)

There is more biological diversity within viruses than in all the rest of the bacterial, plant and animal kingdoms put together. This is the result of the success of viruses in parasitizing all known groups of living organisms, and understanding this diversity is the key to comprehending the interactions of viruses with their hosts. This book deals with 'molecular virology' in a rather broad sense – that is, 'virology at a molecular level', or perhaps even, 'molecules and viruses'. Protein–protein, protein–nucleic acid, and protein–lipid interactions determine the structure of virus particles, the synthesis and expression of virus genomes and the effects of viruses on the host cell. This is virology at a molecular level.

However, before exploring the subject further, it is necessary to understand the nature of viruses. It would also be useful to know something of the history of virology, or more accurately, how virology as a discipline in its own right arose, in order to understand its current concerns and future directions. These are the purposes of this introductory chapter. The principles behind certain techniques mentioned in this chapter may not be familiar to some readers. It may be helpful to use the further reading at the end of this chapter to become conversant with these methods. In this and the subsequent chapters, terms in the text in **bold** print are defined in the glossary (Appendix 2).

Viruses are Distinct from 'Living' Organisms

Viruses are submicroscopic, obligate intracellular parasites. This simple but useful definition goes a long way towards describing and differentiating viruses from all other groups of living organisms. However, this short definition is in itself inadequate. Clearly, there is no problem in delimiting viruses from higher 'macroscopic' organisms. Even within a broad definition of microbiology encompassing prokaryotic organisms and microscopic eukaryotes such as algae, protozoa and fungi, in most cases it will suffice. There are, however, a few groups of prokaryotic organisms that have specialized parasitic life-cycles and which confound the above definition. These are the Rickettsiae and Chlamydiae – obligate intracellular parasitic bacteria which have evolved to be so cell-associated that they can exist outside the cells of their hosts only for a short period of time before losing viability. Therefore it is necessary to add further clauses to the definition of what constitutes a virus:

- Virus particles are produced from the assembly of pre-formed components, whereas other agents 'grow' from an increase in the integrated sum of their components and reproduce by division
- Virus particles (**virions**) themselves do not 'grow' or undergo division
- Viruses lack the genetic information which encodes apparatus necessary for the generation of metabolic energy or for protein synthesis (ribosomes).

No known virus has the biochemical or genetic potential to generate the energy necessary to drive all biological processes (e.g. macromolecular synthesis). They are therefore absolutely dependent on the host cell for this function. It is often asked whether viruses are 'alive' or not. One view is that inside the host cell, viruses are 'alive', whereas outside it they are merely complex assemblages of metabolically inert chemicals. That is not to say that chemical changes do not occur in extracellular virus particles, as will be explained elsewhere, but these are in no sense the 'growth' of a living organism.

Although there will always be some exceptions and uncertainties in the case of organisms which are too small to see and in many cases very difficult to study, for the most part, the above guidelines will suffice to determine the definition of a virus. Unfortunately, molecular investigation of certain infectious agents has revealed further complications. A number of novel, pathogenic entities possess properties which confound the above definition, yet are clearly more similar to viruses than other organisms. These are the entities known as **viroids**, **virusoids**, and **prions**. Viroids are very small (200–400 nucleotides) circular RNA molecules with a rod-like secondary structure. They have no capsid or envelope and are associated with certain plant diseases. Their replication strategy is like that of viruses – they are obligate intracellular parasites. Virusoids are **satellite**, viroid-like molecules, somewhat larger than viroids (e.g. approximately 1000 nucleotides), which are dependent on the presence of virus replication for multiplication (hence 'satellite'); they are packaged into virus capsids as passengers. Prions are infectious agents generally believed to consist of a single type of protein molecule with no nucleic acid component. Confusion arises from the fact that the prion protein and the gene which encodes it are also found in normal 'uninfected' cells. These agents are associated with 'slow virus diseases' such as Creutzfeldt–Jakob disease in humans, scrapie in sheep and bovine spongiform encephalopathy (BSE) in cattle. Chapter 8 deals with these novel infectious agents in more detail. Moreover, molecular genetics has shown that perhaps 5–10% of the eukaryotic cell **genome** is composed of mobile retrovirus-like elements (**retrotransposons**), which may have had a considerable role in shaping these complex genomes (Chapter 3). Furthermore, certain **bacteriophage** genomes closely resemble bacterial plasmids in their structure and in the way they are replicated. Therefore, recent research has revealed that the relationship between viruses and other living organisms is perhaps more complex than was previously thought.

The Origins of Virology

It is easy to regard events which occurred prior to our own personal experience as prehistoric. Thus much has been written about virology as a 'new' discipline in biology and as far as the formal recognition of viruses as distinct from other living organisms is concerned, this is true. However, we now realize that not only were ancient peoples aware of the effects of virus infection, but in some instances also carried out active research into the causes and prevention of virus diseases. Perhaps the first written record of a virus infection consists of a hieroglyph from Memphis, the capital of ancient Egypt, drawn in approximately 1400 BC, which depicts a temple priest showing typical clinical signs of paralytic poliomyelitis. In addition, the Pharoh Ramses V, who died in 1196 BC and who's extraordinarily well-preserved mummified body is now in a Cairo Museum, is believed to have succumbed to smallpox – the comparison between the pustular lesions on the face of the mummy and those of more recent patients is startling.

Smallpox was endemic in China by 1000 BC. In response, the practice of **variolation** was developed. Recognizing that survivors of smallpox outbreaks were protected from subsequent infection, the Chinese inhaled the dried crusts from smallpox lesions like snuff, or in later modifications, inoculated the pus from a lesion into a scratch on the forearm. Variolation was practised for centuries and was shown to be an effective, although risky, method of disease prevention, because the outcome of the inoculation was never certain. Edward Jenner was nearly killed by variolation at the age of seven! Not surprisingly, this experience spurred him on to find a safer alternative treatment. On 14 May 1796, he used cowpox-infected material obtained from the hand of Sarah Nemes, a milkmaid from his home village of Berkeley in Gloucestershire, England, to successfully vaccinate 8-year-old James Phipps. Although initially controversial, **vaccination** against smallpox was almost universally adopted worldwide during the nineteenth century.

This early success, although a triumph of scientific observation and reasoning, was not based on any real understanding of the nature of infectious agents, which arose separately from another line of reasoning. Antony van Leeuwenhoek (1632–1723), a Dutch merchant, constructed the first simple microscopes and with these, identified bacteria as the 'animalcules' he saw in his specimens. However, it was not until Robert Koch and Louis Pasteur in the 1880s jointly proposed the 'germ theory' of disease that the significance of these organisms became apparent. Koch defined the four famous criteria now known as Koch's postulates which are still generally regarded as the proof that an infectious agent is responsible for a specific disease:

(1) The agent must be present in every case of the disease
(2) The agent must be isolated from the host and grown *in vitro*
(3) The disease must be reproduced when a pure culture of the agent is inoculated into a healthy susceptible host
(4) The same agent must be recovered once again from the experimentally infected host.

Subsequently, Pasteur worked extensively on rabies, which he identified as being caused by a 'virus' (from the Latin for 'poison') but despite this, he did not discriminate between bacteria and other agents of disease. In 1892, Dimitri Iwanowski, a Russian botanist, showed that extracts from diseased tobacco plants could transmit disease to other plants after passage through ceramic filters fine enough to retain the smallest known bacteria. Unfortunately, he did not realize the full significance of these results. A few years later (1898), Martinus Beijerinick confirmed and extended Iwanowski's results on tobacco mosaic virus and was the first to develop the modern idea of the virus, which he referred to as *contagium vivum fluidum* ('soluble living germ'). Freidrich Loeffler and Paul Frosch (1898) showed that a similar agent was responsible for foot-and-mouth disease in cattle, but despite the realization that these new-found agents caused disease in animals as well as plants, there was resistance to the idea that they might have anything to do with human diseases. This was finally dispelled by Karl Landsteiner and E. Popper in 1909, who showed that poliomyelitis was caused by a 'filterable agent' – the first human disease to be recognized as having a viral cause.

Frederick Twort (1915) and Felix d'Herelle (1917) were the first to recognize viruses which infect bacteria, which d'Herelle called **bacteriophages** (eaters of bacteria). In the 1930s and subsequent decades, pioneering virologists such as S.E. Luria, M. Delbruck, and many others used these viruses as model systems to investigate many aspects of virology, including virus structure (Chapter 2), genetics (Chapter 3), replication (Chapter 4), etc. These relatively simple agents have since been very important to our understanding of all types of viruses, including those of humans which are much more difficult to propagate and study. The further history of virology is the story of the development of experimental tools and systems with which viruses could be examined and by which whole new areas of biology were opened up, including not only the biology of the viruses themselves, but inevitably also the biology of the host cells on which these agents are entirely dependent.

Living Host Systems

In 1881, Louis Pasteur began to study rabies in animals. Over a number of years, he developed methods of producing **attenuated** virus preparations by progressively drying the spinal cords of rabbits experimentally infected with the agent which, when inoculated into animals, would protect from challenge with virulent virus. This was the first artificially produced virus **vaccine**, since the ancient practice of variolation (above) and Jenner's use of cowpox virus for vaccination relied on naturally occurring viruses. Whole plants have been used to study the effects of plant viruses after infection ever since tobacco mosaic virus was first discovered by Iwanowski; usually the studies involve rubbing preparations containing virus particles into the leaves or stem of the plant.

During the Spanish–American War of the late nineteenth century and the subsequent building of the Panama Canal, American deaths due to yellow fever were colossal. The disease also appeared to be spreading slowly northward into the continental United States. Through experimental transmission to mice Walter Reed, in 1900, demonstrated that yellow fever was caused by a virus, spread by mosquitoes. This discovery eventually enabled Max Theiler in 1937 to propagate the virus in chick embryos and to produce an attenuated vaccine – the 17D strain – which is still in use today.

The success of this approach led many other investigators from the 1930s to the 1950s to develop animal systems to identify and propagate pathogenic viruses. **Eukaryotic** cells can be grown *in vitro* (tissue culture) and viruses can be propagated in these cultures, but these techniques are expensive and technically quite demanding. Some viruses will replicate in the living tissues of developing embryonated hens eggs, such as influenza virus. Egg-adapted strains of influenza virus replicate well in eggs and very high virus **titres** can be obtained. Embryonated hens eggs were first used to propagate viruses in the early decades of this century. This method has proved to be highly effective for the isolation and culture of many viruses, particularly strains of influenza virus and various poxviruses (e.g. vaccinia virus). Counting the 'pocks' on the chorioallantoic membrane of eggs produced by the replication of vaccinia virus was the first quantitative assay for any virus. Animal host systems still have their uses in virology:

- To produce viruses which cannot be propagated *in vitro*, e.g. hepatitis B virus
- To study the pathogenesis of virus infections, e.g. coxsackieviruses
- To test vaccine safety, e.g. oral poliovirus vaccine.

Nevertheless, they are increasingly being discarded for the following reasons:

- Breeding and maintenance of animals infected with pathogenic viruses is expensive
- Whole animals are complex systems, in which it is sometimes difficult to discern events
- Results obtained are not always reproducible, due to host variation
- Unnecessary or wasteful use of experimental animals is morally repugnant
- They are rapidly being overtaken by 'modern science' – cell culture and molecular biology.

The use of whole plants as host organisms does not engender the same moral objections as the use of living animals and continues to play an important part in the study of plant viruses, although such systems are sometimes slow to deliver results and expensive to maintain.

In recent years, an entirely new technology has been employed to study the effects of viruses on host organisms. This involves the creation of **transgenic** animals and plants by inserting all or part of the virus genome into the DNA of the experimental organism, resulting in expression of virus mRNA and proteins in somatic cells (and sometimes in the cells of the germ line). Thus the pathogenic effects of virus proteins, individually and in various combinations, can be studied in living hosts. 'SCID-hu' mice have been constructed from immunodeficient lines of animals transplanted with human tissue. These mice form an intriguing model to study the pathogenesis of HIV since there is no real alternative to study the properties of this important virus *in vivo*. While these techniques often raise the same moral objections as 'old-fashioned' experimental infection of animals by viruses, they are immensely powerful new tools for the study of viral pathogenicity. A growing number of plant and animal viruses genes have been analysed in this way, but the results have not always been as expected, and in many cases it has proved difficult to equate the observations obtained with those gathered from experimental infections. Nevertheless, this method will undoubtedly become much more widely used as more of the technical difficulties associated with the construction of transgenic organisms are solved.

Cell Culture Methods

Cell culture began early this century with whole-organ cultures, then progressed to methods involving individual cells, either **primary cell** cultures (somatic cells from an experimental animal or taken from a human patient which can be maintained for a short period in culture) or **immortalized cell** lines, which, given appropriate conditions, continue to grow in culture indefinitely.

In 1949, J.F. Enders and his colleagues were able to propagate poliovirus in primary human cell cultures. This opened what many regard as the 'Golden Age of Virology' – the identification and isolation during the 1950s and 1960s of many viruses and their association with human diseases, for example, many enteroviruses and respiratory viruses, such as adenoviruses. Widespread virus isolation led to the realization that subclinical virus infections were very common; for example, even in epidemics of the most virulent strains of poliovirus, there are approximately 100 subclinical infections for each paralytic case of poliomyelitis.

Renato Dulbecco in 1952 was the first to quantify accurately animal viruses using a plaque assay. In this technique, dilutions of the virus are used to infect a cultured cell **monolayer**, which is then covered with soft agar to restrict diffusion of the virus, resulting in localized cell killing and the appearance of **plaques** after the monolayer is stained (Figure 1.1). Counting the number of plaques directly determines the number of infectious virus particles applied to the plate. The same technique can also be used biologically to clone a virus, i.e. isolate a pure form from a mixture of types. This technique had been in use for some time to quantify the number of infectious virus particles in bacteriophage suspensions applied to confluent 'lawns' of bacterial cells on agar plates, but its application to eukaryotic viruses enabled rapid advances in the study of virus replication to be made. Plaque assays largely replaced earlier endpoint dilution techniques, such as the $TCID_{50}$ (tissue culture infectious dose) assay, which are statistical means of measuring virus populations in culture. However, endpoint techniques may still be used in certain circumstances, for example, for viruses which do not replicate in culture, or which are not cytopathic and do not produce plaques, e.g. human immunodeficiency virus (HIV).

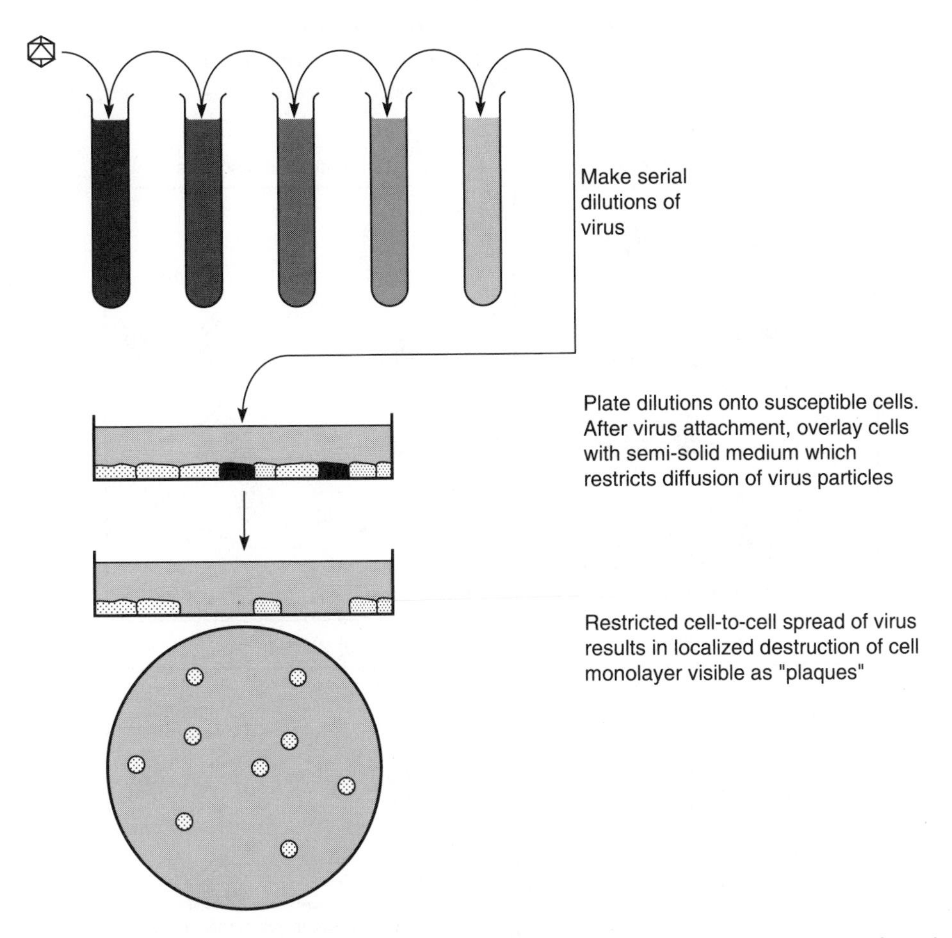

Figure 1.1 Plaque assays are performed by applying a suitable dilution of a virus preparation to a confluent or semi-confluent adherent monolayer of susceptible cells. After allowing time for virus attachment to and infection of the cells, liquid medium is replaced by a semi-solid culture medium containing a polymer such as agarose or carboxymethyl cellulose, which restricts diffusion of virus particles from infected cells. Only direct cell-to-cell spread can occur, resulting in localized destruction of the monolayer. After a suitable period, the medium is usually removed and the cells stained to make the holes in the monolayer (**plaques**) more easily visible. Each plaque therefore results from infection by a single **plaque-forming unit (p.f.u.)**.

Serological/Immunological Methods

As virology was emerging in the early part of this century, the techniques of immunology were also developing and, as with molecular biology in more recent years, the two disciplines have always been very closely linked. Understanding mechanisms of immunity to virus infections has, of course, been very important. Recently, the role that the immune system itself plays in pathogenesis has become known (see Chapter 7). Immunology as a discipline in its own right has contributed many of the classical techniques to virology (Figure 1.2).

George Hirst, in 1941, observed **haemagglutination** of red blood cells by influenza virus (see Chapter 4). This proved to be an important tool not only in the study of influenza, but also with several other groups of viruses, for

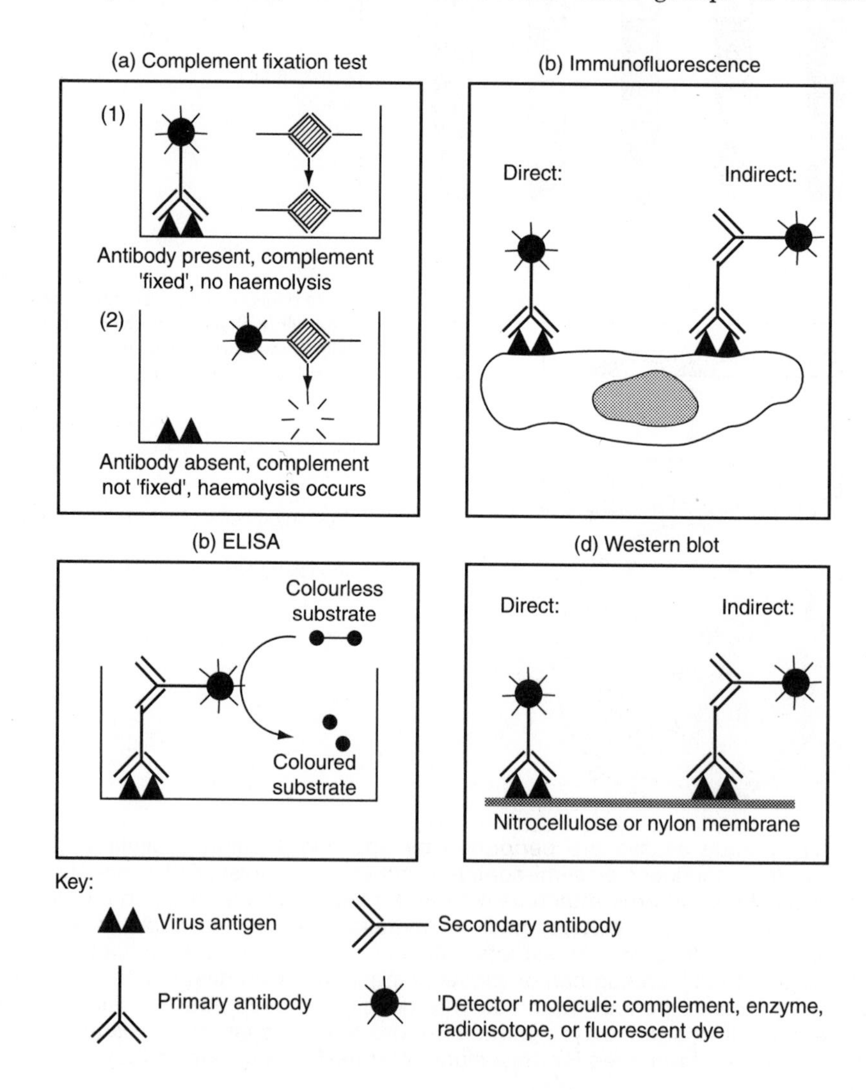

example, rubella virus. In addition to measuring the **titre** (i.e. relative amount) of virus present in any preparation, this technique can also be used to determine the antigenic type of the virus. Haemagglutination will not occur in the presence of antibodies which bind to and block the virus haemagglutinin. If an antiserum is titrated against a given number of haemagglutinating units, the haemagglutination inhibition titre and specificity of the antiserum can be determined. Also, if antisera of known specificity are used to inhibit haemagglutination, the antigenic type of an unknown virus can be determined. In the 1960s and subsequent years, many improved detection methods for viruses were developed, for example:

- Complement fixation tests
- Radioimmunoassays

Figure 1.2 It is hard to overestimate the importance of serological techniques in virology. Four assays illustrated by the diagrams in this figure have been used for many years and are of widespread value. (a) The complement fixation test works on the basis that complement is sequestered by antigen-antibody complexes. 'Sensitized' antibody-coated red blood cells and known amounts of complement plus a particular virus antigen are added to the wells of a multi-well plate, plus the serum to be tested. In the absence of antibodies directed against the virus antigen, some free complement remains and this causes lysis of the sensitized red blood cells (haemolysis). If, however, the test serum contains a sufficiently high titre of anti-viral antibodies, no free complement remains and haemolysis does not occur. Titrating the test serum by means of serial dilutions allows a quantitative measurement of the amount of anti-virus antibody present to be made. (b) Immunofluorescence is performed using derivatized antibodies containing a covalently-linked fluorescent molecule which emits a characteristically coloured light when illuminated by light of a different wavelength, e.g. rhodamine (red) or fluorescein (green). In direct immunofluorescence, the anti-virus antibody itself is conjugated to the fluorescent marker, whereas in indirect immunofluorescence, a second antibody reactive to the anti-virus antibody carries the marker. Immunofluorescence can be used not only to identify virus-infected cells in populations of cells or in tissue sections, but also to determine the sub-cellular localization of particular virus proteins, e.g. in the nucleus or in the cytoplasm. (c) *E*nzyme-*L*inked *I*mmuno-*S*orbent *A*ssays (ELISA) are a rapid and sensitive means of identifying or quantifying small amounts of virus antigens or anti-virus antibodies. Either an antigen (in the case of an ELISA to detect antibodies) or antibody (in the case of an antigen ELISA) are bound to the surface of a multi-well plate. An antibody specific for the test antigen, which has been conjugated with an enzyme molecule (such as alkaline phosphatase or horseradish peroxidase), is then added. As with immunofluorescence, ELISA assays may rely on direct or indirect detection of the test antigen. During a short incubation, a colourless substrate for the enzyme is converted to a coloured product, thus amplifying the signal produced by a very small amount of antigen. The intensity of the product can easily be measured in a specialized spectrophotometer (a 'plate reader'). ELISA assays can be mechanized and are therefore suitable for routine tests on large numbers of clinical samples. (d) Western blotting is used to analyse a specific virus protein from a complex mixture of antigens. Virus antigen-containing preparations (particles, infected cells or clinical materials) are subjected to electrophoresis on a polyacrylamide gel. Proteins from the gel are then transferred to a nitrocellulose or nylon membrane and immobilized in their relative positions from the gel. Specific antigens are detected by allowing the membrane to react with antibodies directed against the antigen of interest. By the use of samples containing proteins of known sizes in known amounts, the apparent molecular weight and relative amounts of antigen in the test samples can be determined.

- Immunofluorescence (direct detection of virus antigens in infected cells or tissue)
- Enzyme-linked immunosorbent assays (ELISAs)
- Radioimmune precipitation
- Western blot assays.

These techniques are sensitive, quick and quantitative.

In 1975, Kohler and Milstein isolated the first monoclonal antibodies from clones of cells selected *in vitro* to produce an antibody of a single specificity directed against a particular antigenic target. This enabled virologists to look not only at the whole virus, but at specific regions – epitopes – of individual virus antigens (Figure 1.3). In recent years, this ability has greatly increased our

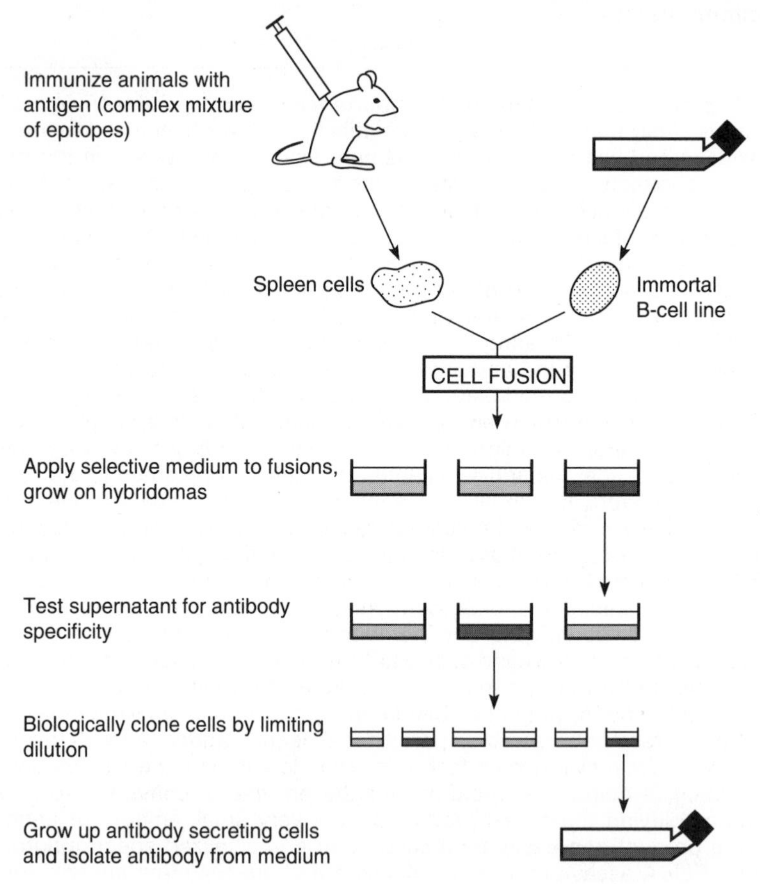

Figure 1.3 Monoclonal antibodies are produced by immunization of an animal with an antigen which usually contains a complex mixture of epitopes. Immature B-cells are later prepared from the spleen of the animal and these are fused with a myeloma cell line, resulting in the formation of **transformed** cells continuously secreting antibodies. A small proportion of these will make a single type of antibody (a monoclonal antibody) against the desired epitope. Recently, *in vitro* molecular techniques have been developed to speed up the selection of monoclonal antibodies, although these have not yet replaced the original approach shown here.

understanding of the function of individual virus proteins. Monoclonal antibodies are also finding increasingly widespread application in other types of serological assay (e.g. ELISAs), to increase their reproducibility, sensitivity, and specificity.

It would be inappropriate here to spend too long on the technical details of what is also a very rapidly expanding field of knowledge. However, I recommend strongly that readers who are not familiar with the techniques mentioned above should familiarize themselves thoroughly with this subject by reading one or more of the texts given in the Further Reading for this chapter. Time spent in this way will repay the reader throughout the rest of this book.

Ultrastructural Studies

These can be considered under three areas: physical methods, chemical methods, and electron microscopy. Physical measurements of virus particles began in the 1930s with the earliest determinations of their proportions by filtration through colloidal membranes of various pore sizes. Experiments of this sort led to the first (rather inaccurate) estimates of the size of virus particles. The accuracy of these was improved greatly by studies of the sedimentation properties of viruses in ultracentrifuges in the 1960s (Figure 1.4). Differential centrifugation proved to be of great use in obtaining purified and highly concentrated preparations of many different viruses, free of contamination from host cell components, which can be subjected to chemical analysis. The relative density of virus particles, measured in solutions of sucrose or CsCl, is also a characteristic feature, revealing information about the proportions of nucleic acid and protein in the particles.

The physical properties of viruses can be determined by spectroscopy, using either ultraviolet light to examine the nucleic acid content of the particle or visible light to determine its light-scattering properties. Electrophoresis of intact virus particles has yielded some limited information, but electrophoretic analysis of individual virion proteins by gel electrophoresis, and particularly also of nucleic acid genomes (Chapter 3), has been far more valuable. However, by far the most important method for the elucidation of virus structures has been the use of X-ray diffraction by crystalline forms of purified virus. This technique permits determination of the structure of virions at an atomic level.

The complete structures of many viruses, representative of many of the major groups, have now been determined at a resolution of a few angstroms (Å) (see Chapter 2). This has advanced our understanding of the functions of the viral particle considerably. However, a number of viruses have proved to be resistant to this type of investigation, a fact which highlights some of the problems inherent in this otherwise powerful technique. One problem is that the virus must first be purified to a high degree, otherwise specific information on the virus cannot be gathered. This presupposes that adequate quantities of the virus can be propagated in culture or obtained from infected tissues or patients and

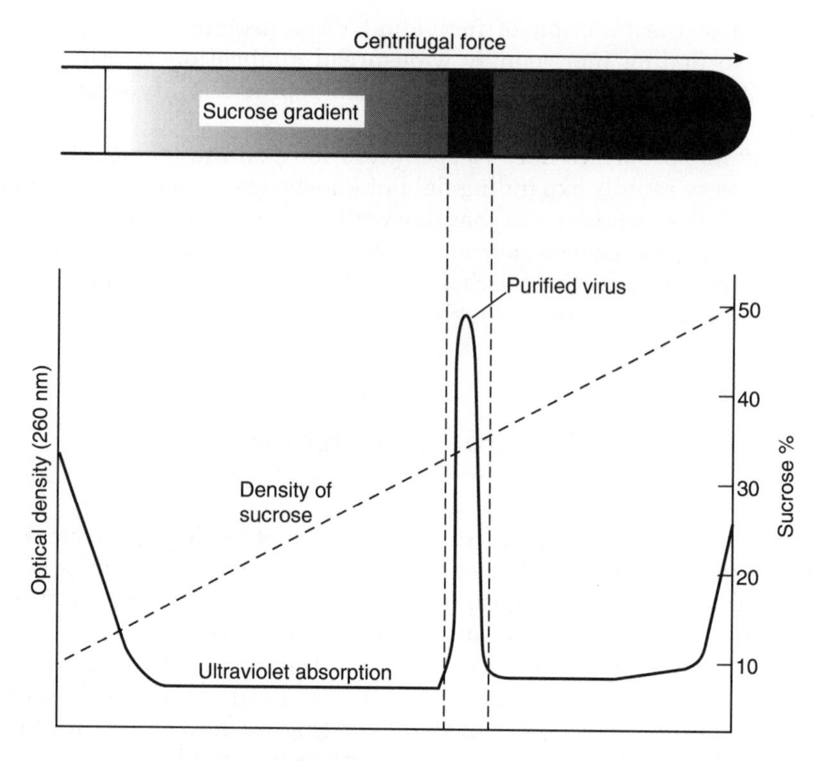

Figure 1.4 A number of different sedimentation techniques can be used to study viruses. In rate zonal centrifugation (shown here), virus particles are applied to the top of a preformed density gradient, i.e. a sucrose or salt solution of increasing density from the top to the bottom of the tube (top of figure). After a period of time in an ultracentrifuge, the gradient is separated into a number of fractions, which are analysed for the presence of virus particles. In the figure, the nucleic acid of the virus genome is detected by its absorption of ultraviolet light (below). This method can be used both to purify virus particles or nucleic acids or to determine their sedimentation characteristics. In equilibrium or isopycnic centrifugation, the sample is present in a homologous mixture containing a dense salt such as caesium chloride. A density gradient forms in the tube during centrifugation and the sample forms a band at a position in the tube equivalent to its own density. This method can thus be used to determine the density of virus particles and is commonly used to purify plasmid DNA.

that a method is available to purify virus particles without loss of structural integrity. In a number of important cases, this requirement rules out further study (e.g. hepatitis C virus). The purified virus must also be able to form paracrystalline arrays large enough to cause significant diffraction of the radiation source. For some viruses, this is relatively straightforward and crystals big enough to see with the naked eye and which diffract strongly are easily formed. This is particularly true for a number of plant viruses, such as tobacco mosaic virus (TMV) (which was first crystallized by W.M. Stanley in 1935) and turnip yellow mosaic virus (TYMV), the structures of which were among the first to be determined during the 1950s. It is significant that these two viruses represent the two fundamental types of virus particle: helical in the case of TMV and isometric for TYMV (see Chapter 2). In many cases, however, only microscopic crystals can be prepared. A partial answer to this problem is to use ever more powerful radiation sources which enable good data to be collected from small crystals. Powerful synchotron sources which generate intense beams of radiation have been built during the last few decades and are now used extensively for this purpose. However, there is a limit beyond which this brute force approach fails to yield further benefit. A number of important viruses steadfastly refuse to crystallize – this is a particularly common problem with irregularly shaped viruses, for example those which have an outer lipid envelope, and to date, no complete high-resolution atomic structure has yet been determined for many viruses of this type (e.g. HIV). Modifications of the basic diffraction technique (such as electron scattering by membrane-associated protein arrays and cryo-electron microscopy) may help to provide more information in the future, but it is unlikely that these variations will solve this problem completely. One further limitation is that some of the largest virus particles, such as poxviruses, contain hundreds of different proteins and are at present too complex to be analysed using these techniques.

Nuclear magnetic resonance (NMR) is increasingly being used to determine the atomic structure of all kinds of molecules, including proteins and nucleic acids. The limitation of this method is that only relatively small molecules can be analysed before the signals obtained become so confusing that they are impossible to decipher with present technology. At present, the upper size limit for this technique restricts its use to molecules with a molecular weight of less than about 30 000–40 000 – considerably less than even the smallest virus particles. Nevertheless, this method may well prove to be of value in the future, certainly for examining isolated viral proteins, if not for intact virions.

Chemical investigation can be used to determine not only the overall composition of viruses and the nature of the nucleic acid which comprises the virus genome, but also the construction of the particle and the way in which individual components relate to each other in the capsid. Many classic studies of virus structure have been based on gradual, stepwise disruption of particles by slow alteration of pH, or gradual addition of protein-denaturing agents such as urea, phenol, or detergents. Under these conditions, valuable information can sometimes be obtained from relatively simple experiments. For example, as urea is gradually added to preparations of purified adenovirus particles, they break down in an ordered, stepwise fashion which releases subviral protein assemblies, revealing how the particles are made up. In the case of TMV, similar studies of capsid organization have been performed by renaturation of the capsid protein under various conditions (Figure 1.5). In simple terms, the reagents used to denature virus capsids can indicate the basis of the stable interactions between its components. Proteins bound together by electrostatic interactions can be eluted by addition of ionic salts or alteration of pH, those bound by non-ionic, hydrophobic interactions can be eluted by reagents such as urea, and proteins which interact with lipid components can be eluted by non-ionic detergents or organic solvents.

In addition to its fundamental structure, progressive denaturation can be used to observe alteration or loss of antigenic sites on the surface of particles and in

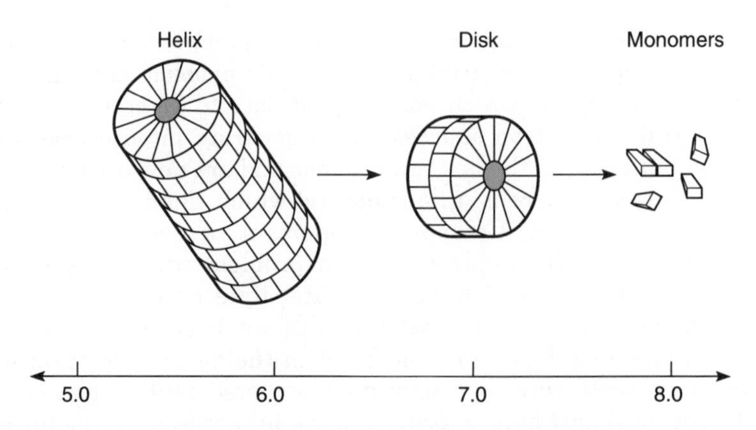

Figure 1.5 The structure and stability of virus particles can be examined by progressive denaturation or renaturation studies. At any particular ionic strength, the purified capsid protein of tobacco mosaic virus (TMV) spontaneously assembles into different structures, dependent on the pH of the solution. At a pH of around 6.0, the particles formed have a helical structure very similar to infectious virus particles. As the pH is increased to about 7.0, disk-like structures are formed. At even higher pH values, individual capsid monomers fail to assemble into more complex structures.

this way, a picture of the physical state of the particle can be built up. Proteins exposed on the surface of viruses can be labelled with various compounds (e.g. iodine) to indicate which parts of the protein are exposed and which are protected inside the particle or by lipid membranes. Cross-linking reagents such as psoralens, or newer synthetic reagents with side-arms of specific lengths are used to determine the spatial relationship of proteins and nucleic acids in intact viruses.

Electron microscopes, developed in the 1930s, overcome the fundamental limitation of light microscopes, i.e. the inability to resolve individual virus particles owing to physical constraints caused by the wavelength of visible light illumination and the optics of the instruments. The first electron micrograph of a virus (TMV) was published in 1939. During subsequent years, techniques were developed which allowed the direct examination of viruses at magnifications of over ×100 000. There are two fundamental types of electron microscope, the transmission electron microscope (TEM) and the scanning electron microscope (SEM) (Figure 1.6). Although beautiful images with the appearance of three dimensions are produced by the SEM, for practical investigations of virus structure, the higher magnifications achievable with the TEM have proved to be of most value. Two fundamental types of information can be obtained by electron microscopy of viruses: the absolute number of virus particles present in any preparation (total count) and the appearance and structure of the virions (see below). Electron microscopy can provide a rapid method of virus detection and diagnosis, but in itself, may give misleading information. Many cellular components, for example ribosomes, can resemble 'virus-like particles,' particularly in crude preparations. This difficulty can be overcome by using antisera specific for particular virus antigens conjugated to electron-dense markers such as the iron-containing protein ferritin, or colloidal gold suspensions. This highly

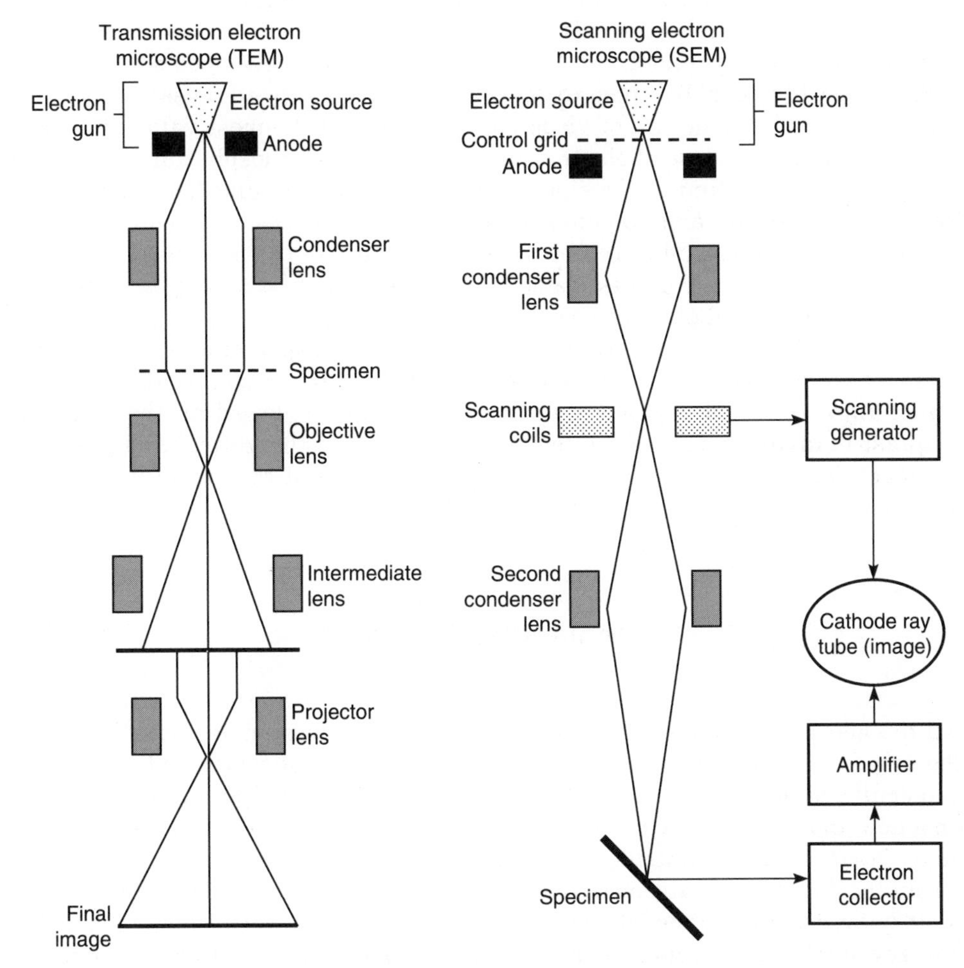

Figure 1.6 Working principles of transmission and scanning electron microscopes.

specific technique, known as immunoelectron microscopy, is gaining ground as a rapid method for diagnosis.

In recent years, developments in electron microscopy have allowed investigation of the structure of fragile viruses which cannot be determined by X-ray crystallography. These include: cryo-electron microscopy, in which the virus particles are maintained at very low temperatures on cooled specimen stages; examination of particles embedded in vitreous ice, which does not disrupt the particles by the formation of ice crystals; low-irradiation electron microscopy, reducing the destructive bombardment of the specimen with electrons; and sophisticated image-analysis and image-reconstruction techniques which permit accurate, three-dimensional images to be formed from multiple images that individually would appear as very poor quality. Conventional electron microscopy can resolve structures down to 50–70 Å in size (a typical atomic diameter is

2–3 Å, a protein α-helix 10 Å and a DNA double helix 20 Å). Using these newer techniques it is possible to resolve structures of 25–30 Å.

In the late 1950s, Brenner and Horne (among others) developed sophisticated techniques which enabled them to use electron microscopy to reveal many of the fine details of the structure of virus particles. One of the most valuable techniques proved to be the use of electron-dense dyes such as phosphotungstic acid or uranyl acetate to examine virus particles by negative staining. The small metal ions in such dyes are able to penetrate the minute crevices between the protein subunits in a viral capsid to reveal the fine structure of the particle. Using such data, Francis Crick and James Watson (1956) were the first to suggest that virus capsids are composed of numerous identical protein subunits arranged either in helical or cubic (icosahedral) symmetry. In 1962, Caspar and Klug extended these observations and elucidated the fundamental principles of symmetry which allow repeated protomers to form virus capsids, based on the principle of **quasi-equivalence** (see Chapter 2). This combined theoretical and practical approach has resulted in our present understanding of the structure of virus particles.

'Molecular Biology'

All the above techniques of investigation are themselves 'molecular biology' in the original sense of the term. However, during the last two decades the term 'molecular biology' ('genetic engineering' or 'genetic manipulation') has taken on a new and different meaning. These new techniques for manipulating nucleic acids *in vitro* (that is, outside living cells or organisms) do not comprise a new discipline, but are an outgrowth of earlier developments in biochemistry and cell biology over the previous 50 years. Despite this, this powerful new technology has revolutionized virology and, to a large extent, has shifted the focus of attention away from the virus particle onto the virus genome. Again, this book is not the place to discuss in detail the technical aspects of these methods, and readers are referred to one of the many texts published in recent years, such as those given at the end of this chapter.

Virus infection has long been used to probe the working of 'normal' (i.e. uninfected) cells, for example, to look at macromolecular synthesis. This is true, for example, of the applications of bacteriophages in bacterial genetics, and in many instances where the study of eukaryotic viruses has revealed fundamental information about the cell biology and genomic organization of higher organisms. In 1970, John Kates first observed that vaccinia virus mRNAs were polyadenylated at their 3′ ends. In the same year, Howard Temin and David Baltimore jointly identified the enzyme reverse transcriptase (RNA-dependent DNA polymerase) in retrovirus-infected cells. This shattered the so-called 'central dogma' of biology, i.e. that there is a one-way flow of information from DNA through RNA into protein, and revealed the plasticity of the eukaryotic genome. Subsequently, the purification of this enzyme from retrovirus particles permitted cDNA cloning, which greatly accelerated the study of viruses with RNA genomes – a good illustration of the catalytic nature of scientific advances. In 1977, Philip Sharp recognized that adenovirus mRNAs were spliced to remove intervening sequences, indicating the similarities between viral and cellular genomes.

Initially at least, the effect of this new technology was to shift the emphasis of investigation from proteins to nucleic acids. As the power of the techniques developed, it quickly became possible to determine the nucleotide sequences of entire virus genomes, beginning with the smallest bacteriophages in the mid-1970s and working up to the largest of all virus genomes, those of the herpesviruses and poxviruses, many of which have now been determined.

This nucleic acid-centred technology, in addition to its 'ultimate' achievement of nucleotide sequencing and the artificial manipulation of virus genomes, also offered significant advances in detection of viruses and virus infections involving nucleic acid hybridization techniques. There are many variants of this basic idea, but essentially, a hybridization probe, labelled in some fashion to facilitate detection, is allowed to react with a crude mixture of nucleic acids. The specific interaction of the probe sequence with complementary virus-encoded sequences, to which it binds by hydrogen-bond formation between the complementary base pairs, reveals the presence of the viral genetic material (Figure 1.7). In the past few years, this approach has been taken a stage further by the development of various *in vitro* nucleic acid amplification procedures, such as the polymerase chain reaction (PCR). This is an even more sensitive technique, capable of detecting just a single molecule of viral nucleic acid (Figure 1.8).

More recently, there has also been renewed interest in virus proteins, based on a new biology which is itself dependent on manipulation of nucleic acids *in vitro* and advances in protein detection arising from immunology. Methods for *in vitro* synthesis and expression of proteins from molecularly cloned DNA have advanced rapidly, and many new analytical techniques are now available. Studies of protein–nucleic acid interactions are proving to be particularly valuable in understanding virus structure and gene expression.

In addition to examining virus proteins, molecular biologists have one further trick up their sleeves. Because of the repetitive, digitized nature of nucleotide sequences, computers are the ideal means of storing and processing this mass of information. Moreover, as computer software becomes more sophisticated, computers are being used increasingly to make predictions based on nucleotide sequences (Figure 1.9). These include detecting the presence of open reading frames, the amino acid sequences of the proteins encoded by them, control regions of genes such as promoters and splice signals and the secondary structure of proteins and nucleic acids. However (particularly in the case of RNA), the secondary structure assumed by molecules is almost as important as the primary nucleotide sequence in determining the biological reactions which the molecule may undergo. Much caution is needed in interpreting such predicted rather than factual information, and the validity of such predictions should not be accepted without question unless confirmed by biochemical and/or genetic data. Although knowing the primary amino acid sequence of a polypeptide is a long way from understanding the subtleties of the protein molecule itself, this is

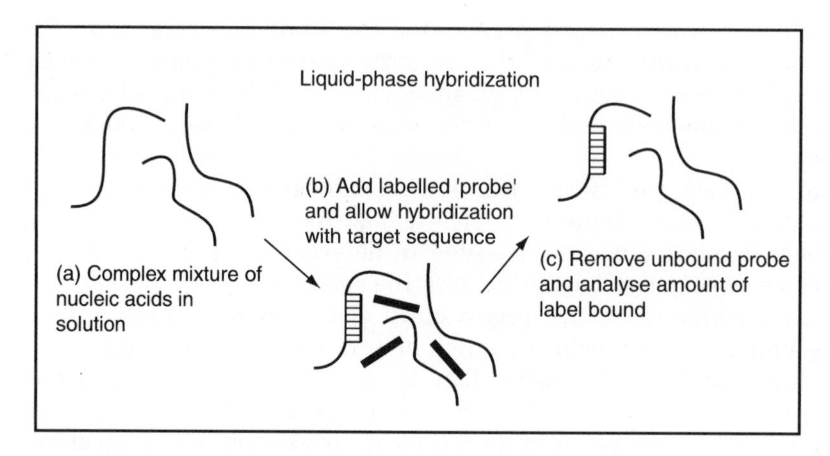

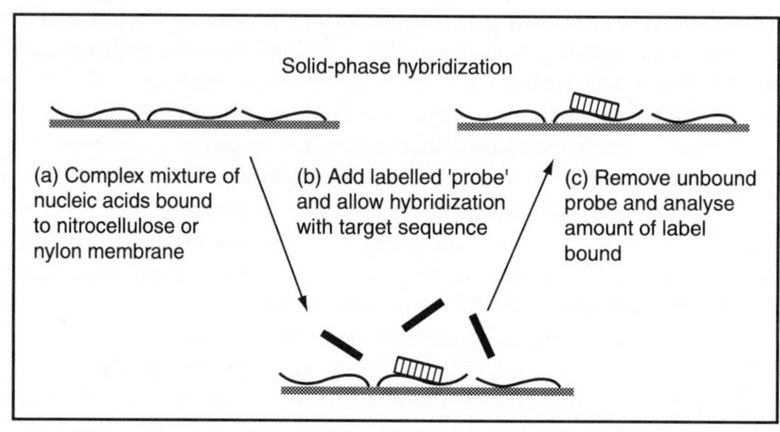

Figure 1.7 Nucleic acid hybridization relies on the specificity of base-pairing which allows a labelled nucleic acid probe to pick out a complementary target sequence from a complex mixture of sequences in the test sample. The label used to identify the probe may be a radio-isotope or a non-isotopic label such as an enzyme or chemiluminescent system. Hybridization may be performed with both the probe and test sequences in the liquid phase (top of figure) or with the test sequences bound to a solid phase, usually a nitrocellulose or nylon membrane (below). Both methods may be used to quantify the amount of the test sequence present, but solid-phase hybridization is also used to locate the position of sequences immobilized on the membrane. Plaque and colony hybridization are used to locate recombinant molecules directly from a mixture of bacterial colonies or bacteriophage plaques on an agar plate. 'Northern' and 'Southern' blotting are used to detect RNA and DNA respectively after transfer of these molecules from gels following separation by electrophoresis (c.f. 'Western' blotting, Fig. 1.2).

still a powerful tool. Vast international databases of nucleotide and protein sequence information have now been compiled, and these can rapidly be consulted by computers to compare newly determined sequences with those whose function may have been studied in great detail.

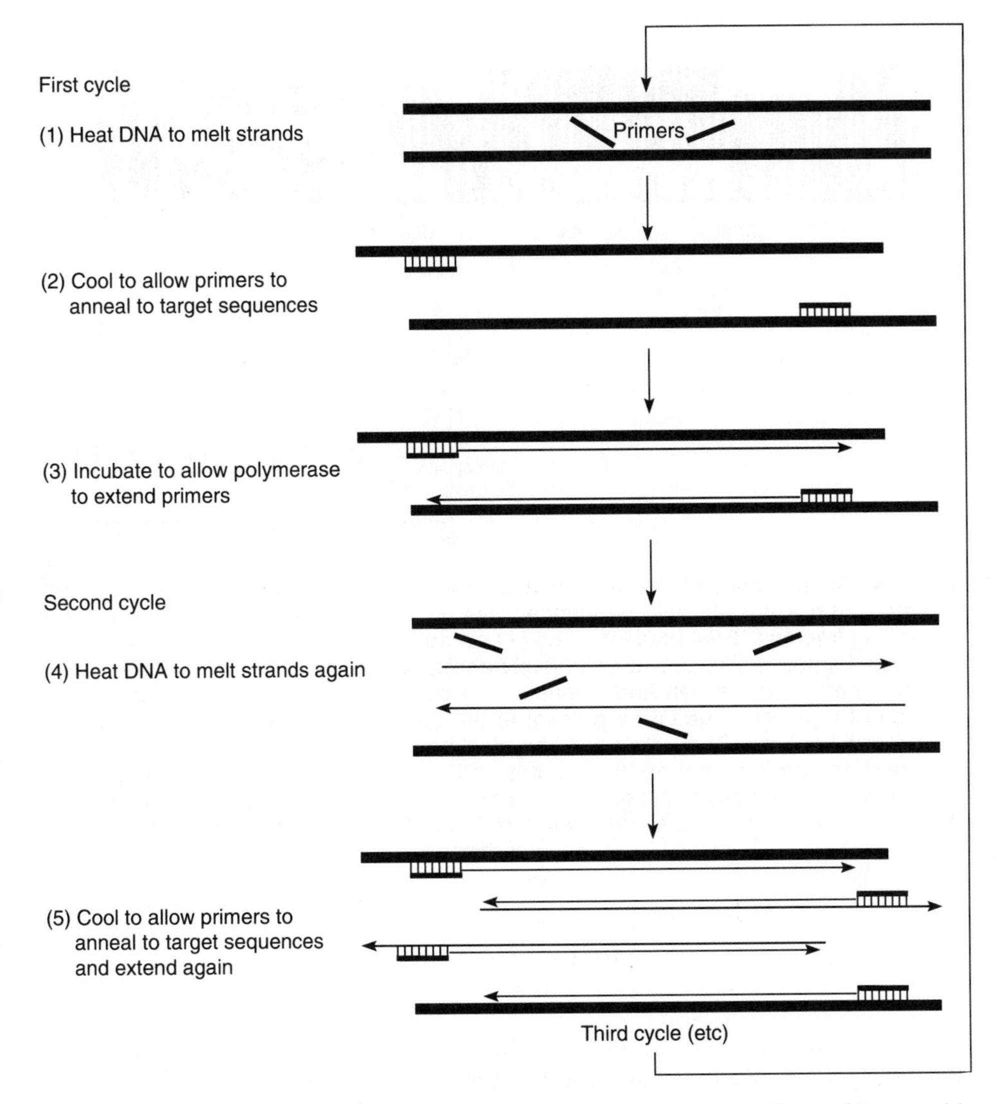

Figure 1.8 The polymerase chain reaction (PCR) relies on the specificity of base-pairing between short synthetic olignucleotide probes and complementary sequences in a complex mixture of nucleic acids to prime DNA synthesis using a thermostable DNA polymerase. Multiple cycles of primer annealing, extension and thermal denaturation are carried out in an automated process, resulting in a massive amplification (2^n-fold increase after n cycles of amplification) of the target sequence located between the two primers.

Thus we have, in a sense, come full circle in our investigations of viruses – from particles via genomes back to proteins again – and have emerged with a far more profound understanding of these organisms. However, the present pace of research in virology tells us that there is still far more that we need to know.

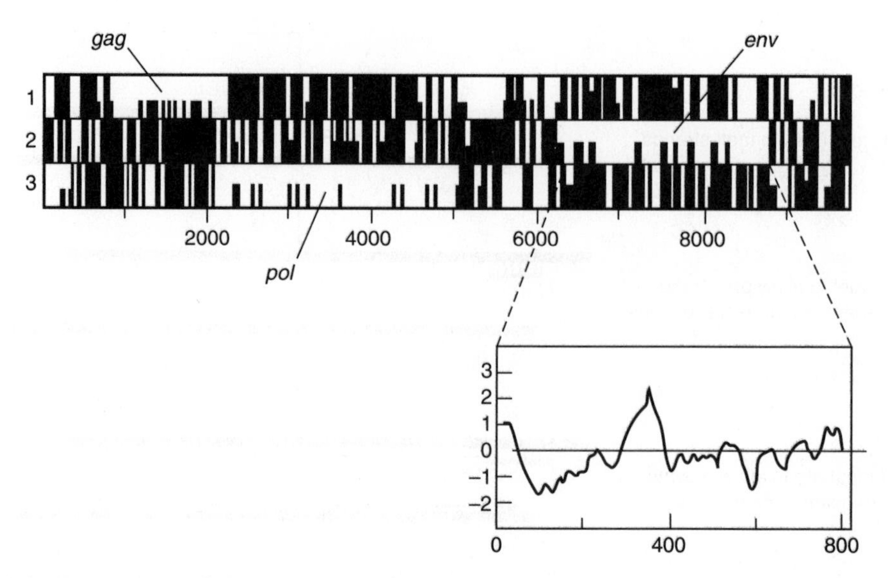

Figure 1.9 An example of the use of a computer to store and process 'digitized' information from a nucleic acid sequence. This figure shows an analysis of all of the open reading frames (ORFs) present in the (+) sense strand of an HIV-1 provirus. In each of the three frames, a line drawn completely across the frame represents a translation termination codon and a half line a methionine codon – a potential initiation site for translation of a protein. The ORFs present in the three main retrovirus genes, *gag*, *pol* and *env*, can be seen. The computer has also been used to predict the amino acid sequence of the protein encoded by the *env* ORF and to analyse the hydrophobicity of the envelope protein (bottom of figure). This complex analysis took only a few minutes to perform using an ordinary personal computer. Manually, the same task may have taken several days.

Further Reading

Alberts, B., Bray, D., Lewis, J., Raff, M., Roberts, K. and Watson J.D. (Eds). (1994). *Molecular Biology of the Cell* (3rd edition). Garland, New York.

Brady, H.J.M. *et al.* (1994). Transgenic mice as models of HIV gene expression and related cellular effects. *J. Gen. Virol.* **75**: 2549–58.

Chisari, F.V., Oldstone, M.B.A. (Eds). (1996). Transgenic models of human viral and immunological disease. *Curr. Top. Microbiol. Immunol.* Vol. **206**.

Levine, A.J. (1992). *Viruses.* W.H. Freeman/Scientific American Library, New York.

McCune, J. *et al.* (1991). The scid-hu mouse: a small animal model for HIV infection and pathogenesis. *Annu. Rev. Immunol.* **9**: 399–429.

Old, R.W. and Primrose, S.B. (1994). *Principles of Gene Manipulation* (5th edition). Blackwell Scientific, Oxford.

Roitt, I. (1994). *Essential Immunology* (8th edition). Blackwell Scientific, Oxford.

Sambrook, J., Fritsch, E.F. and Maniatis, T. (1989). *Molecular Cloning* (2nd edition). Cold Spring Harbor Laboratory Press, Cold Spring Harbor.

Watt, I.M. (1985). *The Principles and Practice of Electron Microscopy.* Cambridge University Press, Cambridge.

Self-Assessment Questions

(1.1) Are the following statements true or false?
- (a) Viruses are sub-microscopic, obligate intracellular parasites.
- (b) Virus particles increase by the integrated sum of their components and reproduce by division.
- (c) Viruses lack the genetic information which encodes apparatus necessary for the generation of metabolic energy.
- (d) Viruses lack the genetic information which encodes apparatus necessary for protein synthesis (ribosomes).
- (e) Prions are infectious agents believed to consist of a single nucleic acid component.

(1.2) Are the following statements true or false?
- (a) Antony van Leeuwenhoek (1632–1723) was the first person to see virus particles under the microscope.
- (b) Edward Jenner first vaccinated a patient against smallpox on 14th May 1796.
- (c) Louis Pasteur invented the term 'virus' in the 1890s.
- (d) Dimitri Iwanowski showed that viruses could pass through filters fine enough to retain the smallest known bacteria.
- (e) Viruses which infect bacteria are called bacteriophiles.

(1.3) These are Koch's Postulates:
- (a) The agent must be present in every case of the disease.
- (b) The agent must be isolated from the host and grown *in vitro*.
- (c) The disease must be reproduced when a pure culture of the agent is inoculated into a healthy susceptible host.
- (d) The same agent must be recovered once again from the experimentally infected host.

(1.4) The following are immunological methods of studying viruses (true or false?):
- (a) Complement fixation
- (b) Northern blot
- (c) Southern blot
- (d) Western blot
- (e) PCR

(1.5) Are the following statements true or false?
- (a) Electron microscopes allow direct examination of viruses at magnifications of up to 1000 times.
- (b) X-ray diffraction by crystalline forms of purified virus can be used to determine their structure.
- (c) The atomic structures of all known viruses have now been determined.
- (d) Viruses are believed to originate by spontaneous generation.
- (e) More than 4000 different viruses have been identified.

Answers to Self-Assessment Questions are given in Appendix 3.

Chapter 2

Particles

The Function and Formation of Virus Particles

Why bother to form a virus particle to contain the **genome**? Indeed, some infectious agents, such as viroids, do not (see Chapter 8). However, the fact that viruses struggle with the genetic and biochemical burden entailed in encoding and assembling the components of a particle indicates that there must be positive benefits to this strategy. At the simplest level, the function of the outer shells of a virus particle is to protect the fragile nucleic acid genome from physical, chemical, or enzymatic damage. After leaving the host cell, the virus enters a hostile environment which would quickly inactivate the unprotected genome. Nucleic acids are susceptible to physical damage such as shearing by mechanical forces and to chemical modification by ultraviolet light (from sunlight). The natural environment is heavily laden with nucleases either derived from dead or leaky cells or deliberately secreted by vertebrates as defence against infection. In viruses with single-stranded genomes, the breaking of a single phosphodiester bond or chemical modification of one nucleotide is sufficient to inactivate that virus particle, making replication of the genome impossible. How is protection against this achieved? The protein subunits in a virus **capsid** are multiply redundant, i.e. present in many copies per particle. Damage to one or more subunits may render that particular subunit non-functional, but rarely does limited damage destroy the infectivity of the whole particle. This makes the capsid an effective barrier.

The outer surface of the virus is also responsible for recognition of and the first interaction with the host cell. Initially, this takes the form of binding of a specific **virus-attachment protein** to a cellular **receptor** molecule. However, the capsid also has a role to play in initiating infection by delivering the genome in a form in which it can interact with the host cell. In some cases this is a simple process which consists only of dumping the genome into the cytoplasm of the cell. In other cases this stage is much more complex; for example, retroviruses carry out extensive modifications to the virus genome while it is still inside the particle, converting two molecules of single-stranded RNA to one molecule of double-stranded DNA before delivering it to the cell nucleus. Hence, the role of the capsid is vital in allowing viruses to establish an infection.

To form infectious particles, viruses must overcome two fundamental problems. First, they must assemble the particle utilizing the information available from the components which make up the particle itself. Second, virus particles form regular geometric shapes, even though the proteins from which they are made are irregularly shaped. How do these simple organisms solve these difficulties? The solution to the first problem lies in the rules of symmetry, as does the answer to the second.

Capsid Symmetry and Virus Architecture

It is possible to imagine a virus particle, the outer shell of which (the **capsid**) consists of a single, hollow protein molecule, which, as it folds to assume its mature conformation, traps the virus genome inside. In practice, this arrangement cannot occur, for the following reason. The triplet nature of the genetic code means that three nucleotides (or base pairs in the case of viruses with double-stranded genomes) are necessary to encode one amino acid. Viruses cannot, of course, utilize an alternative, more economical, genetic code because this could not be deciphered by the host cell. Since the approximate molecular weight of a nucleotide triplet is 1000 and the average molecular weight of a single amino acid is 150, a nucleic acid can only encode a protein that is at most 15% of its own weight. Therefore, virus capsids must be made up of multiple protein molecules (subunit construction) and viruses must overcome the problem of how these subunits are arranged.

In 1955, Fraenkel-Conrat and Williams showed that when mixtures of purified tobacco mosaic virus (TMV) RNA and coat protein were incubated together, virus particles formed. The discovery that virus particles could form spontaneously from purified subunits without any extraneous information indicated that the particle was in the free energy minimum state and was therefore the favoured structure of the components. This stability is an important feature of the virus particle. Although some viruses are very fragile and unable to survive outside the protected host cell environment, many are able to persist for long periods, in some cases for years.

The forces which drive the assembly of virus particles include hydrophobic and electrostatic interactions – only rarely are covalent bonds involved in holding together the multiple subunits. In biological terms, this means that protein–protein, protein–nucleic acid, and protein–lipid interactions are used. It would be fair to say that the subtlety of these interactions is not fully understood for the majority of virus structures, but we now have a good understanding of general principles and repeated structural motifs which appear to govern the construction of diverse, unrelated viruses. These are discussed below under the two main classes of virus structures: helical and icosahedral symmetry.

Helical capsids

Tobacco mosaic virus (TMV) is representative of one of the two major structural classes seen in viruses, those with helical symmetry. The simplest way to arrange multiple, identical protein subunits is to use rotational symmetry and to arrange the irregularly shaped proteins around the circumference of a circle to form a disk. Multiple disks can then be stacked on top of one another to form a cylinder, with the virus genome coated by the protein shell or contained in the hollow centre of the cylinder. Denaturation and phase-transition studies of TMV suggest that this is the form the particle takes (see Chapter 1).

Closer examination of the TMV particle by X-ray crystallography reveals that the structure of the capsid actually consists of a **helix** rather than a pile of stacked disks. A helix can be defined mathematically by two parameters: its amplitude (diameter) and pitch (the distance covered by each complete turn of the helix (Figure 2.1). Helices are rather simple structures formed by stacking repeated components with a constant relationship (amplitude and pitch) to one another. Note that if this simple constraint is broken, a spiral forms rather than a helix and this is quite unsuitable for containing a virus genome. In terms of individual protein subunits, helices are described by the number of subunits per

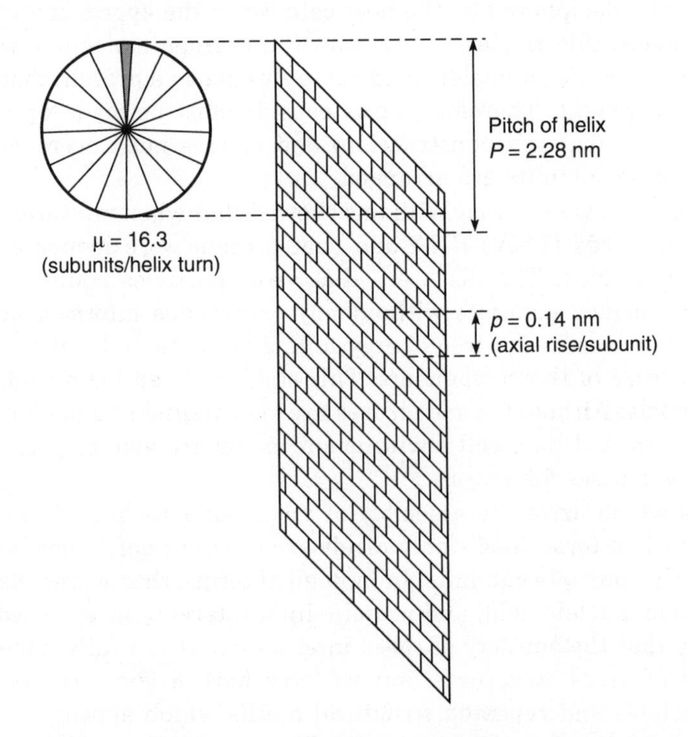

Figure 2.1 Tobacco mosaic virus has a capsid consisting of many molecules of a single coat protein arranged in a constant relationship, forming a helix with a pitch of 2.28 Å.

turn of the helix, μ, and the axial rise per subunit, p. The pitch of the helix, P, is therefore equal to:

$$P = \mu \times p$$

For TMV, $\mu = 16.3$, that is, there are 16.3 coat protein molecules per helix turn, and $p = 0.14$ nm. Therefore, the pitch of the TMV helix is $16.3 \times 0.14 = 2.28$ nm. TMV particles are rigid, rod-like structures, but some helical viruses demonstrate considerable flexibility and longer helical virus particles are often curved or bent. Flexibility is probably an important attribute. Long helical particles are likely to be subject to shear forces and the ability to bend reduces the likelihood of breakage or damage.

That helical symmetry is a useful way of arranging a single protein subunit to form a particle is confirmed by the large number of different types of virus which have evolved with this capsid arrangement. Among the simplest helical capsids are those of the well-known bacteriophages of the family *Inoviridae*, such as M13 and fd (known as Ff phages). These phages are about 900 nm long and 9 nm diameter and the particles contain five proteins (Figure 2.2). The major coat protein is the product of phage gene 8 (g8p) and there are 2700–3000 copies of this protein per particle, together with approximately five copies each of four minor capsid proteins, g3p, g6p, g7p and g9p, located at the ends of the filamentous particle. The primary structure of the major coat protein g8p explains many of the properties of the particle. Mature molecules of g8p consist of approximately 50 amino acid residues (a signal sequence of 23 amino acids is cleaved from the precursor protein during its translocation into the outer membrane of the host bacterium), and is almost entirely α-helical in structure so that the molecule forms a short rod. There are three distinct domains within this rod. There is a negatively charged region at the amino-terminal end which contains acidic amino acid residues which forms the outer, hydrophilic surface of the virus particle, and a basic, positively charged region at the carboxy-terminal end which lines the inside of the protein cylinder adjacent to the negatively charged DNA genome. Between these two, there is a hydrophobic region which is responsible for interactions between the g8p subunits that allow the formation of and stabilize the phage particle (Figure 2.2). Ff phage particles are held together by the hydrophobic interactions between the coat protein subunits and this is demonstrated by the fact that the particles fall apart in the presence of chloroform, even though they do not contain any lipid component. The g8p subunits in successive turns of the helix interlock with the subunits in the turn below and are tilted at an angle of approximately 20° to the long axis of the particle, overlapping one another 'like the scales of a fish'. The value of μ (protein subunits per complete helix turn) is 4.5 and p (axial rise per subunit) = 1.5 nm.

Because the phage DNA is packaged inside the core of the helical particle, the length of the particle is dependent on the length of the genome. In all Ff phage preparations, 'polyphage' (containing more than one genome length of DNA), 'miniphage' (deleted forms containing 0.2–0.5 phage genomes length of DNA), and 'maxiphage' (genetically defective forms but containing more than one phage genome length of DNA) occur. This plastic property of these filamentous particles has been exploited by molecular biologists to develop the M13 genome

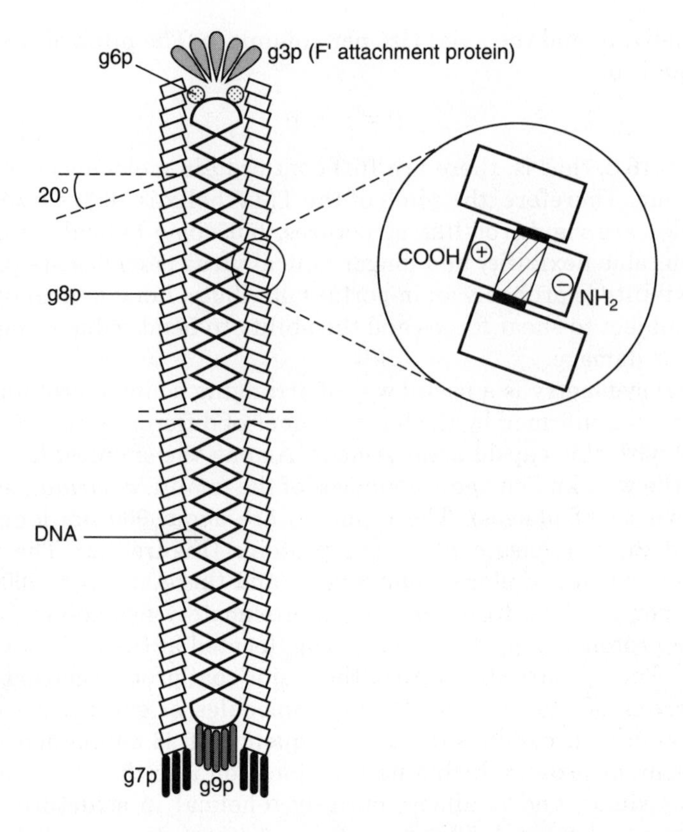

Figure 2.2 Schematic representation of the bacteriophage M13 particle. Major coat protein g8p is arranged helically, the subunits overlapping 'like the scales of a fish'. Other capsid proteins required for the biological activity of the virion are located at either end of the particle. Inset shows the hydrophobic interactions between the g8p monomers (shaded region).

as a cloning vector. Insertion of foreign DNA into the genome results in recombinant phage particles which are longer than the wild-type filaments. Unlike most viruses, there is no sharp cut-off genome length at which the genome can no longer be packaged into the particle. However, as M13 genome size increases, the efficiency of replication declines. While recombinant phage genomes 1–10% longer than the wild-type do not appear to be significantly disadvantaged, those 10–50% longer than the wild-type replicate significantly more slowly. Above a 50% increase over the normal genome length it becomes progressively more difficult to isolate recombinant phage.

The structure of the Ff capsid also explains the events which occur on infection of suitable bacterial host cells. Ff phages are 'male-specific', i.e. they require the F pilus on the surface of *Escherichia coli* for infection. The first event in infection is an interaction between g3p located at one end of the filament together with g6p, and the end of the F pilus. This interaction causes a conformational change in g8p. Initially, its structure changes from 100% α-helix to 85% α-helix, causing the filament to shorten. The end of the particle attached to the F pilus flares open, exposing the phage DNA. Subsequently, a second conformational change in the g8p subunits reduces its α-helical content from 85% to 50%, causing the phage particle to form a hollow spheroid about 40 nm in diameter and expelling the phage DNA, thus initiating the infection of the host cell.

Many plant viruses show helical symmetry. These include members of the *Capillovirus*, *Carlavirus*, *Closterovirus*, *Furovirus*, *Hordeivirus*, *Potexvirus*, *Potyvirus*, *Tobamovirus* and *Tobravirus* groups. These particles vary from approximately 100 nm (*Tobravirus*) to approximately 1000 nm (*Closterovirus*) in length. The best studied example is, as stated above, TMV from the *Tobamovirus* group. Quite why so many groups of plant virus have evolved this structure is not clear, but it may be related either to the biology of the host plant cell, or alternatively, to the way in which they are transmitted between hosts.

Helical, naked (i.e. non-enveloped) animal viruses do not exist. Once again, this probably reflects aspects of host cell biology and virus transmission, but the reasons are not clear. There are a large number of animal viruses based on helical symmetry, but all have the addition of an outer lipid **envelope** (see below). There are too many viruses with this structure to list individually, but this category includes many of the best known human pathogens, such as influenza virus (*Orthomyxoviridae*), mumps and measles viruses (*Paramyxoviridae*), and rabies virus (*Rhabdoviridae*). All possess single-stranded, negative-sense RNA genomes (see Chapter 3). The molecular design of all of these viruses is similar. The viral nucleic acid and a basic, nucleic acid-binding protein condense together in the infected cell to form a helical **nucleocapsid**. This protein-RNA complex serves to protect the fragile virus genome from physical and chemical damage and in some instances also provides other functions associated with virus replication. The envelope and its associated proteins are derived from the membranes of the host cell and are added to the nucleocapsid core of the virus during replication (see Chapter 4).

Some of these helical, enveloped animal viruses are relatively simple in structure, for example, rabies virus and the closely related virus, vesicular stomatitis virus (VSV) (Figure 2.3). These viruses are built up around the negative-sense RNA genome, which in rhabdoviruses is about 11 000 nucleotides (11 kilobases, 11 kb) long. The RNA genome and basic nucleocapsid (N) protein interact to form a helical structure with a pitch of approximately 5 nm, which, together with two non-structural proteins, L and NS (which form the virus polymerase: see Chapter 4), makes up the core of the virus particle. There are 30–35 turns of the nucleoprotein helix in the core, which is about 180 nm long and 80 nm in diameter. The individual N protein monomers are approximately 9 × 5 × 3 nm and each covers about nine nucleotides of the RNA genome. As in the case of the filamentous phage particles described above, the role of the N protein is to stabilize the RNA genome and protect it from chemical, physical and enzymatic damage. In common with most enveloped viruses, the nucleocapsid is surrounded by an amorphous layer which interacts with both the core and the overlying lipid envelope linking them together. This is known as the matrix. The matrix (M) protein is usually the most abundant protein in the virus particle; for example, there are approximately 1800 copies of the M protein and approximately 1250

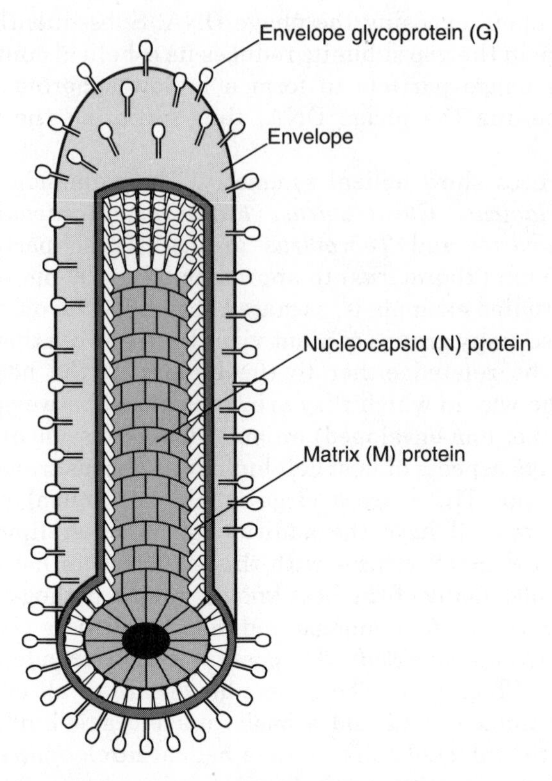

Figure 2.3 Rhabdovirus particles, such as those of vesicular stomatitis virus, have an inner helical nucleocapsid surrounded by an outer lipid envelope and its associated glycoproteins.

copies of the N protein in VSV particles. The lipid envelope and its associated proteins are discussed in more detail below.

It is clear that many different groups of viruses have evolved around helical symmetry. Simple viruses with small genomes use this architecture to provide protection for the genome without the need to encode multiple capsid proteins. More complex virus particles utilize this structure as the basis of the virus particle, but elaborate on it with additional layers of protein and lipid.

Icosahedral (isometric) capsids

An alternative way of building a virus capsid is to arrange protein subunits in the form of a hollow quasispherical structure, enclosing the genome within. The criteria for arranging subunits on the surface of a solid are a little more complex than those for building a helix. In theory, a number of solid shapes can be constructed from repeated subunits, e.g. a tetrahedron (four triangular faces), cube (six square faces), octahedron (eight triangular faces), dodecahedron (12 pentagonal faces) and an **icosahedron**, a solid shape consisting of 20 triangular faces arranged around the surface of a sphere (Figure 2.4).

Early in the 1960s, direct examination of a number of small 'spherical' viruses by electron microscopy revealed that they appeared to have icosahedral symmetry. At first sight, it is not obvious why this pattern should have been 'chosen' by diverse virus groups. However, although in theory it is possible to construct virus capsids based on simpler symmetrical arrangements, such as tetrahedra or cubes, there are practical reasons why this does not occur. As described above, it is more economic in terms of genetic capacity to design a capsid based on a large number of identical, repeated protein subunits rather than fewer, larger subunits. It is unlikely that a simple tetrahedron consisting of four identical protein molecules would be large enough to contain even the smallest virus genome. If it were, it is probable that the gaps between the subunits would be so large that the particle would be leaky, and fail to carry out its primary function of protecting the virus genome.

In order to construct a capsid from repeated subunits, a virus must 'know the rules' which dictate how these are arranged. For an icosahedron the rules are based on the rotational symmetry of the solid, known as 2–3–5 symmetry, which has the following features (Figure 2.4):

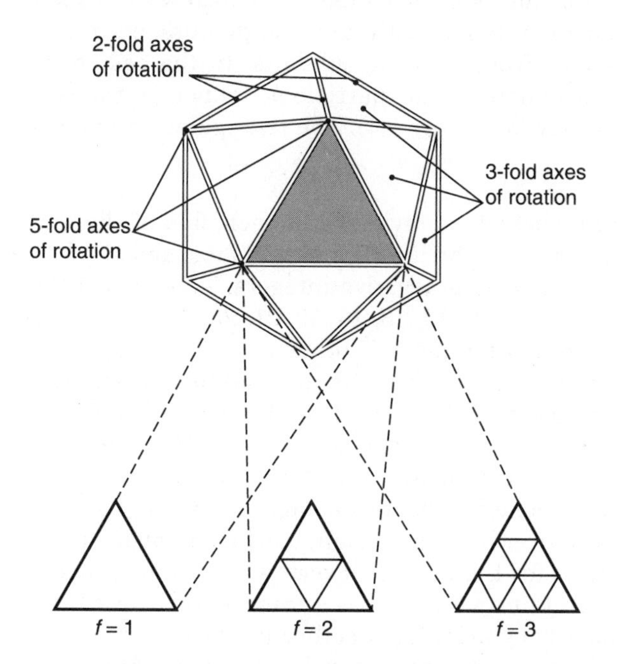

Figure 2.4 Illustration of the 2–3–5 symmetry of an icosahedron. More complex ('higher order') icosahedra can be defined by the triangulation number of the structure, $T = f^2 \times P$. Regular icosahedra have faces consisting of equilateral triangles and are formed when the value of P = 1 or 3. All other values of P give rise to more complex structures with either a left-hand or right-hand skew.

- An axis of twofold rotational symmetry through the centre of each edge
- An axis of threefold rotational symmetry through the centre of each face
- An axis of fivefold rotational symmetry through the centre of each corner (vertex).

Since protein molecules are irregularly shaped and are not regular equilateral triangles, the simplest icosahedral capsids are built up by using three identical subunits to form each triangular face. This means that 60 identical subunits are required to form a complete capsid. A few simple virus particles are constructed in this way; for example, bacteriophages of the family *Microviridae*, such as ϕX174, the complete structure of which has recently been determined by X-ray crystallography.

In most cases, analysis reveals that icosahedral virus capsids contain more than 60 subunits, for the reasons of genetic economy given above. This presents a difficulty. A regular icosahedron composed of 60 identical subunits is a very stable structure because all the subunits are equivalently bonded, i.e. they show the same spacing relative to one another and each occupies the minimum free energy state. With more than 60 subunits it is impossible for them all to be arranged completely symmetrically with exactly equivalent bonds to all their neighbours, since a true regular icosahedron consists of only 20 subunits. To solve this problem, in 1962 Caspar and Klug proposed the idea of **quasi-equivalence**. Their simple idea was that subunits in *nearly* the same local environment form *nearly* equivalent bonds with their neighbours, permitting self-assembly of icosahedral capsids from multiple subunits. In the case of these higher order icosahedra, the symmetry of the particle is defined by the **triangulation number** of the icosahedron (Figure 2.4). The triangulation number, T, is defined by:

$$T = f^2 \times P$$

where f is the number of subdivisions of each side of the triangular face; f^2 is therefore the number of subtriangles on each face; and $P = h^2 + hk + k^2$, where h and k are any distinct, non-negative integers. This means that values of T fall into the series 1, 3, 4, 7, 9, 12, 13, 16, 19, 21, 25, 27, 28, etc. When $P = 1$ or 3, a regular icosahedron is formed. All other values of P give rise to icosahedra of the 'skew' class, where the subtriangles making up the icosahedron are not symmetrically arranged with respect to the edge of each face. Detailed structures of icosahedral virus particles with $T = 1$ (*Microviridae*, e.g. ϕX174), $T = 3$ (many insect, plant, and animal RNA viruses: see below), $T = 4$ (togaviruses) and $T = 7$ (the heads of the tailed bacteriophages such as λ) have all been determined. Virus particles with still larger triangulation numbers use different kinds of subunit assemblies for the faces and vertices of the icosahedron and have internal scaffolding proteins which act as a framework. These direct the assembly of the capsid, typically by bringing together pre-formed subassemblies of proteins. Variations on the theme of icosahedral symmetry occur over and over again in virus particles. For example, Geminivirus particles consist of a fused pair of $T = 1$ icosahedra, hence their name from the twins of Greek mythology, Castor and Pollux. Elements of icosahedral symmetry occur frequently as part of larger assemblies of proteins (see Complex Structures).

The capsids of picornaviruses (*Picornaviridae*) provide a good illustration of the construction of icosahedral virus particles. In recent years, atomic structures of the capsids of a number of different picornaviruses have been determined. These include poliovirus types 1 and 3 (PV1 and PV3), foot-and-mouth disease virus (FMDV) and human rhinovirus type 14 (HRV-14), and a number of others. In fact, the structure of these virus particles is remarkably similar to those of many other unrelated viruses, such as insect viruses of the family *Nodaviridae* and plant viruses from the *Comovirus* group. All these virus groups have icosahedral capsids approximately 30 nm diameter with triangulation number $T = 3$ (Figure 2.5). The capsid is composed of 60 repeated subassemblies of proteins, each containing three major subunits, VP1, VP2, and VP3. This means that there are $60 \times 3 = 180$ surface monomers in the whole picornavirus particle. All three proteins are based on a similar structure, consisting of 150–200 amino acid residues in what has been described as an '8-strand anti-parallel β-barrel' (Figure 2.6). This subunit structure has been found in all $T = 3$ icosahedral RNA virus capsids which have been examined so far (e.g. picornaviruses, comoviruses, nepoviruses), possibly reflecting distant evolutionary relationships between distinct virus families.

Knowledge of the structure of these $T = 3$ capsids also reveals information about the way in which they are assembled and the function of the mature capsid. Picornavirus capsids contain four structural proteins. In addition to the three major proteins VP1–3 (above), there is a small fourth protein, VP4. VP4 is located predominantly on the inside of the capsid and is not exposed at

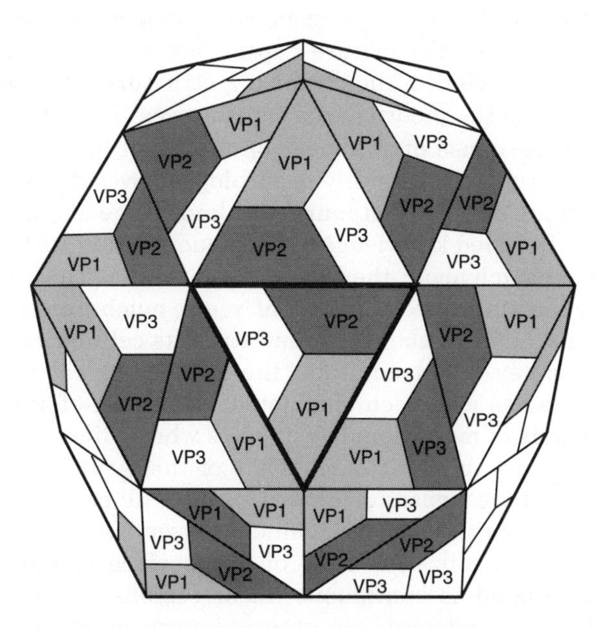

Figure 2.5 Picornavirus particles are icosahedral structures with a triangulation number $T = 3$. Three virus proteins, VP1, 2 and 3, comprise the surface of the particle. A fourth protein, VP4, is not exposed on the surface of the virion but is present in each of the 60 repeated units which make up the capsid.

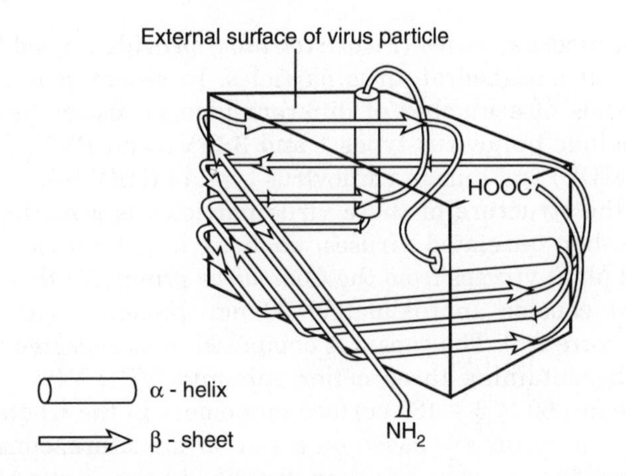

Figure 2.6 The '8-strand anti-parallel β-barrel' subunit structure found in all $T = 3$ icosahedral RNA virus capsids.

the surface of the particle. The way in which the four capsid proteins are processed from the initial **polyprotein** (see Chapter 5) has long been known from biochemical studies of picornavirus-infected cells (Figure 2.7). VP4 is formed from cleavage of the VP0 precursor into VP2+VP4 late in assembly and is myristylated at its amino terminus, i.e. it is modified after translation by the covalent attachment of myristic acid, a 14-carbon unsaturated fatty acid. Five VP4 monomers form a hydrophobic micelle, driving the assembly of a pentameric subassembly. There is biochemical evidence that these pentamers, which form the vertices of the mature capsid, are a major precursor in the assembly of the particle. Hence, the chemistry, structure, and symmetry of the proteins which make up the picornavirus capsid reveal how assembly is driven.

Because they are the cause of a number of important human diseases, picornaviruses have been studied intensively by virologists over the last decade or so. This interest has resulted in an outpouring of knowledge about these structurally simple viruses. Detailed knowledge of the structure and surface geometry of HRV-14 has revealed much about the interaction of rhinoviruses with their host cells and with the immune system. In recent years, much has been learned not only about this virus, but also about the identity of its cellular receptor, ICAM-1 (see below and Chapter 4). In addition, the immunological structure of the picornavirus particle has also been elucidated. A number of studies have been published using panels of monoclonal antibodies whose binding sites have been 'mapped' to the primary amino acid sequence of the virus by examining their reactivity towards mutant viruses or by using synthetic peptides to block binding. The information from these experiments has been used to identify a number of discrete antibody-neutralization sites on the surface of the virus particle. Some of these correspond to contiguous linear regions of the primary amino acid sequence of the capsid proteins, others, known as conformational sites, result from the bringing together of separated stretches of amino acids in the mature virus. With the elucidation of detailed picornavirus capsid structures, these regions have now been physically identified on the surface of the particle.

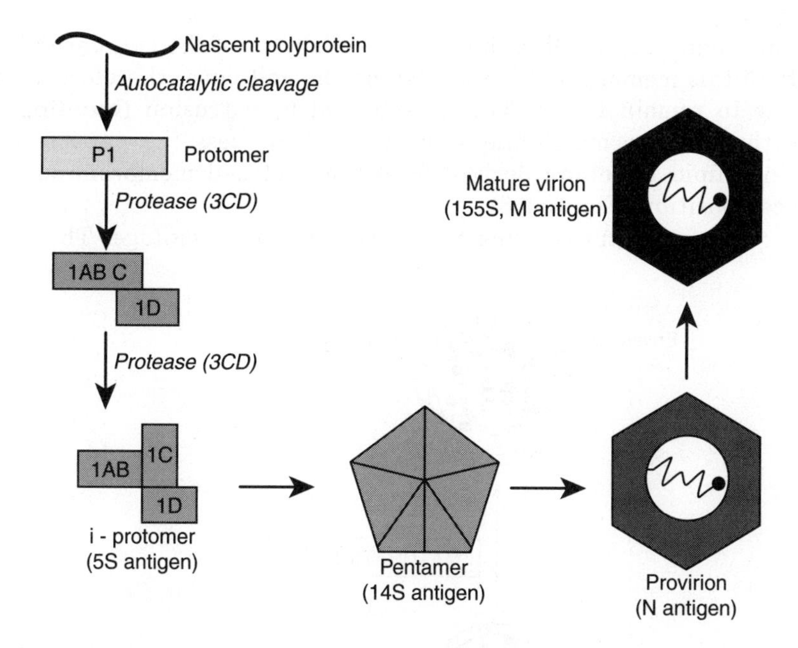

Figure 2.7 Proteolytic processing of picornavirus capsid proteins.

They correspond mostly to hydrophilic, exposed loops of amino acid sequence, readily accessible to antibody binding, which are repeated on each of the pentameric subassemblies of the capsid. Now that the physical constraints on these sites are known, this type of information is being used to artificially manipulate them, even to build 'antigenic chimeras' with the structural properties of one virus but expressing crucial antigenic sites from another. With the application of the rational, computer-aided design tools now available, it is probable that the efficiency with which this type of chimera can be designed and built should increase. Indeed, these compound viruses may prove to be the vaccines of the future.

Enveloped Viruses

So far, this chapter has concentrated on the structure of 'naked' virus particles, i.e. those in which the capsid proteins are exposed to the external environment. Such viruses are produced from infected cells at the end of the replicative cycle when the cell dies, breaks down and lyses, releasing the virions which have been built up internally. This simple strategy has drawbacks. In some circumstances it is wasteful, resulting in the premature death of the cell. It also reduces the possibilities for persistent or latent infections. Therefore, many viruses have devised strategies to effect an exit from the infected cell without its total destruction. The difficulty this presents is that all living cells are covered by a membrane composed of a lipid bilayer. The viability of the cell depends on the integrity of this membrane. Viruses leaving the cell must, therefore, allow this membrane to remain intact. This is achieved by extrusion (**budding**) of the particle through the membrane, during which process the particle becomes coated in a lipid **envelope** derived from the host cell membrane and with a similar composition (Figure 2.8).

Viruses have also turned this necessity into an advantage. The structure

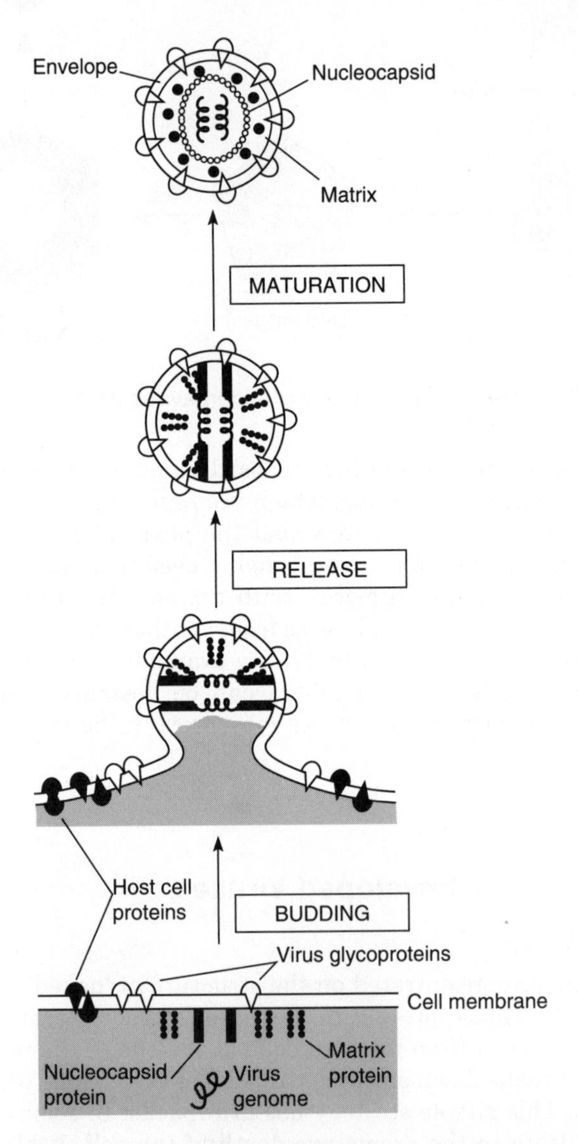

Figure 2.8 Enveloped virus particles are formed by budding through a host cell membrane, during which the particle becomes coated with a lipid bilayer derived from the cell membrane. For some viruses, assembly of the structure of the particle and budding occur simultaneously, whereas in others a preformed core pushes out through the membrane.

underlying the envelope may be based on helical or icosahedral symmetry and may be formed before or as the virus leaves the cell. In the majority of cases, enveloped viruses use cellular membranes as sites allowing them to direct assembly. The formation of the particle inside the cell, maturation and release are in many cases a continuous process. The site of assembly varies for different viruses. Not all use the cell surface membrane; many use cytoplasmic membranes such as the Golgi apparatus, others, such as herpesviruses, which replicate in the nucleus may utilize the nuclear membrane. In these cases, the virus is usually extruded into some form of vacuole, in which it is transported to the cell surface and subsequently released. These points are discussed in more detail in Chapter 4.

If the virus particle became covered in a smooth, unbroken lipid bilayer, this would be its undoing. Such a coating is effectively inert, and although effective as a protective layer preventing desiccation of or enzymatic damage to the particle, it would not permit recognition of receptor molecules on the host cell. Therefore, viruses modify their lipid envelopes by the synthesis of several classes of proteins which are associated in one of three ways with the envelope (Figure 2.9). These can be summarized as follows:

- *Matrix proteins*: These are internal virion proteins whose function is effectively to link the internal nucleocapsid assembly to the envelope. Such proteins are not usually glycosylated and are often very abundant, for example, in retroviruses they comprise approximately 30% of the total weight of the virion. Some matrix proteins contain transmembrane anchor domains, others are associated with the membrane by hydrophobic patches on their surface or by protein–protein interactions with envelope glycoproteins
- *Glycoproteins*: These are transmembrane proteins anchored to the membrane by a hydrophobic domain, and can be subdivided into two types by their function:
  (a) *External glycoproteins* are anchored in the envelope by a single transmembrane domain. Most of the structure of the protein is on the outside of the membrane, with a relatively short internal tail. Often, individual monomers associate to form the 'spikes' visible in electron micrographs on the surface of many enveloped viruses. Such proteins are the major antigens of enveloped viruses. The glycosylation is either *N*- or *O*-linked, and many of these proteins are heavily glycosylated; up to 75% of the protein by weight may consist of sugar groups added post-translationally. These proteins are usually the major antigens of enveloped viruses and provide contact with the external environment, frequently serving a number of important functions; for example, influenza virus haemagglutinin is required for receptor binding, membrane fusion, and haemagglutination
  (b) *Transport channel proteins* contain multiple hydrophobic transmembrane domains, forming a protein-lined channel through the envelope. This enables the virus to alter the permeability of the membrane (e.g. ion channels). Such proteins are often important in modifying the internal environment of the virion, permitting or even driving biochemical

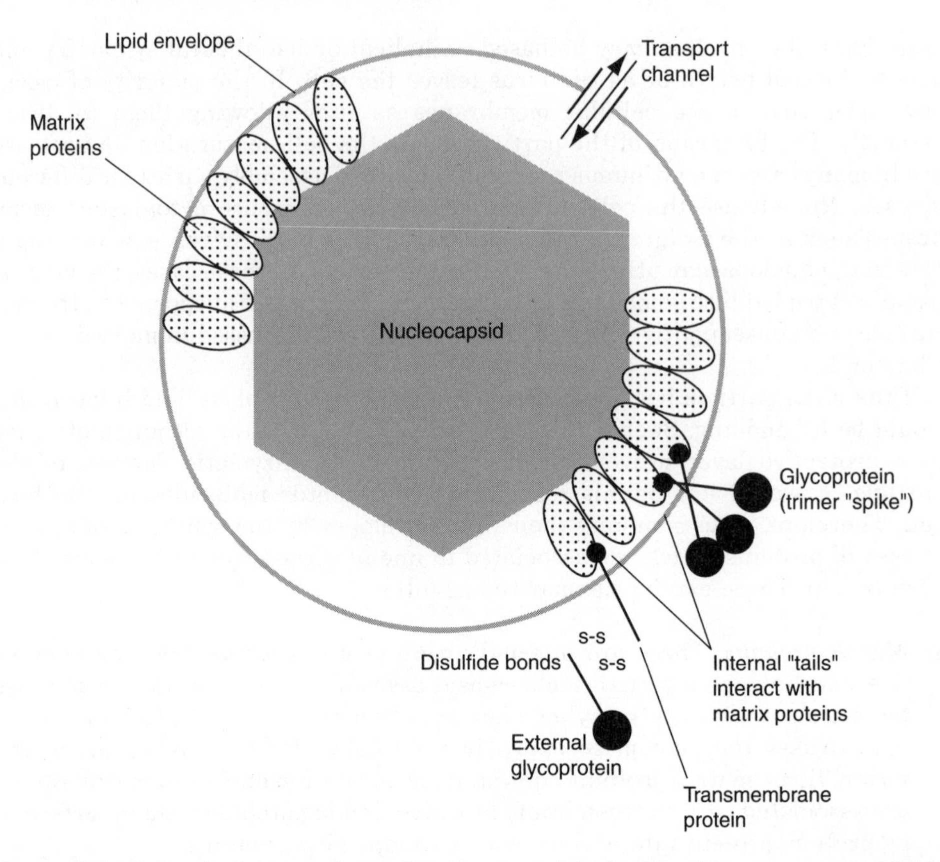

Figure 2.9 Several classes of proteins are associated with virus envelopes. Matrix proteins link the envelope to the core of the particle. Virus-encoded glycoproteins inserted into the envelope serve several functions. External glycoproteins are responsible for receptor recognition and binding, while *trans*-membrane proteins act as transport channels across the envelope. Host-cell derived proteins are also sometimes found to be associated with the envelope, usually in small amounts.

changes necessary for maturation of the particle and the development of infectivity (e.g. influenza virus M2 protein).

While there are many enveloped vertebrate viruses, there are only a few enveloped plant viruses. Most of these belong to the *Rhabdoviridae* family, whose structure has already been discussed (see Helical Symmetry). Except for plant rhabdoviruses, only a few bunyaviruses which infect plants and the members of the *Tospovirus* group have outer lipid envelopes. This relative paucity of enveloped plant viruses probably reflects aspects of host cell biology, in particular the mechanism of release of the virus from the infected cell, which requires a breach in the rigid cell wall. This constraint does not apply to viruses of prokaryotic organisms, where there are a number of enveloped virus families (*Plasmaviridae*, the SSV-1 group, *Lipothrixviridae*, and *Cystoviridae*).

Complex Structures

The majority of viruses can be fitted into one of the three structural classes outlined above, i.e. those with helical symmetry, icosahedral symmetry, or enveloped viruses. However, there are many viruses whose structure is more complex. In these cases, although the general principles of symmetry already described are often used to build part of the virus shell (this term being appropriate here since such viruses often consist of several layers of protein and lipid), the larger and more complex viruses cannot be simply defined by a mathematical equation as can a simple helix or icosahedron. Because of the complexity of some of these viruses, they have defied attempts to determine detailed atomic structures using the techniques described in Chapter 1.

An example of such a group is the *Poxviridae.* These viruses have oval or 'brick-shaped' particles 200–400 nm long. In fact, these particles are so large that they were first observed in vaccine lymph using high-resolution optical microscopes in 1886, and thought at that time to be 'the spores of micrococci'. The external surface of the virion is ridged in parallel rows, sometimes arranged helically. The particles are extremely complex and have been shown to contain more than 100 different proteins (Figure 2.10). During replication, two forms of particle are observed: extracellular forms which contain two membranes and intracellular particles which only have an inner membrane. Thin sections under the electron microscope reveal that the outer surface of the virion is composed of lipid and protein. This surrounds the core, which is biconcave (dumbbell-shaped), and two 'lateral bodies' whose function is unknown. The core is composed of a tightly compressed nucleoprotein and the double-stranded DNA genome is wound around it. Antigenically, poxviruses are very complex, inducing both specific and cross-reacting antibodies – hence the possibility of vaccinating

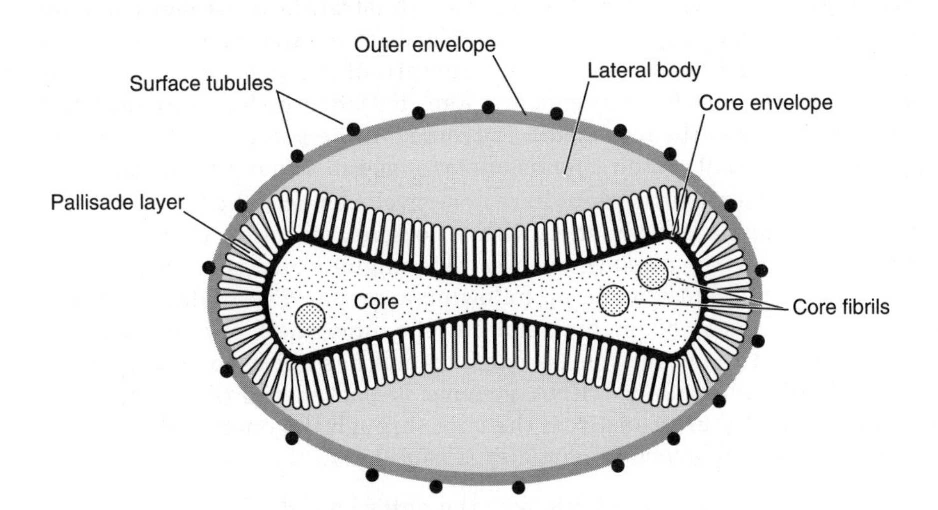

Figure 2.10 Poxvirus particles are the most complex virions known and contain more than 100 virus-encoded proteins, arranged in a variety of internal and external structures.

against one disease with another virus (e.g. the use of vaccinia virus to immunize against smallpox (variola) virus). Poxviruses and a number of other complex viruses also emphasize the true complexity of some viruses – there are at least 10 enzymes present in poxvirus particles, mostly involved in nucleic acid metabolism/genome replication.

Poxviruses form the most complex particles known and even though the complete nucleotide sequence of the genome of several representatives of this family has now been determined (see Chapter 3), a complete elucidation of the structure of these particles has not yet been achieved. This is an extreme case in virology, included here as a counterbalance to the description of some of the simplest viruses given above. In other cases, the particle structure of complex viruses has been much more completely investigated. One of the prime examples of such a group are the tailed phages of *E. coli*. These viruses, members of the families *Myoviridae*, *Siphoviridae* and *Podoviridae*, have been extensively studied for excellent reasons – they are easy to propagate in bacterial cells, they can be obtained in high titres, and they are easily purified, facilitating biochemical and structural studies. The head of the particles consists essentially of an icosahedral shell with $T = 7$ symmetry, attached by a collar to a contractile, helical tail. At the end of the tail is a plate which functions in attachment to the bacterial host and also in penetration of the bacterial cell wall by virtue of lysozyme-like enzymes associated with the plate. In addition to these structures, thin protein fibres are attached to the plate, which, along with the tail plate, are involved in binding to the receptor molecules in the wall of the host cell. The structure of these phages is actually rather more complex than this simple picture; for example, there are several internal proteins and polyamines associated with the genomic DNA in the head and an internal tube structure inside the outer sheath of the helical tail. In the infected bacterial cell, there are separate assembly pathways for the head and tail sections of the particle, which come together at a late stage to make up the virion (Figure 2.11). Therefore, these viruses illustrate how complex particles can be built up from the simple principles outlined above. An even clearer example of this phenomenon is provided by the structure of geminivirus particles, which consist of two twinned $T = 1$ icosahedra. Each icosahedron has one morphological subunit missing and the icosahedra are joined at the point such that the mature particle contains 110 protein monomers arranged in 22 morphological subunits.

Members of the reovirus genus have non-enveloped, icosahedral $T = 13$ capsids composed of double protein shell with a complex structure (Figure 2.12). The structure of orbivirus capsids, such as bluetongue virus, is slightly smaller in diameter but apparently made up following a similar $T = 13$ design. The outer shell is approximately 80 nm diameter and the inner shell, or core, about 60 nm. The structure of the outer shell is complex and has not been completely determined. The double-stranded RNA genome is packed tightly inside the core. Notably, 12 'spikes' protrude from the core through the outer shell. The overall composition of *Orthoreovirus* particles is as follows:

σ1, σ3, μ1c – in the outer capsid
λ1, λ2, λ3, σ2, μ2 – in the virus core
μNS, σNS, σ1s – non-structural proteins

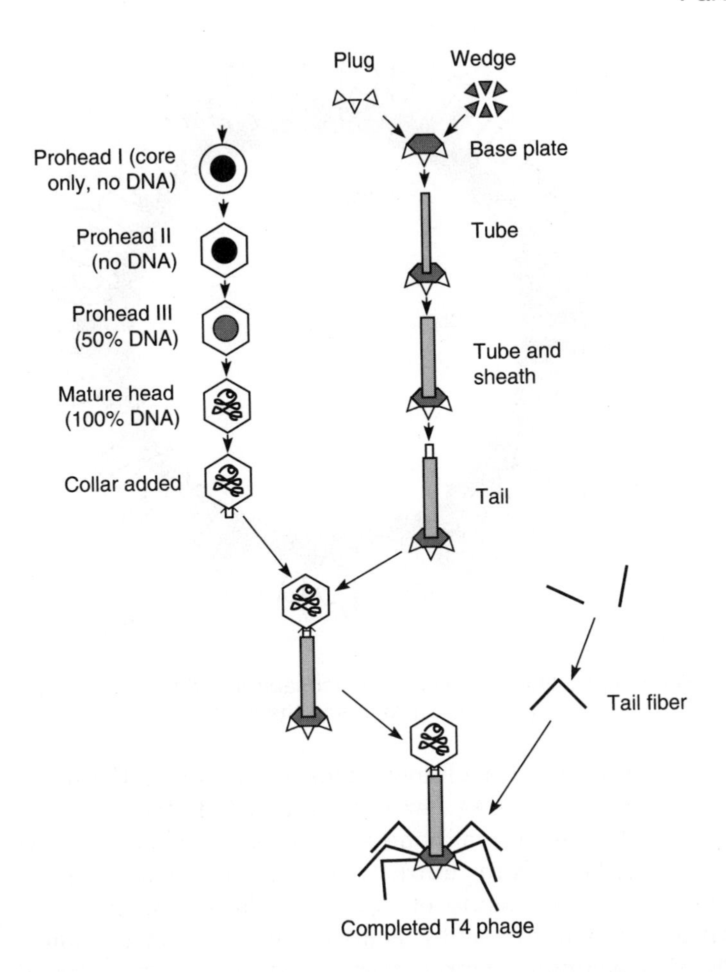

Figure 2.11 Simplified version of the assembly of bacteriophage T4 particles. The head and tail sections are assembled separately and are brought together at a relatively late stage. This complex process was painstakingly worked out by the isolation of phage mutants in each of the virus genes involved. In addition to the major structural proteins, a number of minor 'scaffolding' proteins are involved in guiding the formation of the complex particle.

After many years of study and the isolation of a very large number of mutants, there is now probably more detail known about the structure–function relationships of the reovirus capsid than that of any other virus, with the possible exception of the picornaviruses. These particles are very stable in the environment – this is notably the case with the rotaviruses, which are commonly spread through contaminated drinking water. Rotavirus particles in faecal material stored at room temperature are able to retain their infectivity for 7 months and are not very susceptible to chlorine-containing disinfectants, thus demonstrating the efficiency with which these complex particles protect the fragile RNA genome.

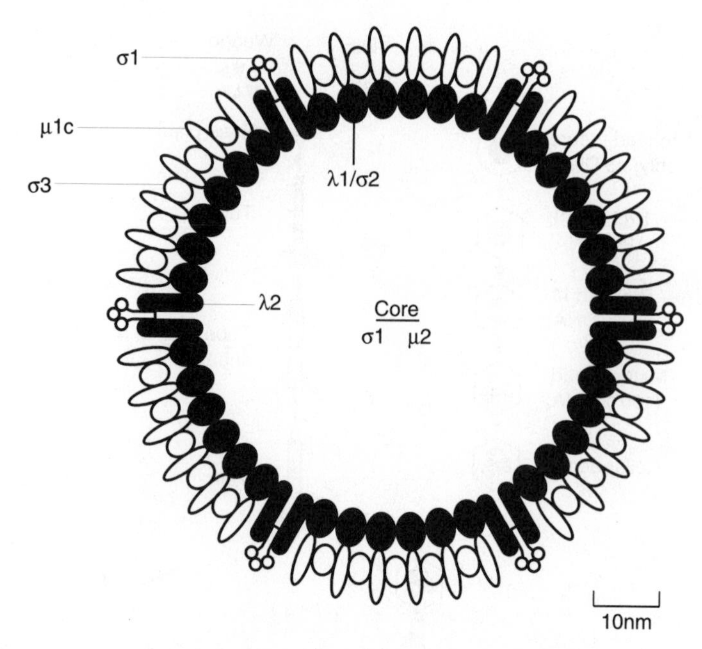

Figure 2.12 Reovirus particles consist of an icosahedral, double-shell arrangement of proteins surrounding the core.

The complex structure of cauliflower mosaic virus (CaMV) has been determined to 3 nm resolution using newer techniques of structural investigation. Although this level of resolution is one order of magnitude less refined than that which can be achieved by X-ray diffraction, this achievement represents a considerable advance. The structure of CaMV has been a long-standing mystery since the virus is very fragile and it is distorted by drying, sectioning, or negative-staining. By examining electron-density maps of frozen–hydrated virus preparations and using sophisticated image reconstruction techniques, the CaMV virion was found to be 54 nm diameter, and to consist of three concentric protein layers with an inner solvent-filled cavity of 27 nm diameter. The outer shell comprises 72 capsomers (12 pentamers and 60 hexamers) arranged with $T = 7$ symmetry, making a total of 420 individual subunits. Interestingly, the 8 kbp double-stranded DNA genome (see Chapter 3) is distributed in the inner layers II and III rather than in the internal cavity.

The final example of complex virus structure to be considered is the baculovirus family (Figure 2.13). In recent years, these viruses have attracted much interest for a number of reasons. They are natural pathogens of arthropods and naturally occurring as well as genetically manipulated baculoviruses are under active investigation as biological control agents for insect pests. In addition, occluded baculoviruses (see below) are increasingly being used as expression vectors to produce large amounts of recombinant proteins. These complex viruses contain 12–30 structural proteins, and consist of a rod-like (hence '*baculo*') nucleocapsid 30–60 nm diameter and 250–300 nm long which contains the 88–160 kbp double-stranded DNA genome. The nucleocapsid is surrounded

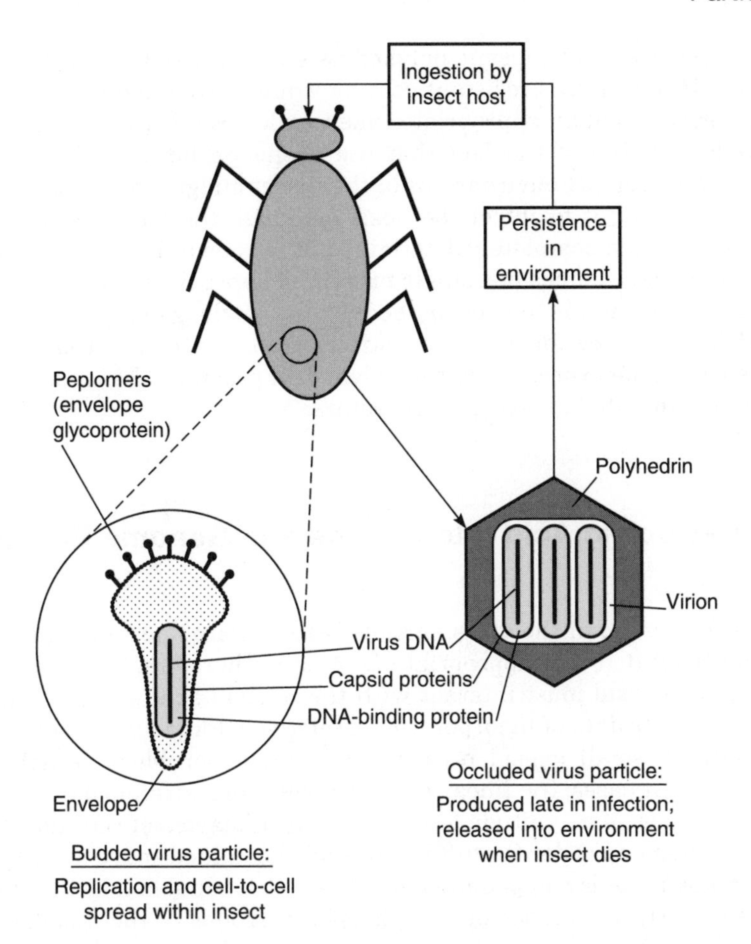

Figure 2.13 Some baculovirus particles exist in two forms, a relatively simple 'budded' form found within the host insect and a crystalline, protein occluded form responsible for environmental persistence.

by an envelope, outside which there may or may not be a crystalline protein matrix. If this outer protein shell is present, the whole assemblage is referred to as an 'occlusion body' and the virus is said to be occluded (Figure 2.13). There are two genera of occluded baculoviruses, those with polyhedral occlusions 1000–15 000 nm diameter which may contain multiple nucleocapsids within the envelope (e.g. nuclear polyhedrosis virus, NPV), and those with ellipsoidal occlusions 200–500 nm diameter (e.g. granulosis viruses, GV). The function of these enormous occlusion bodies is to confer resistance to adverse environmental conditions, which enables the virus to persist in soil or on plant materials for extended periods of time waiting to be ingested by a new host.

The effectiveness of this strategy is such that these viruses can be regarded as being literally armour-plated. Interestingly, the strategy of producing occluded particles appears to have evolved independently in at least three groups of insect viruses. In addition to the baculoviruses, occluded particles are also produced by insect reoviruses (cytoplasmic polyhedrosis viruses) and poxviruses (entomopoxviruses). However, this resistant coating would be the undoing of the virus if it were not removed at an appropriate time to allow replication to proceed. This is achieved by virtue of the fact that the occlusion body is alkali-labile and dissolves in the high pH environment of the insect midgut, releasing the nucleocapsid and allowing it to infect the host. Although the structure of the whole particle has not been completely determined, it is known that the occlusion body is composed of many copies of a single protein of approximately 245 amino acids, polyhedrin. To form the occlusion body, this single gene product is hyper-expressed late in infection by a very strong transcriptional promoter. Cloned foreign genes can be expressed by the polyhedrin promoter, hence baculoviruses have been manipulated as expression vectors.

Protein–Nucleic Acid Interactions and Genome Packaging

The primary function of the virus particle is to contain and protect the genome before delivering it to the appropriate host cell. Therefore, it is clear that the proteins of the capsid must interact with the nucleic acid genome. Once again the physical constraints of incorporating a relatively large nucleic acid molecule into a relatively small capsid present considerable problems which must be overcome. In most cases, the linear virus genome when stretched out in solution is at least an order of magnitude longer than the diameter of the capsid. Merely folding the genome in order to stuff it into such a confined space is quite a feat of topology in itself, but is compounded by the fact that repulsion by the cumulative negative electrostatic charges on the phosphate groups of the nucleotide backbone mean that the genome resists being crammed into a small space. Viruses overcome this difficulty by packaging along with the genome a number of positively charged molecules to counteract this negative charge repulsion. These include small, positively charged ions (Na, Mg, K, etc.), polyamines and various nucleic acid-binding proteins. Some of these latter proteins are virus-encoded and contain amino acids with basic side-chains such as arginine and lysine which interact with the genome. There are many examples of such proteins, for example, retrovirus NC and rhabdovirus N (nucleocapsid) proteins, and influenza virus NP protein (nucleoprotein). Many viruses with double-stranded DNA genomes have basic histone-like molecules closely associated with the DNA. Again, some of these are virus-encoded, e.g. adenovirus polypeptide VII. In other cases, however, the virus may utilize cellular proteins; for example, the polyomavirus genome assumes a chromatin-like structure in association with four cellular histone proteins, H2A, H2B, H3 and H4, similar to that of the host cell genome.

The second problem the virus must overcome is how to achieve the specificity required to select and encapsidate the virus genome from the large background of cellular nucleic acids. In most cases, by the late stages of virus infection when assembly of virus particles occurs (see Chapter 4), transcription of cellular genes has been reduced and a large pool of virus genomes have accumulated. The overproduction of viral nucleic acids eases but does not eliminate the problem of specific genome packaging. Therefore, a specific virus-encoded capsid or nucleocapsid protein is required to achieve this end and many viruses, even those with relatively short, compact genomes such as retroviruses and rhabdoviruses, encode this type of protein.

Viruses with segmented genomes (see next chapter) face further problems: they must not only encapsidate only viral nucleic acid and exclude host cell molecules, but they must also attempt to package one of each of the required genome segments. It is important to realize that during assembly, viruses frequently make mistakes. These can be physically measured by particle:infectivity ratios, i.e. the ratio of the total number of particles in a virus preparation (counted by electron microscopy) to the number of particles able to give rise to infectious progeny (measured by plaque or limiting dilution assays). This value is in some cases found to be several thousand particles to each infectious virion and only rarely approaches a ratio of 1:1. However, a few simple calculations show that viruses such as influenza have far lower particle:infectivity ratios than could be achieved by random packaging of eight distinct genome segments. It is now believed that each influenza virus particle contains more than eight RNA segments. This redundancy may be sufficient to ensure that a reasonable proportion of virus particles in the population will contain at least one of each of the eight segments and will thus be infectious. However, it is not certain that this is the case and it is possible that influenza virus has a mechanism not yet discovered, e.g. incorporation of ribonucleoprotein complexes during morphogenesis, that ensures a complete genetic complement in the majority of particles.

On the other side of the packaging equation are the specific nucleotide sequences in the genome (the **packaging signal**) which permit the virus to select genomic nucleic acids from the cellular background. The packaging signal from a number of virus genomes has been identified. Examples are the 'Ψ' ('psi') signal in murine retrovirus genomes which has been used to package synthetic 'retrovirus vector' genomes into a virus particle, and the sequences responsible for packaging the genomes of several DNA virus genomes (some adenoviruses and herpesviruses) which have been clearly and unambiguously defined. However, it is clear from a number of different approaches that accurate and efficient genome packaging requires information not only from the linear nucleotide sequence of the genome, but also from regions of secondary structure formed by the folding of the genomic nucleic acid into complex forms. In many cases, attempts to find a unique, linear packaging signal in virus genomes have failed. The probable reason for this is that the key to the specificity of genome packaging in most viruses lies in the secondary structure of the genome.

Like many other aspects of virus assembly, the way in which packaging is controlled is, in many cases, not well understood. However, the key must lie in the specific molecular interactions between the genome and the capsid. Until recently, the physical structure of virus genomes within virus particles has been poorly studied, although the genetics of packaging have been extensively investigated in recent years. This is a pity, because it is unlikely that we will be able fully to appreciate this important aspect of virus replication without this information, but is understandable because the techniques used to determine the structure of virus capsids (e.g. X-ray diffraction) only rarely reveal any information about the state of the genome within its protein shell. However, there are some cases where detailed knowledge about the mechanism and specificity of genome encapsidation is now available. These include both viruses with helical symmetry and some with icosahedral symmetry.

Undoubtedly the best understood packaging mechanism is that of a (+) sense RNA helical plant virus, TMV. This is due to the relative simplicity of the virus, which only has a single major coat protein, and will spontaneously assemble from its purified RNA and protein components *in vitro*. In the case of TMV, particle assembly is initiated by association of preformed aggregates of coat protein molecules ('discs') with residues 5444–5518 in the 6.4 kb RNA genome, known as the origin of assembly sequence (OAS) (Figure 2.14). The flat discs have 17 subunits per ring, close to the 16.34 subunits per turn found in the mature virus particle. In fact, the discs are not completely symmetrical, they have a pronounced polarity. Assembly begins when a disc interacts with the OAS in genomic RNA. This converts the discs to a helical 'locked washer' structure, each of which contains 39 coat protein subunits. Further discs add to this structure, switching to the 'locked washer' conformation. RNA is drawn into the assembling structure in what is known as a 'travelling loop', which gives the common name to this mechanism of particle formation. The vRNA is trapped and subsequently buried in the middle of the disc as the helix grows. Extension of the helical structure occurs in both directions but at unequal rates. Growth in the 5′ direction is rapid because a disc can add straight to the protein filament and the travelling loop of RNA is drawn up through it. Growth in the 3′ direction

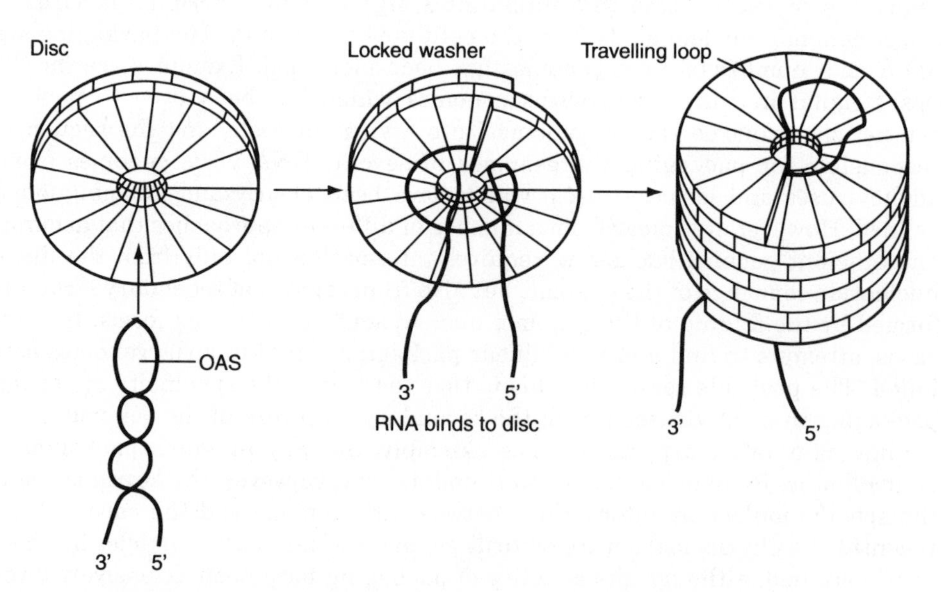

Figure 2.14 Assembly of Tobacco Mosaic Virus (TMV) particles.

is slower because the RNA has to be threaded through the disc before it can add to the structure.

Bacteriophage M13 is another helical virus where protein–nucleic acid interactions in the virus particle are relatively simple to understand (Figure 2.2). The primary sequence of the g8p molecule determines the orientation of the protein in the capsid. In simple terms, the inner surface of the rod-like phage capsid is positively charged and interacts with the negatively charged genome, while the outer surface of the cylindrical capsid is negatively charged. However, the way in which the capsid protein and genome are brought together is a little more complex than this. During replication, the genomic DNA is associated with a non-structural DNA-binding protein, g5p. This is the most abundant of all virus proteins in the infected *E. coli* cell, and coats the newly replicated single-stranded phage DNA, forming an intracellular rod-like structure similar to the mature phage particle, but somewhat longer and thicker (1100 × 16 nm). The function of this protein is to protect the genome from host cell nucleases and to interrupt genome replication, sequestering newly formed strands as substrates for encapsidation. Newly synthesized coat protein monomers (g8p) are associated with the inner (cytoplasmic) membrane of the cell and it is at this site that assembly of the virus particle occurs. The g5p coating is stripped off as the particle passes out through the membrane and is essentially exchanged for the mature g8p coat (plus the accessory proteins). The forces which drive this process are not fully understood, but the protein–nucleic acid interactions which occur appear to be rather simple and involve opposing electrostatic charges and the stacking of the DNA bases between the planar side-chains of the proteins. This is confirmed by the plasticity of the M13 genome and its ability to freely encapsidate extra genetic material.

Protein–nucleic acid interactions in other helical viruses, such as rhabdoviruses, are rather more complex. In most enveloped helical viruses, a nucleoprotein core forms first and this is then coated by matrix proteins, the envelope and its associated glycoproteins (Figure 2.3). The fine structure of the core has not been determined, but appears to show cross-striations 4.5–5.0 nm apart, each of these presumably equating to one turn of the protein–RNA complex (rather like TMV). The matrix protein shows an apparently hexagonal pattern and it is not clear how this is related to the structure of the underlying nucleocapsid, or how the rounded end of the virus particle is formed.

Rather less is known about the arrangement of the genome inside virus particles with icosahedral symmetry. There are, however, some exceptions to this statement. These are the T = 3 icosahedral RNA viruses whose subunits consist largely of the '8-strand anti-parallel β-barrel' structural motif, discussed earlier. In these viruses, positively-charged inward-projecting arms of the capsid proteins interact with the RNA in the centre of the particle. In BPMV, a T = 3 comovirus with a bipartite genome, X-ray crystallography has shown that the RNA is folded in such a way that it assumes icosahedral symmetry, corresponding to that of the capsid surrounding it. The regions which contact the capsid proteins are single-stranded and appear to interact by electrostatic forces rather than covalent bonds. Recently, the atomic structure of ϕX174 has also shown that a portion of the DNA genome interacts with arginine residues exposed on the inner surface of the capsid in a manner similar to BPMV.

Therefore, a consensus about the physical state of nucleic acids within icosahedral virus capsids appears to be emerging. Just as the icosahedral capsids of many genetically unrelated viruses are based on monomers with a common '8-strand anti-parallel β-barrel' structural motif, the genomes inside also appear to display icosahedral symmetry, the vertices of which interact with basic amino acid residues on the inner surface of the capsid. Thus, these common structural motifs may in time explain how viruses selectively package the required genomic nucleic acids and may even offer opportunities to design specific drugs to inhibit these interactions.

Virus Receptors – Recognition and Binding

Cellular receptor molecules used by a number of different viruses from diverse taxonomic groups have now been identified. The interaction of the outer surface of a virus with a cellular receptor is a major event in determining the subsequent events in replication and the outcome of infections. It is this binding event which activates inert extracellular virus particles and initiates the replication cycle. Receptor binding is considered in detail in Chapter 4.

Other Interactions of the Virus Capsid with the Host Cell

As stated at the beginning of this chapter, the function of the virus capsid is not only to protect the genome, but also to deliver it to a suitable host cell and, more specifically, the appropriate compartment of the host cell (in the case of eukaryote hosts) to allow replication to proceed. One example is the nucleocapsid proteins of viruses which replicate in the nucleus of the host cell. These molecules contain within their primary amino acid sequences 'nuclear localization signals' which are responsible for the migration of the virus genome plus its associated proteins into the nucleus where replication can occur. Again, these events are discussed in Chapter 4.

Virions are not inert structures. Many virus particles contain one or more enzymatic activities, although in most cases these are not active outside the biochemical environment of the host cell. All viruses with negative-sense RNA genomes must carry with them a virus-specific RNA-dependent RNA polymerase because uninfected cells have no mechanism for RNA-dependent RNA polymerization and, therefore, genome replication could not occur if this enzyme were not included in the virus particle. Reverse transcription of retrovirus genomes occurs inside a particulate complex and not free in solution. The more complex DNA viruses (e.g. herpesviruses and poxviruses) carry a multiplicity of enzymes, mostly concerned with some aspect of nucleic acid metabolism.

Summary

This chapter is not intended to be a complete list of all the virus structures that are now known, but attempts to illustrate with examples some of the principles that control the assembly of viruses and the difficulties of studying these minute structures. Readers must refer to scientific articles and databases for the fine detail of known virus structures and for the many which will undoubtedly appear in the next few years. Nevertheless, there are a number of repeated structural motifs found in many different virus groups. The most obvious is the division of many virus structures into those based on helical or icosahedral symmetry. More subtly, common protein structures such as the '8-strand anti-parallel β-barrel' structural motif found in many $T = 3$ icosahedral virus capsids and the icosahedral folded RNA genome present inside some of these viruses are beginning to emerge. Virus particles are not inert. Many are armed with a variety of enzymes which carry out a range of complex reactions, most frequently involved in the replication of the genome.

Further Reading

Cheng, R.H., Olson, N.H. and Baker, T.S. (1992). Cauliflower mosaic virus: a 420 subunit (T=7) multilayer structure. *Virology* **186**: 655–68.

Guo P. (Ed.) Virus Assembly. *Sem. Virol.* **5**: No.1 (1994).

Harrison, S.C. (1996). Virus structure. In: Fields B., Knipe D.M., Howley P.M. (Eds) *Field's Virology* (3rd edition). Lippincott-Raven, New York, chap. 3, pp. 59–99.

McKenna, R. *et al.* (1992). Atomic structure of single-stranded DNA bacteriophage φX174 and its functional implications. *Nature* **355**: 137–43.

Nermut, M.V. and Steven, A.C. (Eds) (1978). Animal virus structure. *Perspectives in Medical Virology*, Vol. 3. Elsevier, Amsterdam.

Palmer, E.L. and Martin, M.L. (1982). *An Atlas of Mammalian Viruses.* CRC Press, Boca Raton.

Rasched, I. and Oberer, E. (1986). Ff Coliphages: structural and functional relationships. *Microbiol. Rev.* **50**: 401–27.

Rohrmann, G.F. (1992). Baculovirus structural proteins. *J. Gen. Virol.* **73**: 749–61.

Venkataram Prasad, B.V. *et al.* (1992). Three-dimensional structure of single-shelled bluetongue virus. *J. Virol.* **66**: 2135–42.

Self-Assessment Questions

(2.1) Are the following statements true or false?

(a) Nucleic acids are resistant to physical damage such as shearing by mechanical forces and to chemical modification.

(b) The protein subunits in a virus capsid are multiply redundant, i.e. present in many copies per particle.
(c) The forces which drive the assembly of virus particles include hydrophobic and electrostatic interactions.
(d) The protein subunits in virus capsids are held together by covalent bonds.
(e) The only function of the outer shells of a virus particle is to protect the fragile genome from physical, chemical, or enzymatic damage.

(2.2) Are the following statements true or false?
(a) A helix can be constructed by stacking repeated components with a constant relationship to one another.
(b) A helix can be defined by two parameters: amplitude and pitch.
(c) A helix can be defined mathematically by the equation: $P = \mu \times p^2$
(d) The 'amplitude' of a helix is its diameter.
(e) Bacteriophage M13 particles are 270 nm long.

(2.3) Are the following statements true or false?
(a) The particles of all known plant viruses have helical symmetry.
(b) There are no known non-enveloped helical viruses of animals.
(c) Rhabdovirus particles have a helical nucleocapsid.
(d) Bacteriophage M13 is a member of the 'Ff' coliphage group.
(e) Bacteriophage M13 requires CD4 on the surface of *Escherichia coli* for infection.

(2.4) Are the following statements true or false?
(a) An icosahedron is a solid shape consisting of 30 triangular faces arranged around the surface of a sphere.
(b) An icosahedron has an axis of threefold rotational symmetry through the centre of each edge.
(c) An icosahedron has an axis of fivefold rotational symmetry through the centre of each face.
(d) An icosahedron has an axis of twofold rotational symmetry through the centre of each corner.
(e) Most icosahedral virus capsids contain only 60 subunits.

(2.5) All the following virus groups have icosahedral symmetry (true or false?):
(a) Orthomyxoviruses
(b) Paramyxoviruses
(c) Picornaviruses
(d) Rhabdoviruses
(e) Tobamoviruses

(2.6) Are the following statements true or false?
(a) The lipid composition of a virus envelope is distinct from that of the host cell.
(b) Virus envelopes are acquired during extrusion (budding) of the particle through the host cell membrane.
(c) Budding can only occur from the host cell surface membrane.

(d) All enveloped viruses have underlying helical symmetry.
(e) All virus envelope proteins are encoded by the host cell.

(2.7) All the following virus groups possess lipid envelopes (true or false?):
(a) Orthomyxoviruses
(b) Paramyxoviruses
(c) Picornaviruses
(d) Rhabdoviruses
(e) Tobamoviruses

(2.8) Are the following statements true or false?
(a) Matrix proteins are internal virion proteins that link the internal nucleocapsid assembly to the envelope.
(b) Matrix proteins are often very abundant in the virus particle.
(c) External glycoproteins are usually the major antigens of enveloped viruses.
(d) External glycoproteins are usually the virus attachment proteins of enveloped viruses.
(e) The influenza virus M2 peptide is a transmembrane protein.

(2.9) Are the following statements true or false?
(a) Poxvirus particles have overall helical symmetry.
(b) Poxvirus particles contain more than 100 different proteins
(c) The tailed phages of *E. coli* are released from the host cell by budding.
(d) The tailed phages of *E. coli* have separate assembly pathways for the head and tail sections of the particle.
(e) Reovirus capsids are composed of a double protein shell.

(2.10) Are the following statements about genome packaging true or false?
(a) It is completely specific for virus nucleic acids.
(b) It requires interaction with a specific virus-encoded capsid or nucleocapsid protein.
(c) It requires specific nucleotide sequences in the virus genome.
(d) It requires positively charged groups to overcome negative electrostatic charges on the genome.
(e) In TMV it is initiated by the OAS region of the genome.

Answers to Self-Assessment Questions are given in Appendix 3.

Chapter 3

Genomes

The Structure and Complexity of Virus Genomes

The compositions and structures of virus genomes are more varied than any of those seen in the entire bacterial, plant or animal kingdoms. The nucleic acid comprising the genome may be single-stranded or double-stranded, and in a linear, circular or segmented configuration. Single-stranded virus genomes may be either **positive (+)sense**, i.e. of the same polarity (nucleotide sequence) as mRNA, **negative (−)sense**, or **ambisense** – a mixture of the two. Virus genomes range in size from approximately 3500 nucleotides (**nt**) (e.g. bacteriophages of the family *Leviviridae* such as MS2 and Qβ) to approximately 235 kilobase pairs (**kbp**), i.e. 470 000 nt (e.g. herpesviruses such as *Cytomegalovirus*). Unlike the genomes of all cells, which are composed of DNA, virus genomes may contain their genetic information encoded in either DNA or RNA.

Whatever the particular composition of a virus genome, all must conform to one condition. Since viruses are obligate intracellular parasites only able to replicate inside the appropriate host cells, the genome must contain information encoded in a form which can be recognized and decoded by the particular type of cell parasitized. Thus, the genetic code employed by the virus must match or at least be recognized by the host organism. Similarly, the control signals which direct the expression of virus genes must be appropriate to the host. Chapter 4 describes the means by which virus genomes are replicated and Chapter 5 deals in more detail with the mechanisms that regulate the expression of virus genetic information. The purpose of this chapter is to describe the diversity of virus genomes and to examine how and why this variation may have arisen.

Although molecular biology has developed many techniques for the manipulation of proteins, the power of this technology has concentrated in the main on nucleic acids. There has been a strong and synergistic relationship between advances in virology dependent on this new technology and the opportunities which viruses themselves have afforded to develop new techniques of investigation. Thus, the first complete genome to be sequenced (in the 1970s) was that of a virus, the bacteriophage φX174. This virus was chosen for a number of reasons. First, it has one of the smallest genomes (5386 nt). Second, large amounts of the phage could be propagated in *Escherichia coli*, easily purified and the genomic

DNA extracted. Third, the genome of this phage consists of single-stranded DNA which could be directly sequenced by chain-termination methods. Although methods of sequencing double-stranded DNA were also being developed at this time, ϕX174 proved the utility of bacteriophages with single-stranded genomes as cloning vectors for DNA sequencing and phages such as M13 have subsequently been highly developed for this purpose.

Virus genome structures and nucleotide sequences have been intensively studied in recent years because the power of recombinant DNA technology has focused much attention on the virus genome. It would be wrong to present molecular biology as the only means of addressing unanswered problems in virology, but it would be equally foolish to ignore the opportunities which it offers and the explosion of knowledge which has resulted from it in the last two decades.

Some of the simpler bacteriophages have been cited above as examples of the smallest and least complex genomes known. At the other end of the scale, the genomes of the largest double-stranded DNA viruses such as herpesviruses and poxviruses are sufficiently complex to have escaped complete functional analysis to date (even though the complete nucleotide sequences of the genomes of a growing number of examples are now known). Many of the DNA viruses of eukaryotes closely resemble their host cells in terms of the biology of their genomes:

- Some DNA virus genomes are complexed with cellular histones to form a **chromatin**-like structure inside the virus particle. Once inside the nucleus of the host cell, these genomes behave like miniature satellite chromosomes, following the dictates of cellular enzymes and the cell cycle
- Vaccinia virus mRNAs were found to be polyadenylated at their 3′ ends by Kates in 1970 – the first observation of this phenomenon
- Split genes containing non-coding **introns**, protein coding **exons** and spliced mRNAs were first discovered in adenoviruses by Sharp in 1977.

Introns in prokaryotes were first discovered in the genome of bacteriophage T4 in 1984. Several examples of this phenomenon have now been discovered in T4 and in other phages. All are similar, being of the class I ‘self-splicing’ type. However, this observation raises an important point. The conventional view is that prokaryote genomes are smaller and replicate faster than those of eukaryotes, and hence can be regarded as ‘streamlined’. The genome of phage T4 consists of 160 kbp of double-stranded DNA and is highly compressed; for example, promoters and translation control sequences are nested within the coding regions of overlapping upstream genes. The presence of introns in bacteriophage genomes, which are under constant ruthless pressure to exclude ‘junk sequences’, indicates that these genetic elements must have evolved mechanisms to escape or neutralize this pressure and to persist as parasites within parasites.

All virus genomes experience pressure to minimize their size. For example, viruses with prokaryotic hosts must be able to replicate sufficiently quickly to keep up with their host cells and this is reflected in the compact nature of many (but not all) bacteriophages. Overlapping genes are common and the maximum genetic capacity is compressed into the minimum genome size. In viruses with eukaryotic hosts there is also pressure on genome size. Here, however, the pressure is mainly from the 'packaging size' of the virus particle, i.e. the amount of nucleic acid which can be incorporated into the virion. Therefore, these viruses commonly show tremendous compression of genetic information when compared with the low density of information in the genomes of eukaryotic cells.

As already stated, there are exceptions to this simple 'rule'. Some bacteriophages (e.g. the family *Myoviridae*, such as T4) have relatively large genomes, up to 160 kbp. Among viruses of eukaryotes, herpesviruses and poxviruses also have large genomes, up to 235 kbp. It is notable that in these cases, the virus genomes contain many genes involved in their own replication, particularly enzymes concerned with nucleic acid metabolism. Therefore, these viruses partially escape the restrictions of the biochemistry of the host cell by encoding additional biochemical apparatus. The penalty is that they have to encode all the information necessary for a large and complex particle to package the genome – also an upward pressure on genome size.

Later sections of this chapter contain detailed descriptions of both small, compact and large, complex virus genomes.

Molecular Genetics

As already described, the new techniques of molecular biology have had a major influence in concentrating much attention on the virus genome. It is beyond the scope of this book to give detailed accounts of this technology. Indeed, it is assumed that readers already have a firm grasp of the principles behind these techniques, as well as the jargon involved! However, it is perhaps worth taking some time here to illustrate how some of these techniques have been applied to virology, remembering that these newer techniques are complementary to and do not replace the classical techniques of virology. Initially, any investigation of a virus genome will usually include questions about the following:

- Composition – DNA or RNA, single-stranded or double-stranded, linear, or circular
- Size and number of segments
- Terminal structures
- Nucleotide sequence
- Coding capacity – open reading frames
- Regulatory signals – transcription enhancers, promoters, and terminators.

It is possible to separate the molecular analysis of virus genomes into two types of approach: physical analysis of structure and nucleotide sequence, essentially performed *in vitro*, and a more biological approach to examine the structure–function relationships of intact virus genomes and individual genetic elements, usually involving analysis of the virus phenotype *in vivo*.

The conventional starting point for the physical analysis of virus genomes has been the isolation of nucleic acids from virus preparations of varying degrees of purity. To some extent, this is still true of molecular biology, although the emphasis on extensive purification has declined as techniques of molecular cloning have become more advanced. DNA virus genomes can be analysed directly by restriction endonuclease digestion without resort to molecular cloning and this was achieved for the first time with SV40 DNA in 1971. The first pieces of DNA to be molecularly cloned were restriction fragments of bacteriophage λ DNA which were cloned into the DNA genome of SV40 by Berg and colleagues in 1972. Thus virus genomes were both the first cloning vectors and the first nucleic acids to be analysed by these techniques! As already mentioned, the genome of ϕX174 was the first **replicon** to be completely sequenced. Subsequently, phage genomes such as M13 were highly modified for use as vectors in DNA sequencing. The enzymology of RNA-specific nucleases was comparatively advanced at this time, such that a spectrum of enzymes with specific cleavage sites could be used to analyse and even determine the sequence of RNA virus genomes (the first short nucleotide sequences of tRNAs having been determined in the mid-1960s). However, direct analysis of RNA by these methods was laborious and notoriously difficult. RNA sequence analysis did not begin to advance rapidly until the widespread use of reverse transcriptase (isolated from avian myeloblastosis virus) to convert RNA into cDNA in the 1970s.

In addition to molecular cloning, other techniques of molecular analysis have also been of great value in virology. Direct analysis by electron microscopy, if calibrated with known standards, can be used to estimate the size of nucleic acid molecules. Perhaps the most important single technique has been gel electrophoresis (Figure 3.1). The earliest gel matrix employed for separating molecules was based on starch and gave relatively poor resolution. It is now most common to use agarose gels to separate large nucleic acid molecules (which may be very large indeed – several megabases in the case of techniques such as pulsed-field gel electrophoresis (PFGE)) and polyacrylamide gel electrophoresis (PAGE) to separate smaller pieces (down to sizes of a few nucleotides). Apart from the fact that sequencing depends on the ability to separate molecules that differ from each other by only one nucleotide in length, gel electrophoresis has been of great value in analysing intact virus genomes, and particularly the analysis of viruses with segmented genomes (below). Hybridization of complementary nucleotide sequences can also be used in a number of ways to analyse virus genomes (Chapter 1).

Phenotypic analysis of virus populations has long been a standard technique of virology. Examination of variant viruses and naturally occurring spontaneous mutants has been a longstanding method of determining the function of virus genes. Molecular biology has added to this the ability to design and create specific mutations, deletions and recombinants *in vitro*; it is a very powerful tool indeed. Although coding capacity and to some extent protein properties can be determined *in vitro* by the use of cell-free extracts to translate mRNAs, complete functional analysis of virus genomes can only be performed on intact viruses. Fortunately, the relative simplicity of virus genomes (compared with even the simplest cell) offers a major advantage here – the ability to 'rescue'

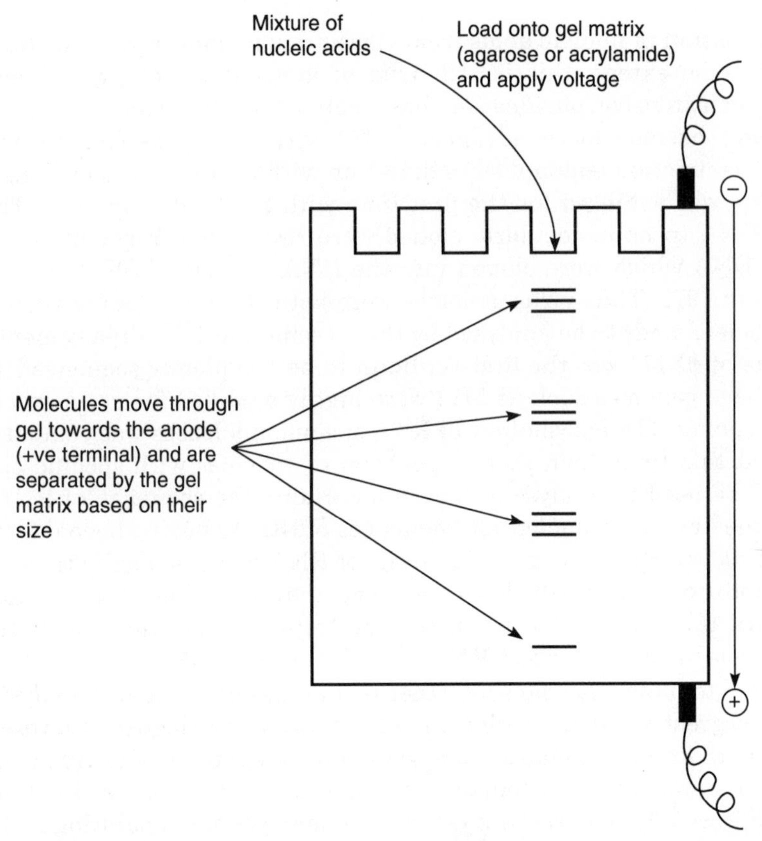

Figure 3.1 In gel electrophoresis, a mixture of nucleic acids (or proteins) are applied to a gel and move through the gel matrix when an electric field is applied. The net negative charge due to the phosphate groups in the backbone of nucleic acid molecules results in their movement away from the cathode and towards the anode. Smaller molecules are able to slip though the gel matrix more easily and thus migrate further than larger molecules, which are retarded, resulting in a net separation based on the size of the molecules.

infectious virus from purified or cloned nucleic acids. Infection of cells caused by nucleic acid alone is referred to as **transfection**.

Virus genomes which consist of (+)sense RNA are infectious when the purified RNA (vRNA) is applied to cells in the absence of any virus proteins. This is because (+)sense vRNA is essentially mRNA and the first event in a normally infected cell is to translate the vRNA to make the virus proteins responsible for genome replication. In this case, direct introduction of RNA into cells circumvents the earliest stages of the replicative cycle (Chapter 4). Virus genomes which are composed of double-stranded DNA are also infectious. The events which occur here are a little more complex, since the virus genome must first be transcribed by host polymerases to produce mRNA. This is relatively simple for phage genomes introduced into prokaryotes, but for viruses which replicate in the nucleus of eukaryotic cells, such as herpesviruses, the DNA must first find its way to the appropriate cellular compartment. Most of the DNA which is introduced into cells by transfection is degraded by cellular nucleases. However, irrespective of its sequence, a small proportion of the newly introduced DNA finds its way into the nucleus, where it is transcribed by cellular polymerases. Unexpectedly, cloned cDNA genomes of (+)sense RNA viruses (e.g. picornaviruses), are also infectious, although less efficient at infecting cells than the vRNA. This is presumably because the DNA is transcribed by cellular enzymes to make RNA. Using these techniques, virus can be rescued from cloned genomes, including those which have been manipulated *in vitro*.

Until recently, this type of approach was not possible for analysis of viruses with (−)sense genomes. This is because such virus particles all contain a virus-specific polymerase. The first event when these virus genomes enter the cell is that the (−)sense genome is copied by the polymerase, forming either (+)sense transcripts which are used directly as mRNA, or a double-stranded molecule, known either as the replicative intermediate (RI) or replicative form (RF), which serves as a template for further rounds of mRNA synthesis. Therefore, since purified (−)sense genomes cannot be directly translated and are not replicated in the absence of the virus polymerase, these genomes are inherently non-infectious. Systems have recently been developed which permit the rescue of viruses with (−)sense genomes from purified or cloned nucleic acids. Such experiments are frequently referred to as 'reverse genetics', i.e. the manipulation of an RNA genome via a DNA intermediate:

- *Influenza virus*: Virus genome segments have been reverse transcribed, cloned as DNA and manipulated *in vitro*. RNA is then produced by *in vitro* transcription and mixed with purified influenza polymerase proteins. The resulting ribonucleoprotein complex is **transfected** into cells which are then **superinfected** with a 'helper' influenza virus strain. In this way, genetically manipulated genome segments can be rescued as 'transfectant virus'
- *Paramyxoviruses*: Genetically manipulated paramyxovirus genomes have been rescued by transfection of naked *in vitro* RNA transcripts into helper virus infected cells. The success of both this and the above approach rely on a strong selection pressure for the manipulated virus, e.g. antibody ablation of the helper virus phenotype
- *Rhabdoviruses*: **Defective-interfering (D.I.) particles** of vesicular stomatitis virus (VSV) have been rescued by co-transfection of plasmids into cells. One plasmid expresses a deleted (D.I.) form of VSV RNA, the other viral proteins which replicate the D.I. RNA and package it into virus particles. Recently, infectious rabies virus has been rescued from cloned cDNA. The feature of these systems is that they are free of helper virus.

Such developments open up possibilities for genetic investigation of negative-strand viruses that have not previously existed.

Virus Genetics

Conventional genetic analysis of animal viruses is based largely on isolation and analysis of mutants. This is usually achieved using plaque-purification techniques. In the case of viruses for which no such systems exist (either because they are not cytopathic or do not replicate in culture), little genetic analysis was possible before the development of molecular genetics. However, certain tricks make it possible to extend standard genetic techniques to non-cytopathic viruses:

- *Biochemical analysis*: The use of metabolic inhibitors to construct genetic 'maps'. Inhibitors of translation (such as puromycin and cycloheximide) and transcription (actinomycin D) can be used to decipher genetic regulatory mechanisms
- *Focal immunoassays*: Replication of non-cytopathic viruses can be visualized by immune-staining to produce visual foci (e.g. human immunodeficiency virus)
- *Molecular biology*: e.g. nucleotide sequencing
- *Physical analysis*: The use of high-resolution electrophoresis to identify genetic polymorphisms of viral proteins or nucleic acids
- Transformed foci: Production of **transformed** 'foci' of cells by non-cytopathic 'focus-forming' viruses (e.g. DNA and RNA tumour viruses).

Various types of genetic map can be derived:

- *Recombination maps*: These are an ordered sequence of mutations derived from the probability of recombination between two genetic markers, which is proportional to the distance between them – a classical genetic technique. This method works for viruses with non-segmented genomes (DNA or RNA)
- *Reassortment maps (or groups)*: In viruses with segmented genomes, the assignment of mutations to particular genome segments results in the identification of genetically linked reassortment groups equivalent to individual genome segments.

Other types of map which can be constructed include the following:

- *Physical maps*: Mutations or other features can be assigned to physical locations on a virus genome using the 'rescue' of mutant genomes by small pieces of the wild-type genome (e.g. heteroduplex formation between mutant and wild-type DNA), after transfection of susceptible cells. Alternatively, cells can be co-transfected with the mutant genome plus individual restriction fragments to localize the mutation. Similarly, various polymorphisms (such as electrophoretic mobility of proteins) can be used to determine the genetic structure of a virus
- *Restriction maps*: Site-specific cleavage of DNA by restriction endonucleases can be used to determine the structure of virus genomes. RNA genomes can be analysed in this way after cDNA cloning

- *Transcription maps*: Maps of regions encoding various mRNAs can be determined as follows:
  Hybridization of mRNA species to specific genome fragments (e.g. restriction fragments). The precise start/finish of mRNAs can be determined by single strand-specific nuclease digestion of radiolabelled probes. Proteins encoded by individual mRNAs can be determined by translation *in vitro*.
  Ultraviolet (UV) irradiation of RNA virus genomes can be used to determine the position of open reading frames because those farthest from the translation start are the least likely to be expressed by *in vitro* translation after partial degradation of the virus RNA by UV light
- *Translation maps*: Pactamycin (which inhibits the initiation of translation) has been used to map protein-coding regions of enteroviruses. Pulse labelling results in incorporation of radioactivity only into proteins initiated before addition of the drug. Proteins nearest the 3′ end of the genome are the most heavily labelled, those at the 5′ end of the genome the least.

Virus mutants

'Mutant', 'strain', 'type', 'variant' and even 'isolate' are all terms used rather loosely by virologists to differentiate particular viruses from each other and from the original 'parental', 'wild-type' or 'street' isolates of that virus. More accurately, these terms are generally applied as follows:

- *Strain*: Different lines or isolates of the same virus, e.g. from different geographical locations or patients
- *Type*: Different serotypes of the same virus (e.g. various antibody neutralization phenotypes)
- *Variant*: A virus whose phenotype differs from the original wild-type strain but where the genetic basis for the difference is not known.

The origin of mutant viruses

Spontaneous mutations In some viruses, mutation rates may be as high as 10^{-3}–10^{-4} per incorporated nucleotide (e.g. in retroviruses such as HIV), whereas in others, they may be as low as 10^{-8}–10^{-11} (e.g. in herpesviruses), which is equivalent to the mutation rates seen in cellular DNA. These differences are due to the mechanism of genome replication, with error rates in RNA-dependent RNA polymerases generally being higher than DNA-dependent DNA polymerases. Some RNA virus polymerases do have proofreading functions, but in general, mutation rates are considerably higher in most RNA viruses than in DNA viruses. For a virus, mutations are a mixed blessing. The ability to generate antigenic variants which can escape the immune response is a clear advantage, but mutation also results in many defective particles, since most mutations are deleterious. In the most extreme cases, e.g. HIV, the error rate is 10^{-3}–10^{-4} per nucleotide incorporated. The HIV genome is approximately 9.7 kb long; therefore, there will be 0.9–9.7 mutations in every genome copied. Hence in this case, the 'wild-type' virus actually consists of a fleeting majority type which dominates the dynamic equilibrium (i.e. the population of genomes) present in all cultures of the virus. These mixtures of molecular variants are known as **quasispecies** and also occur in other RNA viruses, e.g. picornaviruses. However, the majority of these types will be non-infectious or seriously disadvantaged and are therefore rapidly weeded out of a replicating population. This mechanism is an important force in virus evolution (see 'Evolution and Epidemiology').

Induced mutations Historically, most genetic analysis of viruses has been performed on viral mutants isolated from mutagen-treated populations. Mutagens can be divided into two types:

(1) *In vitro* mutagens chemically modify nucleic acids and do not require replication for their activity. Examples include nitrous acid, hydroxylamine and alkylating agents (e.g. nitrosoguanidine)
(2) *In vivo* mutagens require metabolically active, i.e. replicating, nucleic acid for their activity. These compounds are incorporated into newly replicated nucleic acids and cause mutations to be introduced during subsequent rounds of replication. Examples include: base analogues, which cause transitions (i.e. purine–purine and pyrimidine–pyrimidine: AT–GC and CG–TA) and transversions (i.e. purine–pyrimidine and pyrimidine–purine: AT–TA and GC–CG) due to faulty base-pairing when they are incorporated into nucleic acid; intercalating agents (e.g. acridine dyes), which stack between bases, causing insertions or deletions; u.v. irradiation, which causes the formation of pyrimidine dimers which are excised from DNA by repair mechanisms that are much more error-prone than the usual enzymes used in DNA replication. This technique also works on RNA, although the mechanism of action is not known in this case.

Experiments involving chemical mutagens suffer from a number of drawbacks:

- Safety – mutagens are usually carcinogens and are also frequently highly toxic. They are very unpleasant compounds to work with
- The dose of mutagen used must be chosen carefully to give an average of <1 mutation per genome, otherwise the resultant viruses will contain multiple mutations which can complicate interpretation of the phenotype. Therefore, most of the viruses which result will not contain any mutations, which is inefficient since screening for mutants can be very laborious
- There is no control over where mutations occur and it is sometimes difficult or impossible to isolate mutations in a particular gene or region of interest.

For these reasons, site-specific molecular biological methods such as oligonucleotide-directed mutagenesis or PCR-based mutagenesis are now much more commonly used. Together with techniques such as enzyme digestion (to create deletions) and linker scanning (to create insertions), it is now possible to introduce almost any type of mutation precisely and safely at any specific site in a virus genome.

Types of mutant virus

The phenotype of a mutant virus depends on the type of mutation(s) it has and also upon the location of the mutation(s) within the genome. Each of the classes of mutations below can occur naturally in viruses or may be artificially induced for experimental purposes:

- *Biochemical markers*: This category includes drug resistance mutations; specific mutations which result in altered virulence; polymorphisms resulting in altered electrophoretic mobility of proteins or nucleic acids; and altered sensitivity to inactivating agents
- *Deletions*: These are similar in some ways to nonsense mutants (below), but may include one or more virus genes and involve non-coding control regions of the genome (promoters, etc.). Spontaneous deletion mutants often accumulate in virus populations as **defective-interfering (D.I.) particles**. These non-infectious but not necessarily genetically inert genomes are thought to be important in establishing the course and pathogenesis of certain virus infections (see Chapter 6). Genetic deletions can only revert to wild-type by recombination, which usually occurs at comparatively low frequencies. Deletion mutants are very useful for assigning structure–function relationships to virus genomes, since they are easily mapped by physical analysis
- *Host range*: This term can refer either to whole animal hosts or to permissive cell types *in vitro.* **Conditional mutants** of this class have been isolated using amber-suppressor cells (mostly for phages, but also for animal viruses using *in vitro* systems)
- *Nonsense*: These result from alteration of coding sequence of a protein to one of three translation stop codons (UAG, amber; UAA, ochre; UGA, opal). Translation is terminated, resulting in the production of an amino-terminal fragment of the protein. The phenotype of these mutations can be suppressed by propagation of virus in a cell (bacterial or, more recently, animal) with altered suppressor tRNAs. Nonsense mutations are rarely 'leaky' (i.e. the normal function of the protein is completely obliterated) and can only revert to wild type at the original site (see below). They therefore usually show a low reversion frequency
- *Plaque morphology*: Mutants may be either large-plaque mutants which replicate more rapidly than the wild-type, or small-plaque mutants, which are the opposite. Plaque size is often related to a temperature-sensitive (t.s.) phenotype (see below). These mutants are often useful as unselected markers in multifactorial crosses
- *Temperature-sensitive (t.s.)*: These are really 'heat-sensitive' mutants. This type of mutation is very useful as it allows the isolation of **conditional-lethal mutations**, a powerful means of examining virus genes which are essential for replication and whose function cannot otherwise be interrupted. T.s. mutations usually result from mis-sense mutations in proteins (i.e. amino

acid substitutions), resulting in proteins of full size with subtly altered conformation which can function at (permissive) low temperatures but not at (non-permissive) higher ones. Generally, the mutant proteins are immunologically unaltered, which is frequently a useful attribute. These mutations are usually 'leaky', i.e. some of the normal activity is retained even at non-permissive temperatures. Conversely, protein function is often impaired, even at the permissive temperature. Therefore, a high frequency of reversion is often a problem with this type of mutation, because the wild-type virus replicates faster than the mutant. In some viruses (e.g. reoviruses, influenza virus), very many t.s. mutants have been derived over the years for every virus gene, which permitted complete genetic analysis of these genomes before the advent of molecular biology

- *Cold-sensitive (c.s.)*: These mutants are the opposite of t.s. mutants. These mutants are very useful in bacteriophages and plant viruses as their host cells can be propagated at low temperatures, but less useful for animal viruses because their host cells generally will not grow at significantly lower temperatures than normal. It is possible that some c.s. mutants of respiratory viruses, e.g. influenza virus, may be attenuated (i.e. have reduced virulence) and therefore of interest as potential vaccine strains. This is because influenza virus replicates in the cells of the upper respiratory tract, which are at a slightly lower temperature than those at the body core. Attractive though this idea is, it has not proved possible to demonstrate this phenomenon consistently
- *Revertants*: Reverse mutation is a valid type of mutation in its own right. Most of the above classes can undergo reverse mutations, which may be either simple 'back mutations' (i.e. correction of the original mutation) or second site 'compensatory mutations,' which may be physically distant from the original mutation, and not even necessarily in the same gene as the original mutation.

Suppression

Suppression is the inhibition of a mutant phenotype by a second suppressor mutation, which may be either in the virus genome or in that of the host cell. It is important to point out that this mechanism of suppression is not the same as the suppression of chain-terminating amber mutations by host-encoded suppressor tRNAs (above), which could be called 'informational suppression'. Genetic suppression results in an apparently wild-type phenotype from a virus which is still genetically mutant – a **pseudorevertant**. This phenomenon has been best characterized in prokaryotic systems. More recently, some examples have been discovered in animal viruses, for example, reoviruses, vaccinia, influenza (where it has been observed in an attenuated vaccine, leading to an apparently virulent virus – this could therefore be medically important). Suppression may also be important biologically, in allowing viruses to overcome the deleterious effects of mutations, and would therefore be positively selected. Mutant viruses can therefore appear to revert to their original phenotype by three pathways:

(1) Back mutation of the original mutation to give a wild-type genotype/phenotype (true reversion)
(2) A second, compensatory mutation may occur in the same gene as the original mutation, correcting it, e.g. a second frameshift mutation restoring the original reading frame (intragenic suppression)
(3) A suppressor mutation in a different virus gene or a host gene (extragenic suppression).

Genetic interactions between viruses

Genetic interactions between viruses often occur naturally, since host organisms are frequently infected with more than one virus. However, these situations are generally too complicated to be analysed successfully. Experimentally, genetic interactions can be analysed by mixed infection (**superinfection**) of cells in culture. Two types of information can be obtained from such experiments:

- The assignment of mutants to functional groups known as complementation groups
- The ordering of mutants into a linear genetic map by analysis of recombination frequencies.

Complementation can be defined as the interaction(s) of virus gene products during superinfection which results in production of one or both of the parental viruses being increased while both viruses remain unchanged genetically. In this situation, one of the viruses in a mixed infection provides a functional gene product for another virus which is defective for that function (Figure 3.2). If both mutants are defective in the same function, enhancement of replication does not occur and the two mutants are said to be in the same complementation group. The importance of this test is that it allows functional analysis of unknown mutations, if the biochemical basis of any one of the mutations in a particular complementation group is known. In theory, the number of complementation groups is equal to the number of genes in the virus genome. In practice, there are usually fewer complementation groups than genes, since mutations in some genes are always lethal and other genes are non-essential and therefore cannot be scored in this type of test. There are two possible types of complementation:

- Allelic (intragenic) complementation occurs where different mutants have complementing defects in the same protein, e.g. in different functional domains, or in different subunits of a multimeric protein (although this is rare)
- Non-allelic (intergenic) complementation results from mutants with defects in different genes. This is the more common type.

Complementation can be asymmetric, i.e. only one of the (mutant) parental viruses will replicate. This can be an absolute or a partial restriction. When

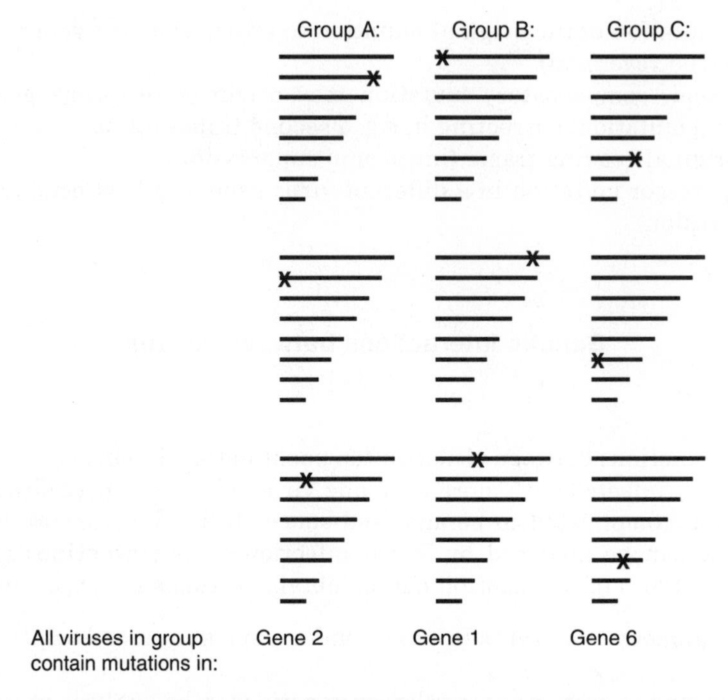

Figure 3.2 Complementation groups of influenza (or other) viruses all contain a mutation in the same virus gene, preventing the rescue of another mutant virus genome from the same complementation group.

complementation occurs naturally, it is usually the case that a replication-competent wild-type virus 'rescues' a replication-defective mutant. In these cases, the wild-type is referred to as a 'helper virus', for example, as in the case of defective transforming retroviruses containing oncogenes (Figure 3.3, Chapter 7).

Recombination is the physical interaction of virus genomes during superinfection resulting in gene combinations not present in either parent. There are three mechanisms by which this can occur, depending on the organization of the virus genome:

(1) *Intramolecular recombination by strand breakage and re-ligation*: This process occurs in all DNA viruses and RNA viruses which replicate via a DNA intermediate. It is believed to be mediated by cellular enzymes, since no virus mutants with specific recombination defects have been isolated
(2) *Intramolecular recombination by 'copy-choice'*: This process occurs in RNA viruses (it has been known in picornaviruses since the 1960s, but has been recognized in other virus groups, such as coronaviruses, only recently), probably by a mechanism in which the virus polymerase switches template strands during genome synthesis. The molecular details of this process are not well understood. There are cellular enzymes which could be involved (e.g. splicing enzymes), but this is unlikely and the process is thought to

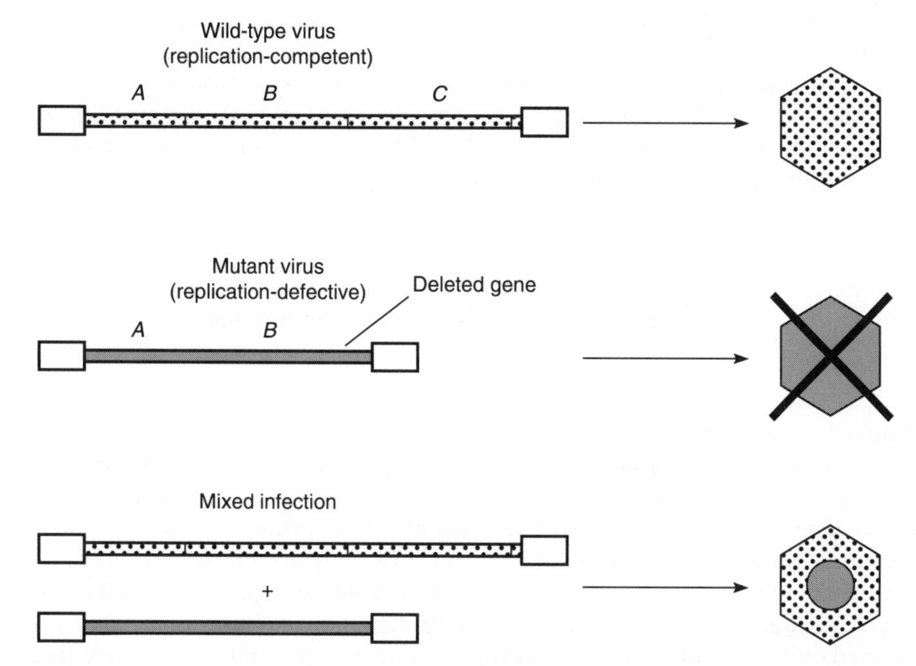

Figure 3.3 Helper viruses are replication-competent viruses which are capable of rescuing replication-defective genomes in a mixed infection, permitting their multiplication and spread.

occur essentially as a random event. **Defective interfering (D.I.)** particles in RNA virus infections are frequently generated in this way (see Chapter 6)

(3) *Reassortment*: In viruses with segmented genomes, the genome segments can be randomly shuffled during superinfection. Progeny viruses receive (at least) one of each of the genome segments, but probably not from a single parent. For example, influenza virus has eight genome segments, therefore in a mixed infection, there could be 2^8 = 256 possible progeny viruses. Packaging mechanisms in these viruses are not understood (see Chapter 2) but may be involved in generating reassortants.

In intramolecular recombination, the probability that breakage-reunion or strand-switching will occur between two markers (resulting in recombination) is proportional to the physical distance between them. Therefore, pairs of markers can be arranged on a linear genetic map, with distances measured in 'map units', i.e. percentage recombination frequency. In reassortment, the frequency of recombination between two markers is either very high (indicating that the markers are on two different genome segments) or comparatively low (which means that they are on the same segment). This is because the frequency of reassortment usually swamps the lower background frequency which is due to intermolecular recombination between strands.

Reactivation is the generation of infectious (recombinant) progeny from non-infectious parental virus genomes. This process has been demonstrated *in vitro* and may be important *in vivo*. For example, it has been suggested that the rescue of defective, long-dormant HIV proviruses during the long clinical course of AIDS may result in increased antigenic diversity and contribute to the pathogenesis of the disease. Recombination occurs frequently in nature, e.g. influenza virus reassortment has resulted in worldwide epidemics (**pandemics**) which have killed millions of people (Chapter 6). This makes these genetic interactions of considerable practical interest and not merely a dry academic matter.

Non-genetic interactions between viruses

A number of non-genetic interactions between viruses occur which can affect the outcome and interpretation of the results of genetic crosses. Eukaryotic cells have a diploid genome with two copies of each chromosome, each bearing its own allele of the same gene. The two chromosomes may differ in allelic markers at many loci. Among viruses, only retroviruses are truly diploid with two complete copies of the entire genome, but some DNA viruses, such as herpesviruses, have repeated sequences and are therefore partially heterozygous. In a few (mostly enveloped) viruses, aberrant packaging of multiple genomes may occasionally result in multiploid particles which are heterozygous (e.g. up to 10% of Newcastle disease virus particles). This process is known as **heterozygosis** and can contribute to the genetic complexity of virus populations.

Another non-genetic interaction which is commonly seen between viruses is **interference**. This process results from the resistance to superinfection by a virus seen in cells already infected by another virus. Homologous interference (i.e. against the same virus) often results from the presence of D.I. particles which compete for essential cell components and block replication. However, interference can also result from other types of mutation (e.g. dominant t.s. mutations) or by sequestration of virus **receptors** due to the production of **virus-attachment proteins** by viruses already present within the cell (e.g. in the case of avian retroviruses).

Phenotypic mixing can vary from extreme cases, where the genome of one virus is completely enclosed within the capsid or envelope of another (**pseudotyping**), to more subtle cases where the capsid/envelope of the progeny contains a mixture of proteins from both viruses. This mixing gives the progeny virus the phenotypic properties (e.g. cell tropism) dependent on the proteins incorporated into the particle, *without any genetic change*. Subsequent generations of viruses inherit and display the original parental phenotypes. This process can occur easily in viruses with naked capsids (non-enveloped) which are closely related (e.g. different strains of enteroviruses), or in enveloped viruses which need not be related to one another (Figure 3.4). In this latter case, the phenomenon is due to the non-specific incorporation of different virus glycoproteins into the envelope, resulting in a mixed phenotype. Rescue of replication-defective transforming retroviruses by helper virus is a form of pseudotyping. Phenotypic mixing has proved to be a very useful tool to examine biological properties of viruses.

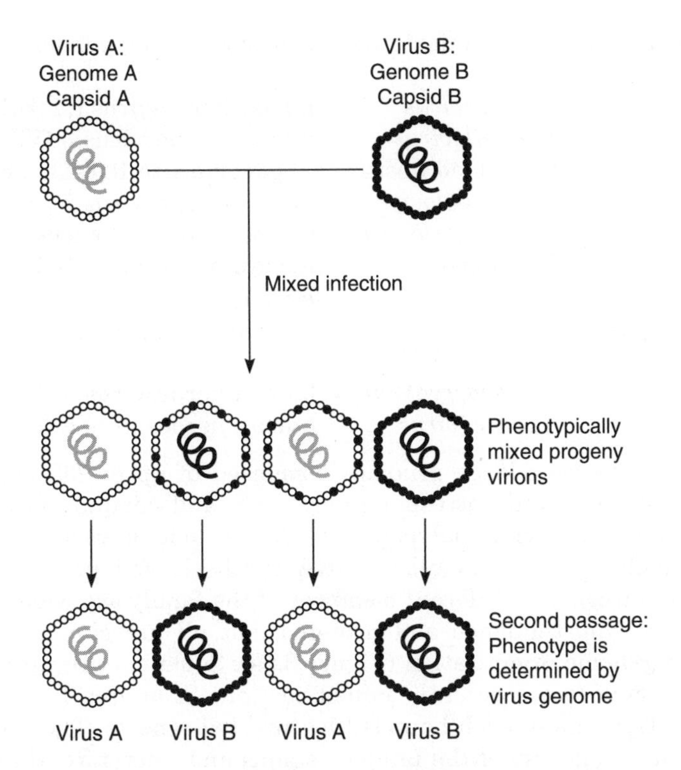

Figure 3.4 Phenotypic mixing occurs in mixed infections, resulting in genetically unaltered virus particles which have some of the properties of the other parental type.

Vesicular stomatitis virus readily forms pseudotypes containing retrovirus envelope glycoproteins, giving a plaque-forming virus with the properties of VSV but with the cell tropism of the retrovirus. This trick has been used in recent years to study the cell **tropism** of HIV and other retroviruses.

'Large' DNA Genomes

There are a number of virus groups which have double-stranded DNA genomes of considerable size and complexity. In many respects, these viruses are genetically very similar to the host cells which they infect. Two examples are the members of the *Adenoviridae* and *Herpesviridae* families.

The *Herpesviridae* are a large family containing more than 100 different members, at least one for most animal species which have been examined to date. There are eight human herpesviruses, all of which share a common overall genome structure, but which differ in the fine details of genome organization and at the level of nucleotide sequence. The family is divided into three subfamilies, based on their biological properties:

Alphaherpesvirinae:	Latent infections in sensory ganglia; genome size 120–180 kbp:	
	Simplexvirus	human herpesvirus 1, 2 (HSV-1, HSV-2)
	Varicellovirus	human herpesvirus 3 (VZV)
Betaherpesvirinae:	Restricted host range; genome size 180–235 kbp:	
	Cytomegalovirus	human herpesvirus 5 (CMV)
	Muromegalovirus	mouse cytomegalovirus 1
	Roseolovirus	human herpesvirus 6, 7 (HHV-6, HSV-7)
Gammaherpesvirinae:	Infection of lymphoblastoid cells; genome size 105–170 kbp:	
	Lymphocryptovirus	human herpesvirus 4 (EBV)
	Rhadinovirus	human herpesvirus 8 (HHV-8)

Herpesviruses have very large genomes composed of up to 235 kbp of linear, double-stranded DNA and correspondingly large and complex virus particles containing about 35 virion polypeptides. All encode a variety of enzymes involved in nucleic acid metabolism, DNA synthesis and protein processing (e.g. protein kinases). The different members of the family are widely separated in terms of genomic sequence and proteins, but all are similar in terms of structure and genome organization (Figure 3.5a). Some but not all herpesvirus genomes consist of two covalently joined sections, a unique long (U_L) and a unique short (U_S) region, each bounded by inverted repeats. The repeats allow structural rearrangements of the unique regions and, therefore, these genomes exist as a mixture of four isomers, all of which are functionally equivalent (Figure 3.5b). Herpesvirus genomes also contain multiple repeated sequences

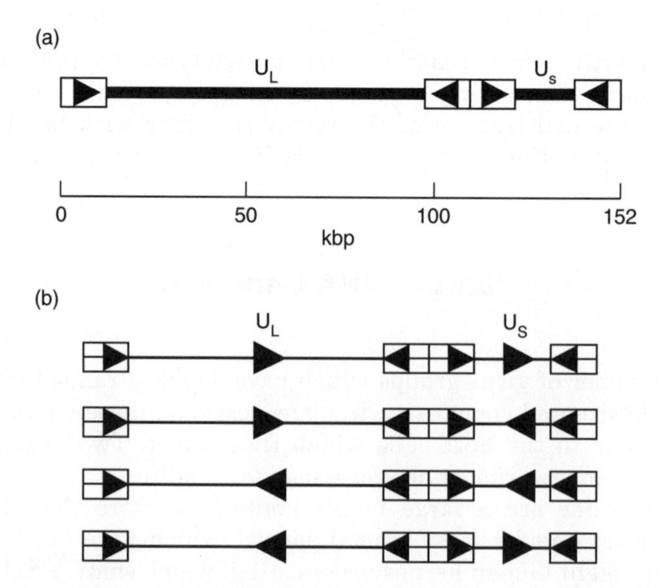

Figure 3.5 (a) Some herpesvirus genomes (e.g. herpes simplex virus) consist of two covalently joined sections, U_L and U_S, each bounded by inverted repeats. (b) This organization permits the formation of four different isomeric forms of the genome.

and, depending on the number of these, the genome size of various isolates of a particular virus can vary by up to 10 kbp.

The prototype member of the family is herpes simplex virus (HSV), whose genome consists of approximately 152 kbp of double-stranded DNA, the complete nucleotide sequence of which has now been determined. This virus contains about 80 genes, densely packed and with overlapping reading frames. However, each gene is expressed from its own promoter (c.f. adenoviruses below). Most of the eight human herpesvirus genomes have now been completely sequenced. Nucleotide sequences are increasingly used as a major criterion in classification of herpesviruses, e.g. in the case of the recently discovered HHV-8 (Chapter 8). Before the development of nucleotide sequencing, the HSV genome had already been extensively mapped by 'conventional' genetic analysis, including the study of a very large number of t.s. mutants. HSV is perhaps the most intensively studied complex virus genome.

In contrast to herpesviruses, the genomes of adenoviruses consist of linear, double-stranded DNA of 30–38 kbp, the precise size of which varies between groups. These virus genomes contain 30–40 genes (Figure 3.6). The terminal sequence of each DNA strand is an inverted repeat of 100–140 bp and, therefore, the denatured single strands can form 'pan-handle' structures. These structures are important in DNA replication, as is a 55 kDa protein known as the terminal protein covalently attached to the 5′ end of each strand. During genome replication, this protein acts as a primer, initiating the synthesis of new DNA strands. Although adenovirus genomes are considerably smaller than those of herpesviruses, the expression of the genetic information is rather more complex. Clusters of genes are expressed from a limited number of shared promoters. Multiply spliced mRNAs and alternative splicing patterns are used to express a variety of polypeptides from each promoter (see Chapter 5).

Human adenoviruses are divided into six groups (subgenera) and in this case, genome structure (determined by cross-hybridization and restriction enzyme mapping) is an additional character which has been used to assign viruses to groups. There is 70–95% nucleotide sequence similarity across the entire genome within groups and 5–20% similarity between different groups.

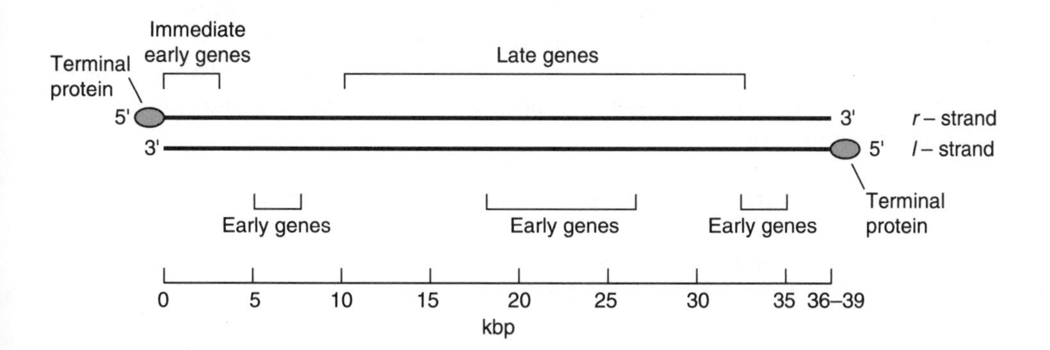

Figure 3.6 Organization of the adenovirus genome.

'Small' DNA Genomes

Bacteriophage M13 has already been mentioned in Chapter 2. The genome of this phage consists of 6.4 kb of single-stranded, (+)sense, circular DNA and encodes 10 genes. Unlike most icosahedral virions, the filamentous M13 capsid can be expanded by the addition of further protein subunits. Hence the genome size can also be increased by the addition of extra sequences in the non-essential intergenic region without becoming incapable of being packaged into the capsid. In other bacteriophages, the packaging constraints are much more rigid; for example, in phage λ, only DNA of between approximately 95 and 110% (approximately 46–54 kbp) of the normal genome size (49 kbp) can be packaged into the virus particle. Not all bacteriophages have such simple genomes as M13, for example, the genome of phage λ is approximately 49 kbp and that of phage T4 is about 160 kbp double-stranded DNA. These last two bacteriophages also illustrate another common feature of linear virus genomes – the importance of the sequences present at the ends of the genome.

In the case of phage λ, the substrate which is packaged into the phage heads during assembly consists of long concatemers of phage DNA which are produced during the later stages of vegetative replication. The DNA is apparently 'reeled in' by the phage head and when a complete genome has been incorporated, the DNA is cleaved at a specific sequence by a phage-coded endonuclease (Figure 3.7). This enzyme leaves a 12 bp 5′ overhang on the end of each of the cleaved strands, known as the *cos* site. Hydrogen bond formation between these 'sticky ends' can result in the formation of a circular molecule. In a newly infected cell, the gaps on either side of the *cos* site are closed by DNA ligase and it is this circular DNA which undergoes vegetative replication or integration into the bacterial chromosome.

Bacteriophage T4 and other phages of *E. coli* illustrate another molecular feature of certain linear virus genomes – **terminal redundancy**. Replication of the T4 genome also produces long concatemers of DNA. These are cleaved by a specific endonuclease, but unlike the λ genome, the lengths of DNA incorporated into the particle are somewhat longer than a complete genome length (Figure 3.8). Therefore, some genes are repeated at each end of the genome, and the DNA packaged into the phage particles contains reiterated information. Bacteriophage genomes are neither necessarily small nor simple!

As further examples of small DNA genomes, consider those of two groups of animal viruses, the parvoviruses and polyomaviruses. The *Parvoviridae* family contains three genera:

(1) *Parvovirus*: Autonomous (replication-competent) pathogenic mammalian viruses
(2) *Dependovirus*: Replication-defective adenovirus-associated viruses (AAV)
(3) *Densovirus*: Autonomous insect viruses.

Parvovirus genomes are linear, non-segmented, single-stranded DNA of about 5 kb. Most of the strands packaged into virions are (−)sense, but AAVs package

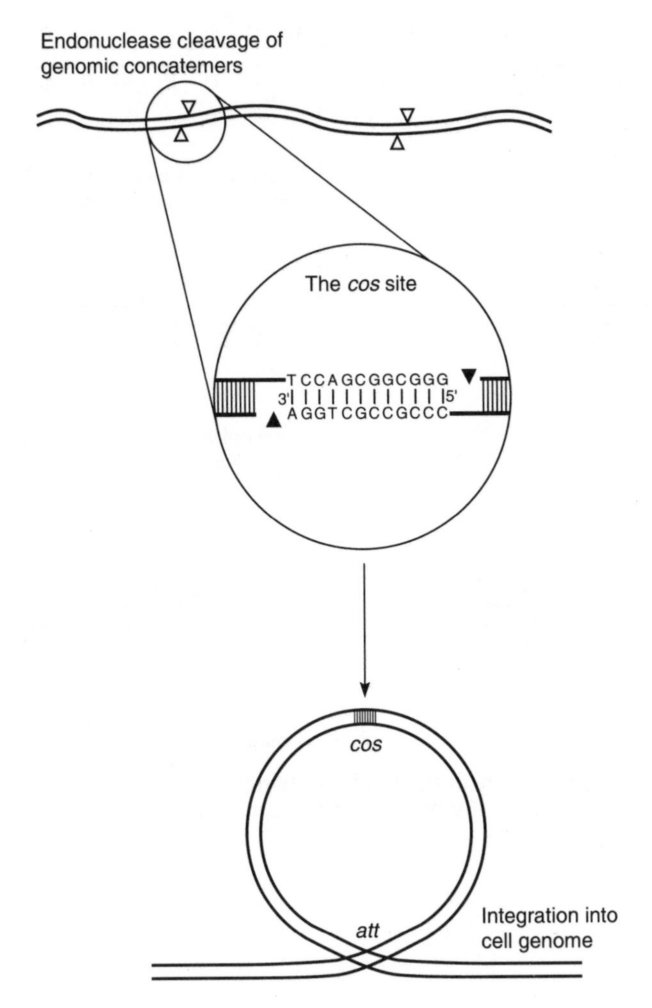

Figure 3.7 The cohesive 'sticky-ends' of the *cos* site in the bacteriophage λ genome are ligated together in newly infected cells, forming a circular molecule. Integration of this circular form into the *Escherichia coli* chromosome occurs by specific recognition and cleavage of the *att* site in the phage genome.

equal amounts of (+) and (−) strands, and all appear to package at least a proportion of (+)sense strands. These are very small genomes, and even the replication-competent parvoviruses contain only two genes: *rep*, which encodes proteins involved in transcription and *cap*, which encodes the coat proteins. However, the expression of these genes is rather complex, resembling the pattern seen in adenoviruses, with multiple splicing patterns seen for each gene (Chapter 5). The ends of the genome have palindromic sequences of about 115 nt, which form 'hairpins' (Figure 3.9). These structures are essential for the initiation of genome replication, again emphasizing the importance of the sequences at the ends of the genome.

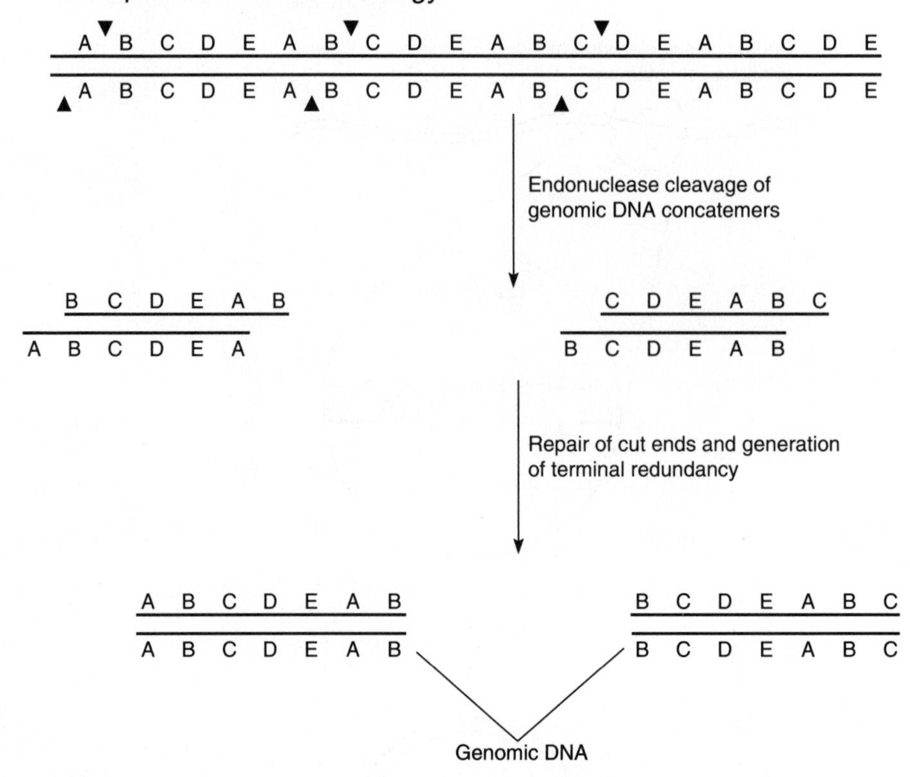

Figure 3.8 Terminal redundancy in the bacteriophage T4 genome results in the reiteration of some genetic information.

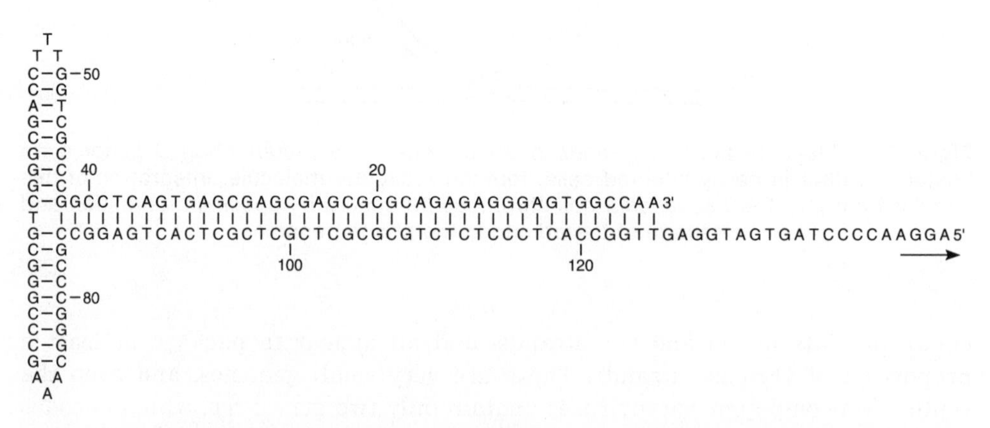

Figure 3.9 Palindromic sequences at the ends of parvovirus genomes result in the formation of hairpin structures involved in the initiation of replication.

The classification of parvoviruses into genera is made on the functional capabilities of the virus genome. *Dependovirus* was the first group to be discovered. This genus contains four types of adeno-associated viruses (AAV 1–4) which are entirely dependent on adenoviruses (or herpesviruses) for essential helper functions required for their replication. The adenovirus genes necessary for *Dependovirus* replication encode early (regulatory) proteins rather than late (structural) adenovirus proteins, but it has recently been shown that treatment of cells with u.v. light, cycloheximide, or some carcinogens can replace the requirement for helper virus. Therefore, the *Dependovirus* requirement appears to be for a modification of the cellular environment rather than a specific virus protein, probably affecting transcription of the defective virus genome.

The genomes of polyomaviruses consist of double-stranded, circular DNA molecules of approximately 5 kbp. The entire nucleotide sequence of all the viruses in the genus is known and the architecture of the polyomavirus genome (i.e. number and arrangement of genes and function of the regulatory signals and systems) has been studied in great detail at a molecular level. Within the particles, the virus DNA assumes a supercoiled form and is associated with four cellular histones H2A, H2B, H3 and H4 (see Chapter 2). The genomic organization of these viruses has evolved to pack the maximum information (six genes) into minimal space (5 kbp). This has been achieved by use of both strands of the genome DNA and overlapping genes (Figure 3.10). VP1 is encoded by a dedicated open reading frame (**ORF**), but the VP2 and VP3 genes overlap so that VP3 is contained within VP2. The origin of replication is surrounded by non-coding regions which control transcription. Polyomaviruses also encode 'T-antigens', which are proteins that can be detected by sera from animals bearing polyomavirus-induced tumours. These proteins bind to the origin of replication

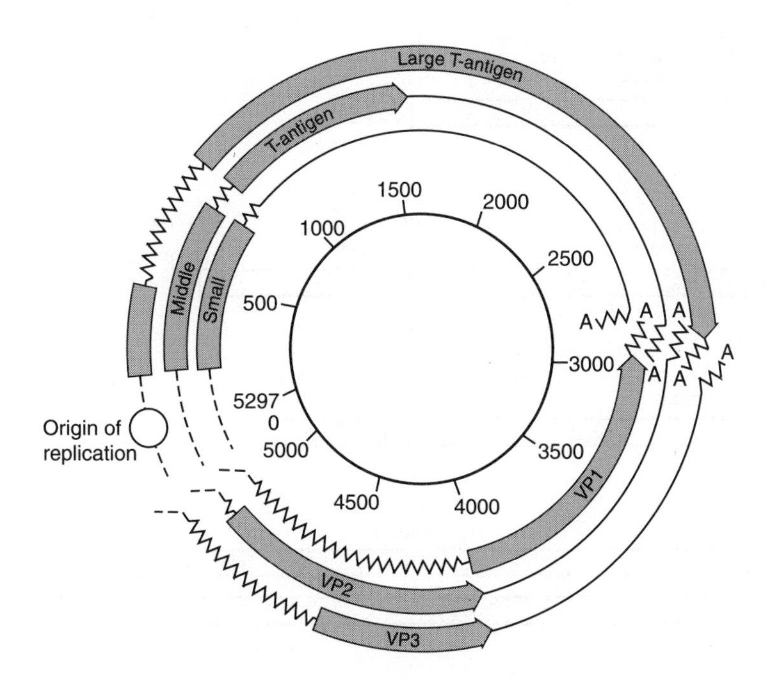

Figure 3.10 The complex organization of the polyomavirus genome results in the compression of much genetic information into a relatively short sequence.

and show complex activities, in that they are involved both in DNA replication and in the transcription of virus genes. This is discussed further in Chapter 7.

Positive-Strand RNA Viruses

The ultimate size of single-stranded RNA genomes is limited by the fragility of RNA and the tendency of long strands to break. In addition, RNA genomes tend to have higher mutation rates than those composed of DNA because they are copied less accurately. This has also tended to drive RNA viruses towards smaller genomes. Single-stranded RNA genomes vary in size from those of coronaviruses, which are approximately 30 kb long, to those of bacteriophages such as MS2 and Qβ, at about 3.5 kb. Although members of distinct families, most (+)sense RNA viruses of vertebrates share common features in terms of the biology of their genomes. In particular, purified (+)sense virus RNA is directly infectious when applied to susceptible host cells in the absence of any virus proteins (although it is about one million times less infectious than virus particles). On examining the features of these families, although the details of genomic organization vary, repeated themes emerge (Figure 3.11).

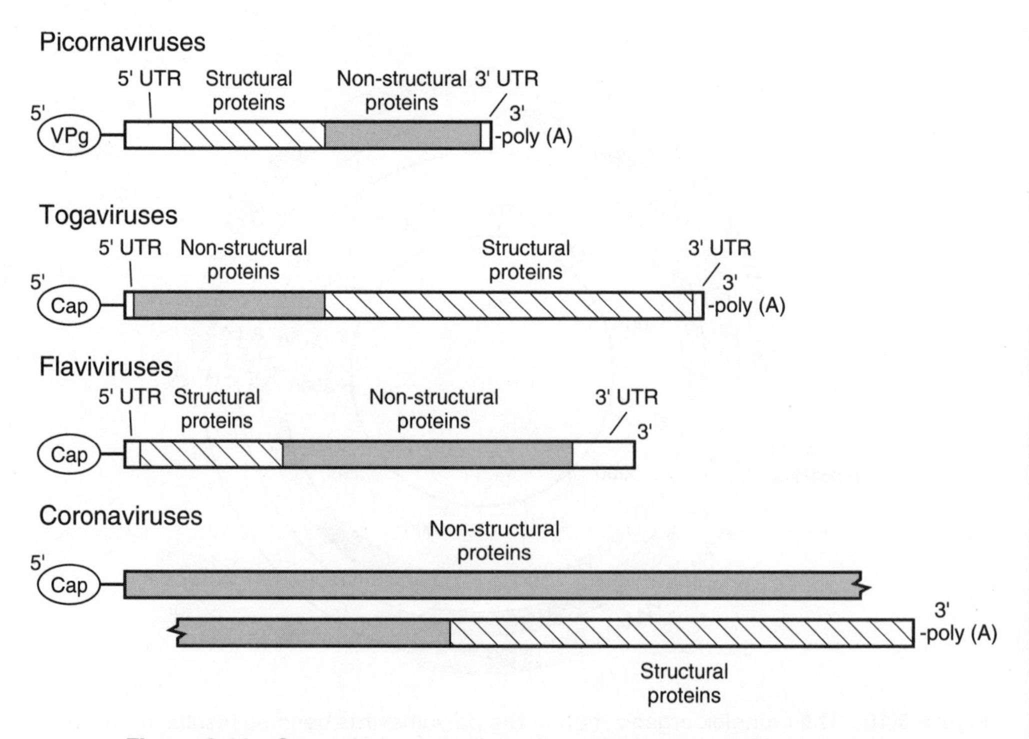

Figure 3.11 Genomic organization of positive-stranded RNA viruses.

Picornaviruses

The picornavirus genome consists of one single-stranded, (+)sense RNA molecule of between 7.2 kb in human rhinoviruses (HRV) to 8.5 kb in foot-and-mouth disease viruses (FMDV), containing a number of features conserved in all picornaviruses:

- There is a long (600–1200 nt) untranslated region (UTR) at the 5′ end which is important in translation, virulence, and possibly encapsidation and a shorter 3′ untranslated region (50–100 nt) necessary for (−)strand synthesis during replication
- The 5′ UTR contains a 'clover-leaf' secondary structure known as the **IRES (Internal Ribosome Entry Site)** (Chapter 5)
- The rest of the genome encodes a single **polyprotein** of between 2100 and 2400 amino acids
- Both ends of the genome are modified, the 5′ end by a covalently attached small, basic protein VPg (23 amino acids), the 3′ end by polyadenylation. The polyadenylic acid sequences are not genetically coded – there is a 'polyadenylation signal' upstream of the 3′ end, as in eukaryotic mRNAs.

Togaviruses

The togavirus genome comprises single-stranded, (+)sense, non-segmented RNA of approximately 11.7 kb. It has the following features:

- It resembles cellular mRNAs in that it has a 5′ methylated cap and 3′ poly(A) sequences
- Expression is achieved by two rounds of translation, producing first non-structural proteins encoded in the 5′ part of the genome and later structural proteins from the 3′ part.

Flaviviruses

The flavivirus genome comprises one single-stranded, (+)sense RNA molecule of about 10.5 kb with the following features:

- It has a 5′ methylated cap but in most cases the RNA is not polyadenylated at the 3′ end
- Genetic organization differs from that of the togaviruses (above) in that the structural proteins are encoded in the 5′ part of the genome and non-structural proteins in the 3′ part
- Expression is similar to that of the picornaviruses, involving the production of a polyprotein.

Coronaviruses

The coronavirus genome consists of non-segmented, single-stranded, (+)sense RNA, approximately 27–30 kb long, which is the longest of any RNA virus. It also has the following features:

- It has a 5′ methylated cap and 3′ poly(A) and the vRNA functions directly as mRNA
- The 5′ 20 kb segment of the genome is translated first to produce a viral polymerase, which then produces a full-length (−)sense strand. This is used as a template to produce mRNA as a 'nested set' of transcripts, all with an identical 5′ non-translated leader sequence of 72 nt and coincident 3′ polyadenylated ends
- Each mRNA is monocistronic, the genes at the 5′ end being translated from the longest mRNA and so on. These unusual cytoplasmic structures are produced not by splicing (post-transcriptional modification) but by the polymerase during transcription.

(+)sense RNA plant viruses

The majority (but not all) of plant virus families have (+)sense RNA genomes. The genome of the Tobamovirus tobacco mosaic virus (TMV) is a well-studied example (Figure 3.12):

- The TMV genome is a 6.4 kb RNA molecule which encodes four genes
- There is a 5′ methylated cap and 3′ end of the genome contains extensive secondary structure, but no poly(A) sequences
- Expression is somewhat reminiscent but distinct from that of Togaviruses, producing non-structural proteins by direct translation of the open reading

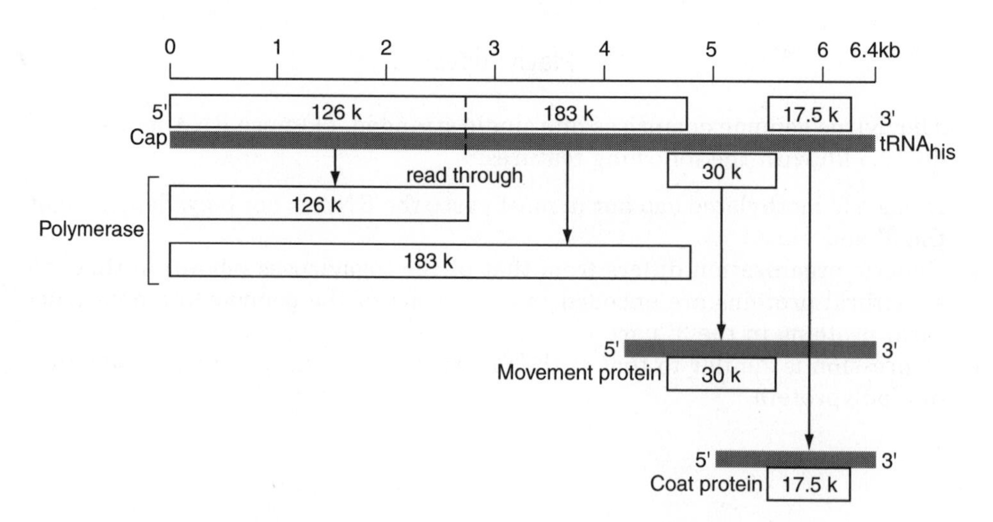

Figure 3.12 Organization of the TMV genome.

frame encoded in the 5′ part of the genome and the virus coat protein and further non-structural proteins from two subgenomic RNAs encoded by the 3′ part.

The similarities and differences between genomes in this class will be considered further in the discussion of virus evolution below and in Chapter 5.

Negative-Strand RNA Viruses

Viruses with negative-sense RNA genomes are a little more diverse than the positive-stranded viruses discussed above. Possibly because of the difficulties of expression, they tend to have larger genomes encoding more genetic information. Because of this, segmentation is a common, although not universal, feature of such viruses (Figure 3.13). None of these genomes are infectious as purified RNA. Because uninfected cells do not contain RNA-dependent RNA polymerase activity, and since the (−)sense genome cannot be translated as mRNA without the virus polymerase packaged in each particle, these genomes are effectively inert. Some of the viruses described in this section are not strictly 'negative-sense' but **ambisense**, since they are part (−)sense and part (+)sense. Ambisense coding strategies occur in both plant (e.g. the *Tospovirus* genus of the Bunyaviruses, and Tenuiviruses such as rice stripe virus) and animal viruses (the *Phlebovirus* genus of the Bunyaviruses, and Arenaviruses).

Bunyaviruses

Members of the *Bunyaviridae* family have single-stranded, (−)sense, segmented RNA. The genome has the following features:

- The genome comprises three molecules: L (8.5 kb), M (5.7 kb) and S (0.9 kb)
- All three RNA species are linear, but in the virion they appear circular because the ends are held together by base-pairing. The three segments are not present in virus preparations in equimolar amounts
- In common with all (−)sense RNAs, the 5′ ends are not capped and the 3′ ends are not polyadenylated
- The members of the *Phlebovirus* and *Tospovirus* genera differ from the other three genera in the family (*Bunyavirus, Nairovirus* and *Hantavirus*) in that genome segment S is rather larger and the overall genome organization is different – **ambisense** (i.e. the 5′ end of each segment is (+)sense, but the 3′ end is (−)sense). The *Tospovirus* genus also has an ambisense coding strategy in the M segment of the genome.

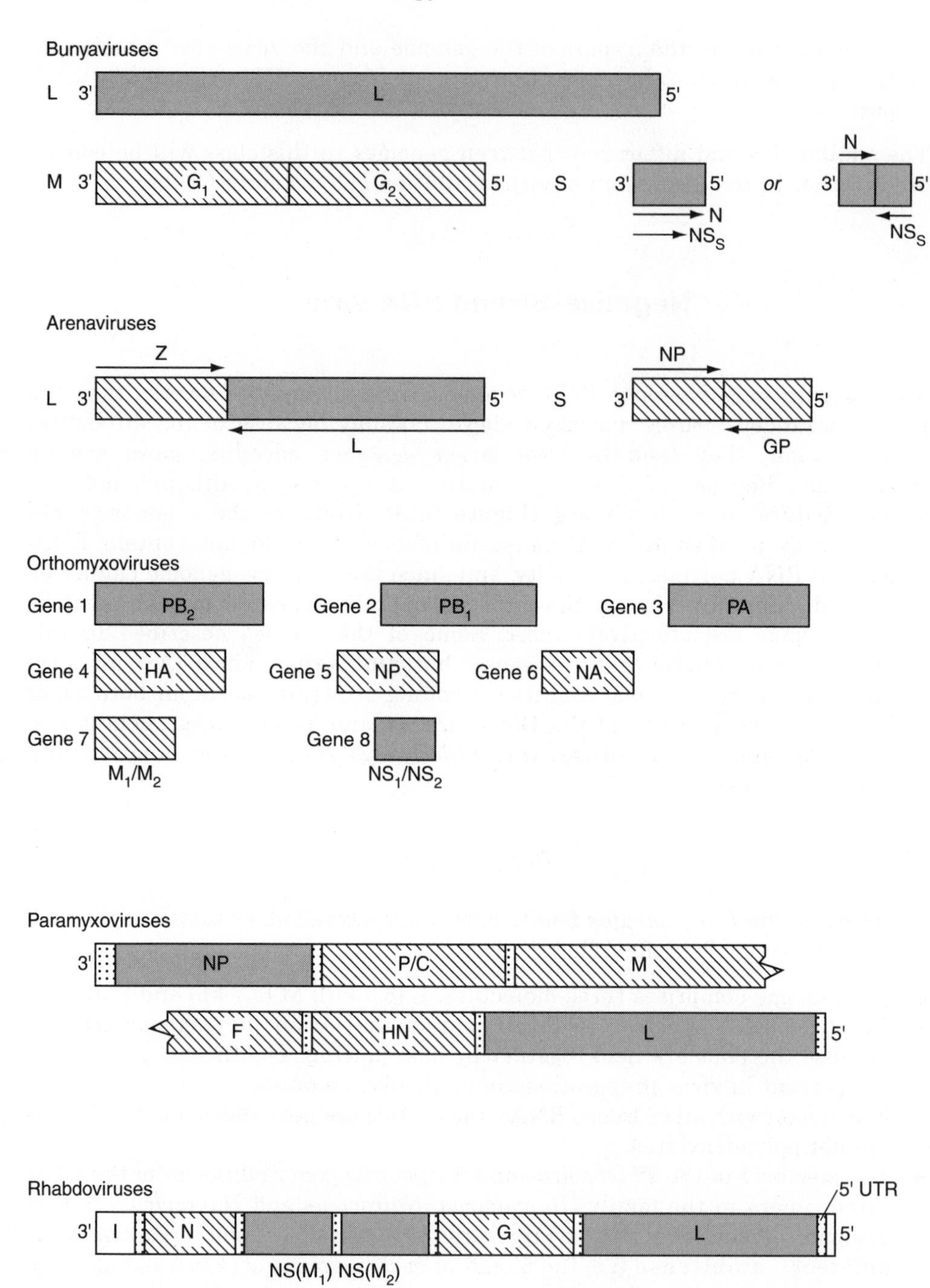

Figure 3.13 Genomic organization of negative-stranded RNA viruses.

Arenaviruses

Arenavirus genomes consist of linear, single-stranded RNA. There are two genome segments: L (5.7 kb) and S (2.8 kb). Both have an ambisense organization, as above.

Orthomyxoviruses

See discussion of segmented genomes (below).

Paramyxoviruses

Members of the *Paramyxoviridae* family have non-segmented (−)sense RNA of 15–16 kb. Typically, six genes are organized in a linear arrangement (3′-NP-P/C/V-M-F-HN-L-5′) separated by repeated sequences: a polyadenylation signal at the end of the gene, an intergenic sequence (GAA), and a translation start signal at the beginning of the next gene.

Rhabdoviruses

Viruses of the *Rhabdoviridae* family have non-segmented, (−)sense RNA of approximately 11 kb. There is a leader region of approximately 50 nt at the 3′ end of the genome and a 60 nt untranslated region (UTR) at the 5′ end of the vRNA. Overall, the genetic arrangement is similar to that of paramyxoviruses, with a conserved polyadenylation signal at the end of each gene and short intergenic regions between the five genes.

Segmented and Multipartite Virus Genomes

There is often some confusion over these two categories of genome structure. Segmented virus genomes are those which are divided into two or more physically separate molecules of nucleic acid, all of which are then packaged into a single virus particle. In contrast, although multipartite genomes are also segmented, each genome segment is packaged into a separate virus particle. These discrete particles are structurally similar and may contain the same component proteins, but often differ in size depending on the length of the genome segment packaged. In one sense, multipartite genomes are, of course, segmented, but this is not the strict meaning of these terms as they will be used here. Virus families with segmented and multipartite genomes are listed in Table 3.1.

Segmentation of the virus genome has a number of advantages and disadvantages. There is an upper limit to the size of a non-segmented virus genome which results from the physical properties of nucleic acids, particularly the tendency of

Table 3.1 Segmented and multipartite virus families

Segmented			Multipartite		
Family	Segments	Host	Family	Segments	Host
Arenavirus (ambisense RNA)	2	Vertebrates	*Geminivirus* (group III) (single-stranded DNA)	Bipartite	Plants
Bunyavirus ((−)/ambisense RNA)	3	Vertebrates/Plants	*Comovirus* (single-stranded RNA)	Bipartite	Plants
Orthomyxovirus ((−)sense RNA)	8 (7[a])	Vertebrates	*Furovirus* (single-stranded RNA)	Bipartite	Plants
Reovirus (double-stranded RNA)	10 (11[b])	Vertebrates/Plants	*Tobravirus* (single-stranded RNA)	Bipartite	Plants
Tenuivirus (ambisense RNA)	5	Plants	*Partitiviridae* (double-stranded RNA)	Bipartite	Plants/Fungi
			Bromoviridae (single-stranded RNA)	Tripartite	Plants
			Hordeivirus (single-stranded RNA)	Tripartite	Plants

[a] Seven segments in influenza C viruses.
[b] Eleven segments in rotaviruses.

long molecules to break due to shear forces (and for each particular virus, the length of nucleic acid which can be packaged into the capsid). The problem of strand breakage is particularly relevant for single-stranded RNA, which is much more chemically labile than double-stranded DNA. The longest single-stranded RNA genomes are those of the coronaviruses, at approximately 30 kb, but the longest double-stranded DNA virus genomes are considerably longer, e.g. cytomegalovirus at approximately 235 kbp. Physical breakage of the genome results in its biological inactivation, since it cannot be completely transcribed, translated, or replicated. Segmentation means that the virus avoids 'having all its eggs in one basket' and also reduces the probability of breakages due to shearing, thus increasing the total potential coding capacity of the entire genome. However, the disadvantage of this strategy is that all the individual genome segments must be packaged into each virus particle, or the virus will be defective as a result of loss of genetic information. In general, it is not understood how this control of packaging is achieved.

Separating the genome segments into different particles (the multipartite strategy) removes the requirement for accurate sorting, but introduces a new problem in that all the discrete virus particles must be taken up by a single host cell to establish a productive infection. This is perhaps the reason multipartite viruses are only found in plants. Many of the sources of infection by plant viruses, such as inoculation by sap-sucking insects or after physical damage to tissues, result in a large inoculum of infectious virus particles, providing opportunities for infection of an initial cell by more than one particle.

The genetics of segmented genomes are essentially the same as those of non-segmented genomes, with the addition of the reassortment of segments, as discussed above. Reassortment can occur whether the segments are packaged into a single particle or are in a multipartite configuration. Reassortment is a powerful means of achieving rapid generation of genetic diversity; this could be another possible reason for its evolution. Segmentation of the genome also has implications for the partition of genetic information and the way in which it is expressed; this will be considered further in Chapter 5.

To understand the complexity of these genomes, consider the organization of a segmented virus genome (orthomyxovirus – influenza virus) and a multipartite genome (geminivirus). The influenza virus genome is composed of eight segments (in influenza A and B strains, seven in influenza C) of single-stranded, (−)sense RNA (Table 3.2). The identity of the proteins encoded by each genome segment were determined originally by genetic analysis of the electrophoretic mobility of the individual segments from reassortant viruses and by analysis of a large number of mutants covering all eight segments. The eight segments have common nucleotide sequences at the 5′ and 3′ ends (Figure 3.14) which are necessary for replication of the genome (Chapter 4). These sequences are complementary to one another and, inside the particle, the ends of the genome segments are held together by base-pairing, forming a 'pan-handle' structure, which again is believed to be involved in replication. The RNA genome segments are not packaged as naked nucleic acid, but in association with the gene *5* product, the nucleoprotein, and are visible in electron micrographs as helical structures. Here, there is a paradox. Biochemically and genetically, each genome segment

Table 3.2 Influenza virus genome segments

Segment	Size(nt)	Polypeptide(s)	Function
1	2341	PB_2	Transcriptase: cap binding
2	2341	PB_1	Transcriptase: elongation
3	2233	PA	Transcriptase: (?)
4	1778	HA	Haemagglutinin
5	1565	NP	Nucleoprotein: RNA binding; part of transcriptase complex
6	1413	NA	Neuraminidase
7	1027	M_1	Matrix protein: major component of virion
		M_2	Integral membrane protein–ion channel
8	890	NS_1	Non-structural (nucleus) function unknown
		NS_2	Non-structural (nucleus+cytoplasm) function unknown

behaves as an individual, discrete entity. However, in electron micrographs of influenza virus particles disrupted with non-ionic detergents, the nucleocapsid has the physical appearance of a single, long helix. Clearly, there is some interaction between the genome segments and it is this which explains the ability of

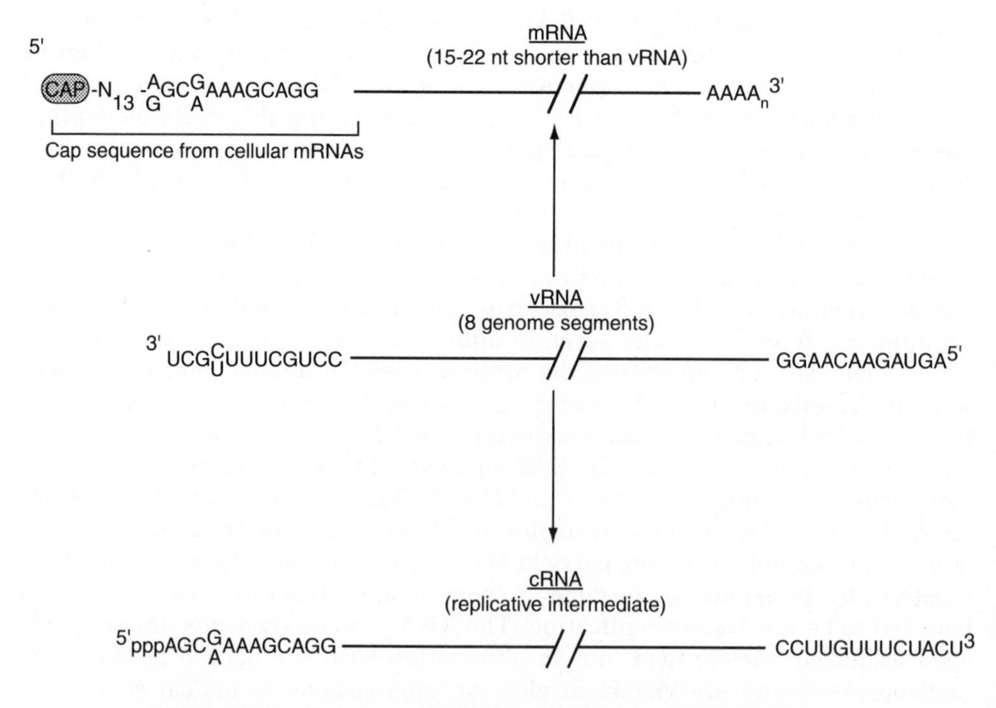

Figure 3.14 Common terminal sequences of influenza RNAs.

influenza virus particles to select and package the genome segments within each particle with a surprisingly low error rate, considering the difficulty of the task (Chapter 2). The genome sequences and protein interaction which operate this subtle mechanism have not yet been defined.

In many tropical and subtropical parts of the world, Geminiviruses are important plant pathogens. Geminiviruses are divided into three taxonomic groups based on their host plants (monocotyledons or dicotyledons) and insect vectors (leafhoppers or whiteflies). In groups I and II, the genome consists of a single-stranded DNA molecule of approximately 2.7 kb. The DNA packaged into these virions has been arbitrarily designated as (+)sense, although both the (+)sense and (−)sense strands found in infected cells contain protein coding sequences. The genome of the group III geminiviruses is bipartite and consists of two circular, single-stranded DNA molecules, each of which is packaged into a discrete particle (Figure 3.15). Both of the strands comprising the genome are approximately 2.7 kb long and differ from one another completely in nucleotide sequence, except for a shared 200 nt non-coding sequence involved in DNA replication. The two genomic DNAs are packaged into entirely separate capsids. Because establishment of a productive infection requires both parts of the genome it is necessary for a minimum of two virus particles bearing one copy of each of the genome segments to infect a new host cell. Although geminiviruses do not multiply in the tissues of their insect vectors (**non-propagative transmission**), a sufficiently large amount of virus is ingested and subsequently deposited onto a new host plant to favour such superinfections.

Both of these examples show a high density of coding information. In influenza virus, genes 7 and 8 both encode two proteins in overlapping reading frames. In geminiviruses, both strands of the virus DNA found in infected cells contain coding information, some of which is present in overlapping reading frames. It is possible that this high density of genetic information is the reason these viruses have resorted to divided genomes, in order to regulate the expression of this information (see Chapter 5).

Reverse Transcription and Transposition

The first successes of molecular biology were the discovery of the double-helix structure of DNA and the elucidation of the language of the genetic code. The importance of these findings does not lie in the mere facts, but in their importance in allowing predictions to be made about the fundamental nature of living organisms. The confidence which flowed from these early triumphs resulted in the development of a grand universal theory, which was called the 'central dogma of molecular biology', namely that all cells (and hence viruses) worked on a simple organizing principle – the unidirectional flow of information from DNA, through RNA, into proteins. In the mid-1960s, there were rumblings that life might not be so simple.

In 1963, Howard Temin showed that the replication of retroviruses, whose

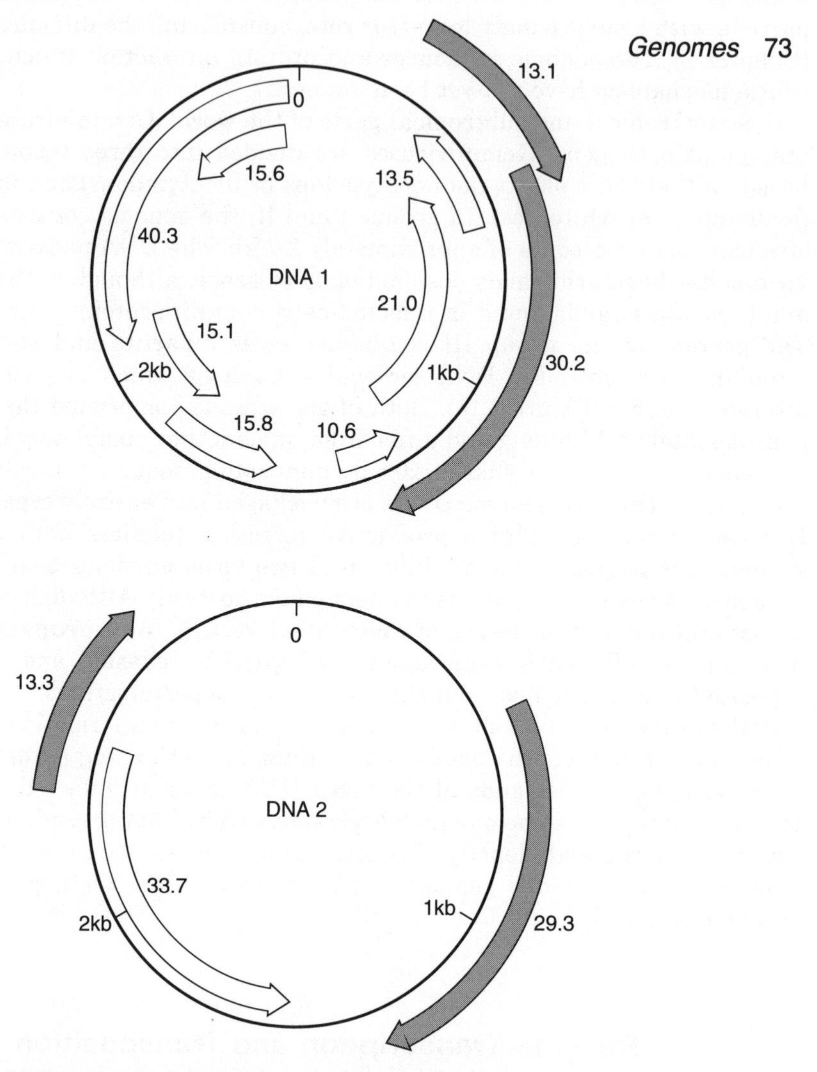

Figure 3.15 Organization and protein-coding potential of the bipartite geminivirus genome.

particles contain RNA genomes, was inhibited by actinomycin D, an antibiotic which binds only to DNA. The replication of other RNA viruses is not inhibited by this drug. So pleased was the scientific community with an all-embracing dogma that these facts were largely ignored until 1970, when Temin and David

Baltimore simultaneously published the observation that retrovirus particles contain an RNA-dependent DNA polymerase – reverse transcriptase. This finding alone was important enough, but like the earlier conclusions of molecular biology, has subsequently had reverberations for the genomes of all organisms, and not merely a few virus families. It is now known that retrotransposons with striking similarities to retrovirus genomes form a substantial part of the genomes of all higher organisms, including man. Earlier ideas of genomes as constant, stable structures have been replaced with the realization that they are, in fact, dynamic and rather fluid entities.

The concept of transposable genetic elements – specific sequences that are able to move from one position in the genome to another – was put forward by Barbara McClintock in the 1940s. Such **transposons** fall into two groups:

(1) Simple transposons, which do not undergo reverse transcription and are found in prokaryotes (e.g. the genome of bacteriophage Mu)
(2) Retrotransposons, which closely resemble retrovirus genomes and are bounded by long direct repeats (long terminal repeats, LTRs). These move by means of a transcription/reverse transcription/integration mechanism and are found in eukaryotes (e.g. the yeast Ty element and retrovirus genomes).

Both types show a number of similar properties:

- They are believed to be responsible for a high proportion of apparently 'spontaneous' mutations
- They promote a wide range of genetic rearrangements in host cell genomes, such as deletions, inversions, duplications, and translocations of the neighbouring cellular DNA
- The mechanism of insertion generates a short (3–13 bp) duplication of the DNA sequence on either side of the inserted element
- The ends of the transposable element consist of inverted repeats, 2–50 bp long
- Transposition is often accompanied by replication of the element – necessarily so in the case of retrotransposons, but this also often occurs with prokaryotic transposition
- Transposons control their own transposition functions, encoding proteins which act on the element in ***cis*** or in ***trans***.

Bacteriophage Mu infects *E. coli* and consists of a complex, tailed particle containing a linear, double-stranded DNA genome of about 40 kb, with host cell-derived sequences of between 0.5 and 3.0 kbp attached to the right-hand end of the genome (Figure 3.16). Mu is a **temperate bacteriophage** whose replication can proceed through two pathways; one involves integration of the genome into that of the host cell and results in **lysogeny**, and the other is **lytic** replication which results in the death of the cell (see Chapter 5). Integration of the phage genome into that of the host bacterium occurs at random sites in the cell genome. Integrated phage genomes are known as **prophage** and integration is essential for the establishment of lysogeny. At intervals in bacterial cells lysogenic for Mu, the prophage undergoes transposition to a different site in the host genome. The mechanism leading to transposition is different from that responsible for the initial integration of the phage genome (which is conservative, i.e.

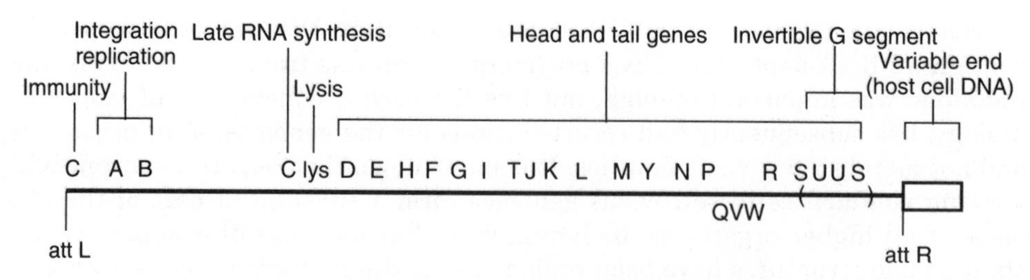

Figure 3.16 Organization of the bacteriophage Mu genome.

does not involve replication), and is a complex process requiring numerous phage-encoded and host cell proteins. Transposition is tightly linked to replication of the phage genome and results in the formation of a 'co-integrate', i.e. a duplicated copy of the phage genome flanking a target sequence in which insertion has occurred. The original Mu genome remains in the same location where it first integrated and is joined by a second integrated genome at another site. (NB Not all prokaryotic transposons use this process. Some, such as TN10, are not replicated during transposition but are excised from the original integration site and integrate elsewhere.) There are two consequences of such a transposition: first the phage genome is replicated during this process (advantageous for the virus) and second, the sequences flanked by the two phage genomes (which form repeated sequences) are at risk of secondary rearrangements, including deletions, inversions, duplications, and translocations (possibly but not necessarily deleterious for the host cell).

The Ty element is representative of a class of sequences found in yeast and other eukaryotes known as retrotransposons. Unlike bacteriophage Mu, such elements are not true viruses, but do bear striking similarities to retroviruses. The genomes of most strains of *Saccharomyces cerevisiae* contain 30–35 copies of the Ty elements, which are around 6 kbp long and contain direct repeats of 245–371bp at each end (Figure 3.17). Within this repeat sequence is a promoter which results in the transcription of a terminally redundant 5.6 kb mRNA. This contains two genes, *TyA*, which has homology to the *gag* gene of retroviruses, and *TyB*, which is homologous to the *pol* gene. The protein encoded by *TyA* is capable of forming a roughly spherical, 60 nm diameter 'virus-like particle' (VLP). The 5.6 kb RNA transcript can be incorporated into such particles, resulting in the formation of intracellular structures known as Ty-VLPs. Unlike true viruses these particles are not infectious for yeast cells, but if accidentally taken up by a cell, can carry out reverse transcription of their RNA content to form a double-stranded DNA Ty element which can then integrate into the host cell genome (see below).

The most significant difference between retrotransposons such as *Ty*, *copia* (a similar element found in *Drosophila melanogaster*), and retroviruses proper is the presence of an additional gene in retroviruses, *env*, which encodes an envelope glycoprotein (see Chapter 2). The envelope protein is responsible for receptor binding and has allowed retroviruses to escape the intracellular lifestyle of retrotransposons, to form a true virus particle and propagate themselves widely by infection of other cells (Figure 3.17). Retrovirus genomes have three unique features:

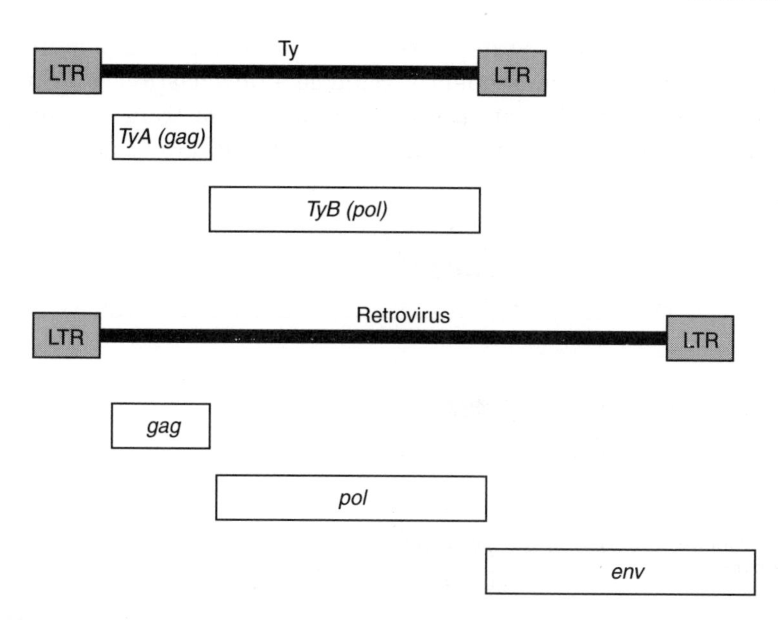

Figure 3.17 The genetic organization of retrotransposons such as Ty (above) and retrovirus genomes (below) shows a number of similarities, inclusing the presence of direct long terminal repeats (LTRs) at either end.

(1) They are the only viruses which are diploid.
(2) They are the only RNA viruses whose genome is produced by cellular transcriptional machinery (without any participation by a virus-encoded polymerase).
(3) They are the only (+)sense RNA viruses whose genome does not serve directly as mRNA immediately after infection.

During the process of reverse transcription (Figure 3.18), the two single-stranded (+)sense RNA molecules which comprise the virus genome are converted into a double-stranded DNA molecule somewhat longer than the RNA templates owing to the duplication of direct repeat sequences at each end – the long terminal repeats (LTRs) (Figure 3.19). Some of the steps in reverse transcription have remained mysterious; for example, the apparent jumps which the polymerase makes from one end of the template strand to the other. In fact, these steps can be explained by the observation that complete conversion of retrovirus RNA into double-stranded DNA only occurs in a partially uncoated core particle and cannot be duplicated accurately *in vitro* with the reagents free in solution. This indicates that the conformation of the two RNAs inside the retrovirus nucleocapsid dictates the course of reverse transcription – the 'jumps' are nothing of the sort, since the ends of the strands are probably held adjacent to one another inside the core.

Reverse transcription has important consequences for retrovirus genetics. Firstly, it is a highly error-prone process, because reverse transcriptase does not carry out the proofreading functions performed by cellular DNA-dependent

DNA polymerases. This results in the introduction of many mutations into retrovirus genomes and, consequently, rapid genetic variation (see Spontaneous Mutations, above). In addition, the process of reverse transcription promotes genetic recombination. Since two RNAs are packaged into each virion and used

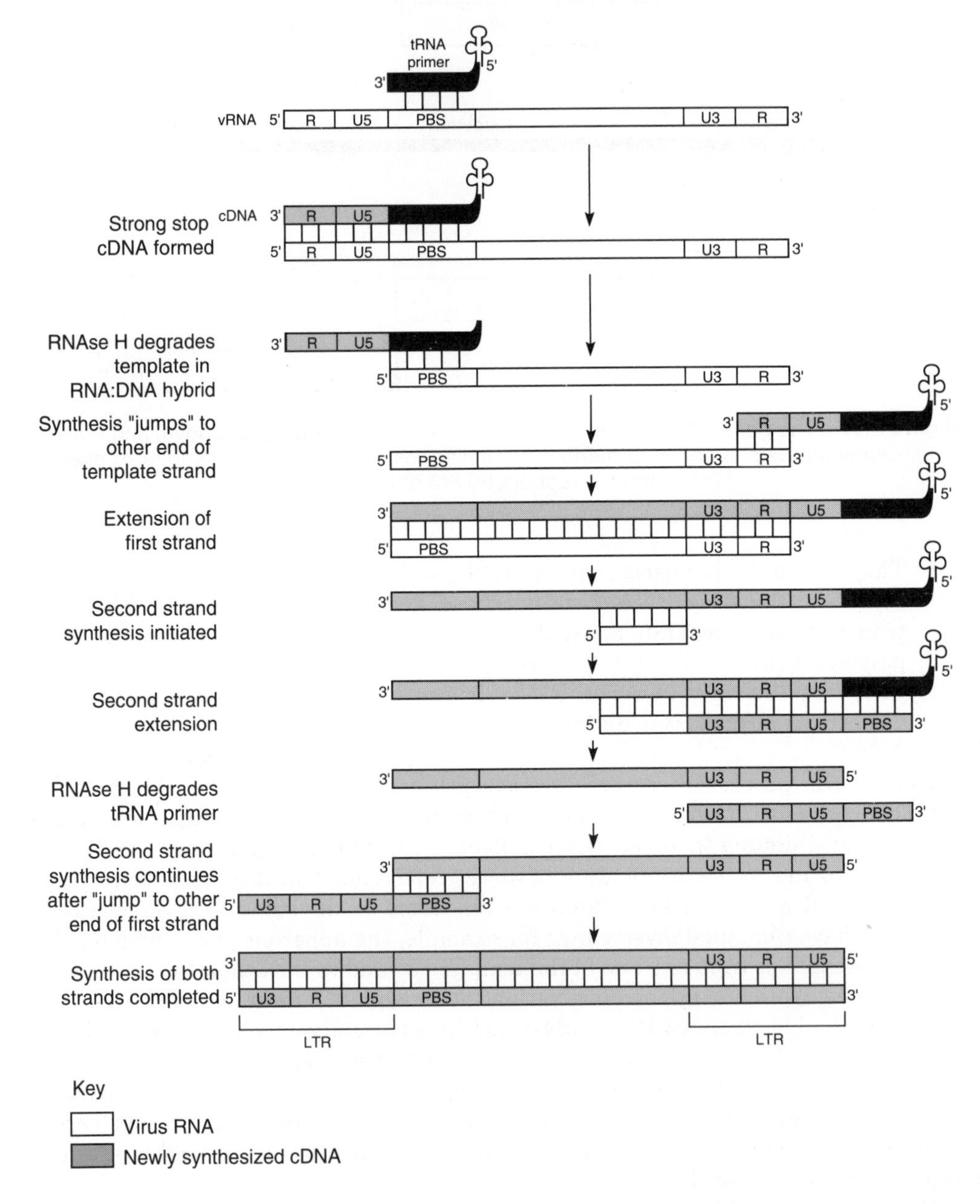

Figure 3.18 Mechanism of reverse transcription of retrovirus RNA genomes, in which two molecules of RNA are converted into a single (terminally redundant) double-stranded DNA provirus.

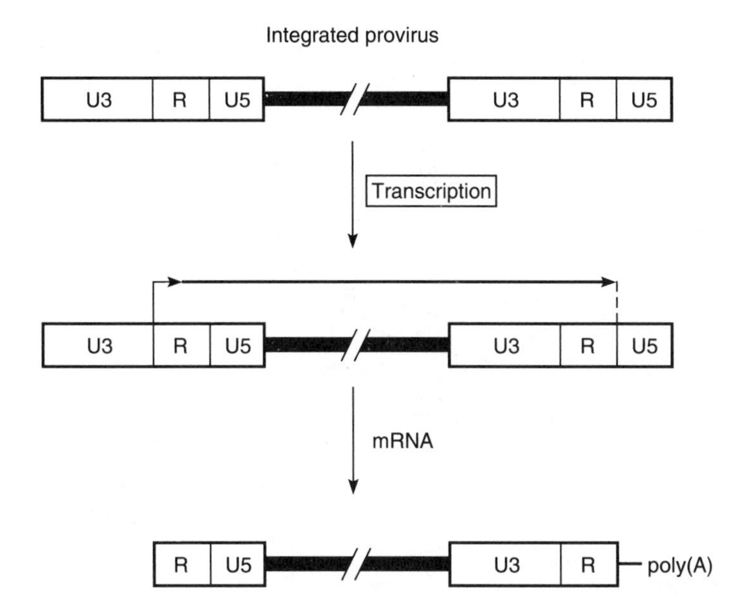

Figure 3.19 Generation of repeated information in retrovirus long terminal repeats. In addition to their role in reverse transcription, these sequences contain important control elements involved in the expression of the virus genome. including a transcriptional promoter in the U3 region and polyadenylation signal in the R region.

as the template for reverse transcription, recombination can and does occur between the two strands. Although the mechanism responsible for this is not clear, if one of the RNA strands differs from the other, for example, by the presence of a mutation, and recombination occurs, then the resulting virus will be genetically distinct from either of the parental viruses.

After reverse transcription is complete, the double-stranded DNA migrates into the nucleus, still in association with virus proteins. The mature products of the *pol* gene are, in fact, a complex of polypeptides which include three distinct enzymatic activities: reverse transcriptase and RNAse H, which are involved in reverse transcription, and integrase, which catalyses integration of virus DNA into the host cell chromatin, following which it is known as the **provirus** (Figure 3.20). Three forms of double-stranded DNA are found in retrovirus-infected cells following reverse transcription: linear DNA and two circular forms which contain either one or two LTRs. From the structure at the ends of the provirus, it was previously believed that the two-LTR circle was the form used for integration. In recent years, systems which have been developed to study the integration of retrovirus DNA *in vitro* show that it is the linear form which integrates. This discrepancy can be resolved by a model in which the ends of the two LTRs are held in close proximity by the reverse transcriptase–integrase complex. The net result of integration is that 1–2 bp are lost from the end of each LTR and 4–6 bp of cellular DNA are duplicated on either side of the provirus. It is unclear whether there is any specificity regarding the site of integration into the cell

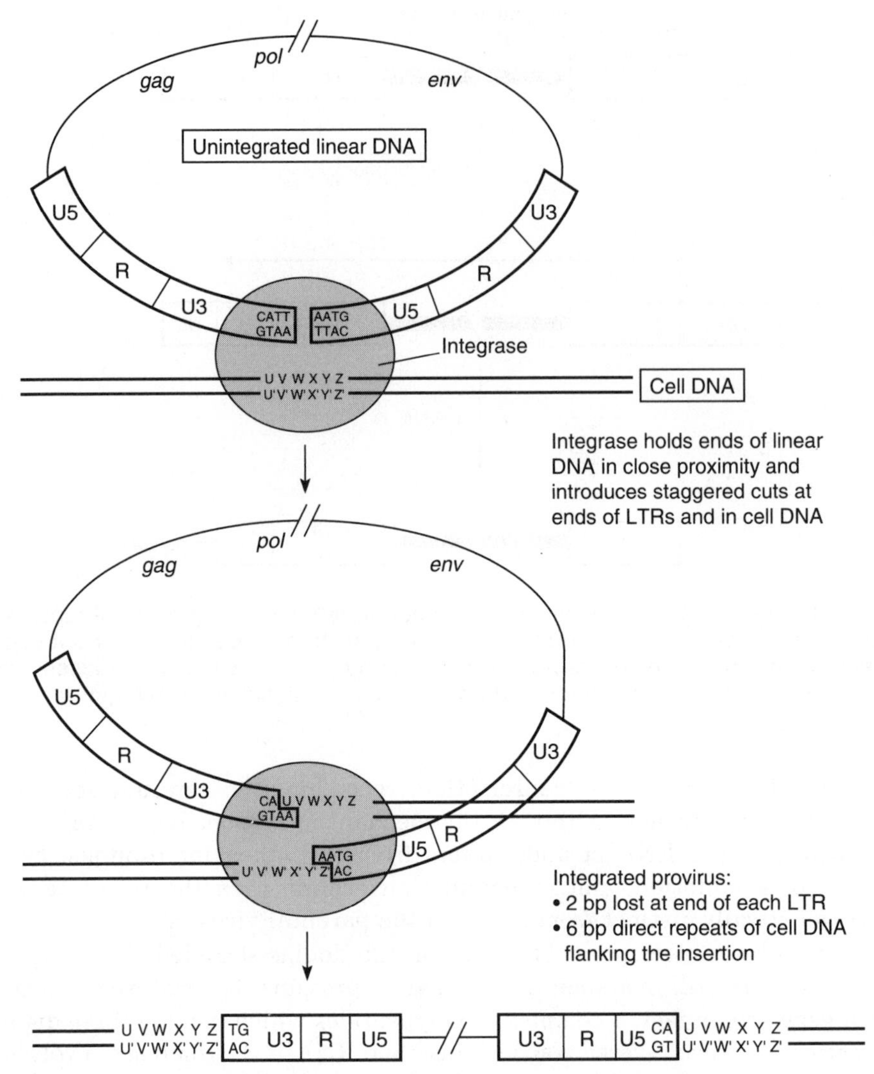

Figure 3.20 Mechanism of integration of retrovirus genomes into the host cell chromatin.

genome. What is obvious is that there is no simple target sequence, but it is possible that there may be (numerous) regions or sites in the eukaryotic genome which are more likely to be integration sites than others.

Following integration, the DNA provirus genome becomes essentially a collection of cellular genes and is at the mercy of the cell for expression. There is no mechanism for the precise excision of integrated proviruses, some of which are known to have been 'fossilized' in primate genomes through millions of years of evolution, although proviruses may sometimes be lost or altered by modifications of the cell genome. The only way out for the virus is transcription, forming what is essentially a full-length mRNA (minus the terminally redundant sequences from the LTRs). This RNA is the vRNA and two copies are packaged into virions (Figure 3.19).

There are, however, two different genome strategies used by viruses which involve reverse transcription. It is at this point that the difference between them becomes obvious. One strategy, as used by retroviruses and described above, culminates in the packaging of RNA into virions as the virus genome. The other, used by hepadnaviruses and caulimoviruses, switches the RNA and DNA phases of replication and results in DNA virus genomes. This is achieved by utilizing reverse transcription not as an early event in replication as retroviruses do, but as a late step during the formation of the virus particle.

Hepatitis B virus (HBV) is the prototype member of the family *Hepadnaviridae*. HBV virions are spherical, lipid-containing particles, 42–47 nm diameter, which contain a partially double-stranded ('gapped') DNA genome, plus an RNA-dependent DNA polymerase (i.e. reverse transcriptase) (Figure 3.21). Hepadnaviruses have very small genomes consisting of a (−)sense strand of 3.0–3.3 kb (varies between different hepadnaviruses) and a (+)sense strand of 1.7–2.8kb (varies between different particles). On infection of cells, three major genome transcripts are produced: 3.5, 2.4, and 2.1 kb mRNAs. All have the same polarity (i.e. are transcribed from the same strand of the virus genome) and the same 3′ ends, but have different 5′ ends (i.e. initiation sites). These transcripts are heterogeneous in size and it is not completely clear which proteins each transcript encodes, but there are four known genes in the virus:

- C encodes the core protein
- P encodes the polymerase

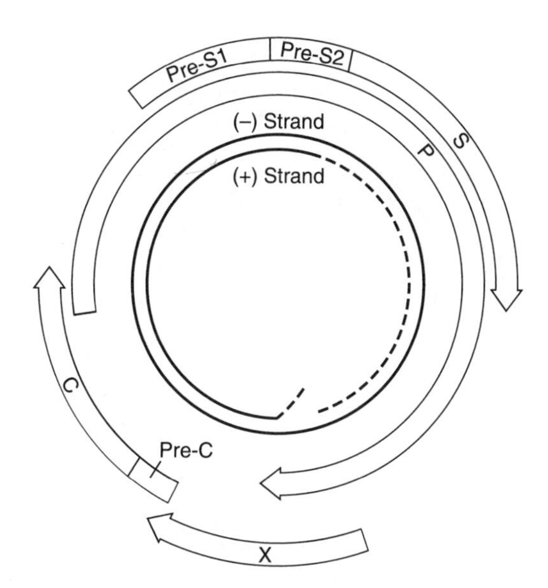

Figure 3.21 Structure, organization and protein-coding potential of the hepatitis B virus genome.

- S encodes the three polypeptides of the surface antigen, pre-S1, pre-S2, and S (which are derived from alternative start sites)
- X encodes a transactivator of viral transcription (and possibly cellular genes?).

Closed circular DNA is found soon after infection in the nucleus of the cell and is probably the source of the above transcripts. This DNA is produced by repair of the gapped virus genome as follows:

(1) Completion of the (+)sense strand.
(2) Removal of a protein primer from the (−)sense strand and an oligoribonucleotide primer from the (+)sense strand.
(3) Elimination of terminal redundancy at the ends of the (−)sense strand.
(4) Ligation of the ends of the two strands.

It is not known how or by which proteins (viral or cellular) these events are carried out. The 3.5kb RNA transcript, core antigen and polymerase form core particles and the polymerase converts the RNA to DNA in the particles as they form in the cytoplasm.

The genome structure and replication of cauliflower mosaic virus (CaMV), the prototype member of the *Caulimovirus* group, is reminiscent of that of hepadnaviruses, although there are differences between them. The CaMV genome consists of a gapped, circular double-stranded DNA molecule of about 8 kbp, one strand of which is known as the α strand and contains a single gap, and a complementary strand which contains two gaps (Figure 3.22). There are eight genes encoded in this genome, although not all eight products have been

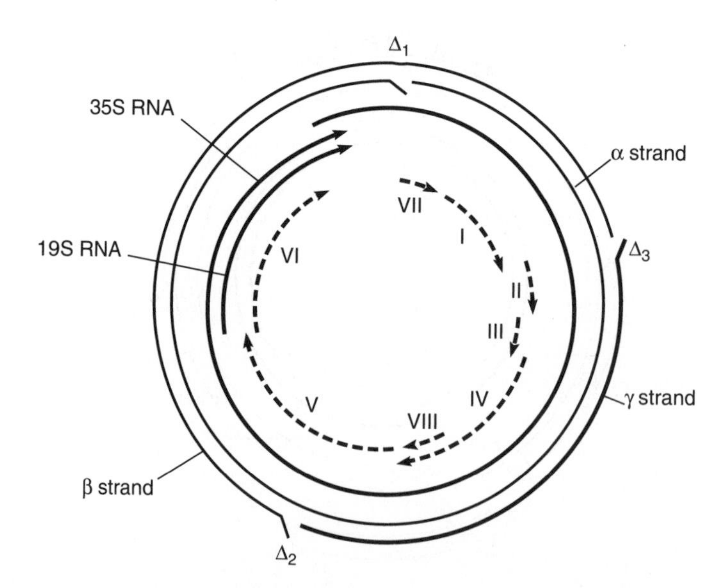

Figure 3.22 Structure, organization and protein-coding potential of the cauliflower mosaic virus genome.

Table 3.3 Reverse transcription of virus genomes

Features	Caulimoviruses	Hepadnaviruses	Retroviruses
Genome	DNA	DNA	RNA
Primer for (−)strand synthesis	tRNA	Protein	tRNA
Terminal repeats (LTRs)	No	No	Yes
Specific integration of virus genome	No	No	Yes

detected in infected cells. Replication of the CaMV genome is similar to that of HBV. The first stage is the migration of the gapped virus DNA to the nucleus of the infected cell where it is repaired to form a covalently closed circle. This DNA is transcribed to produce two polyadenylated transcripts, one long (35S) and one shorter one (19S). In the cytoplasm, the 19S mRNA is translated to produce a protein which forms large **inclusion bodies** in the cytoplasm of infected cells and it is in these sites that the second phase of replication occurs. In these replication complexes, some copies of the 35S mRNA are translated while others are reverse transcribed and packaged into virions as they form. The differences between reverse transcription of these virus genomes and those of retroviruses are summarized in Table 3.3.

Evolution and Epidemiology

Epidemiology is concerned with the distribution of disease and the derivation of strategies to reduce or prevent it. Virus infections present considerable difficulties for this process. Except for epidemics where acute symptoms are obvious, the major evidence of virus infection available to the epidemiologist is the presence of anti-viral antibodies in patients. This information frequently provides an incomplete picture, and it is often difficult to assess whether a virus infection occurred recently or at some time in the past. Alternative techniques, such as the isolation of viruses in experimental plants or animals, are laborious and impossible to apply to large populations. Molecular biology provides sensitive, rapid, and sophisticated techniques to detect and analyse the genetic information stored in virus genomes and has resulted in a new area of investigation, molecular epidemiology.

One drawback of molecular genetic analysis is that some knowledge of the nature of a virus genome is necessary before it can be investigated. However, it should be obvious from this chapter that we now possess a great amount of information about the structure and nucleotide sequences of at least a few representative members of most of the known virus groups. This information allows virologists to look in two directions: back to where viruses came from and forward to chart the course of future epidemics and diseases. Sensitive detection of nucleic acids by amplification techniques such as the polymerase chain reaction is already having a major impact on this type of epidemiological investigation.

There are (at least!) three theories which seek to explain the origin of viruses:

- *Regressive evolution*: This theory states that viruses are degenerate life-forms which have lost many functions that other organisms possess and have only retained the genetic information essential to their parasitic way of life
- *Cellular origins*: In this theory viruses are thought to be subcellular, functional assemblies of macromolecules which have escaped their origins inside cells
- *Independent entities*: This says that viruses evolved on a parallel course to cellular organisms from the self-replicating molecules believed to have existed in the primitive prebiotic 'RNA world'.

While each of these theories has its devotees and this subject provokes fierce disagreements, the fact is that viruses exist, and we are all infected with them. The practical importance of the origin of viruses is that this issue may have implications for virology here and now. Genetic and nucleotide sequence relationships between viruses can reveal the origins not only of individual viruses, but also of whole families and possible 'superfamilies' (Figure 3.23). In a number of groups of viruses previously thought to be unrelated, nucleotide sequencing has revealed that functional regions appear to be grouped together in a similar way. The extent to which there is any sequence similarity between these regions in different viruses varies, although clearly, the active sites of enzymes such as viral replicases are strongly conserved. The emphasis in these groupings is more on functional and organizational similarities. The official classification scheme for viruses does not presently recognize a higher level grouping than the 'family' (see Appendix 1). The one exception to this is the grouping of negative-stranded RNA virus families, the *Mononegvirales*, which embraces the three families of non-segmented negative-stranded viruses, the *Filoviridae, Paramyxoviridae*, and *Rhabdoviridae*. Superfamilies are thus equivalent to the 'orders' of formal biological nomenclature.

Such ideas may allow us to predict the properties and behaviour of new viruses, or to develop new drugs based on what is already known about existing viruses. Whether these shared patterns suggest the descent of present-day viruses from a limited number of primitive ancestors or are evidence of convergent evolution between different virus families is uncertain. Although it is tempting to speculate on events which may have occurred before the origins of 'life' as it is presently recognized, it would be unwise to discount the pressures which might result in viruses with diverse origins assuming common genetic solutions to common problems of storing, replicating, and expressing genetic information. This is particularly true now that we appreciate the plasticity of virus and cellular genomes and the mobility of genetic information from virus to virus, cell to virus, and virus to cell. There is no reason to believe that virus evolution has stopped, and it would be perilous to do so. The practical consequences of ongoing evolution and the concept of **emergent viruses** is described in Chapter 7.

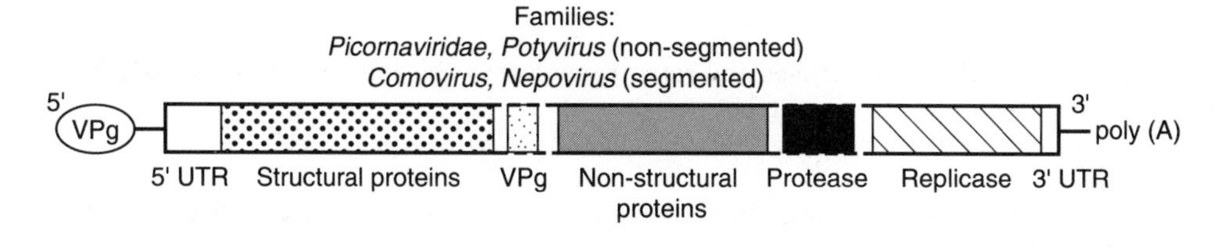

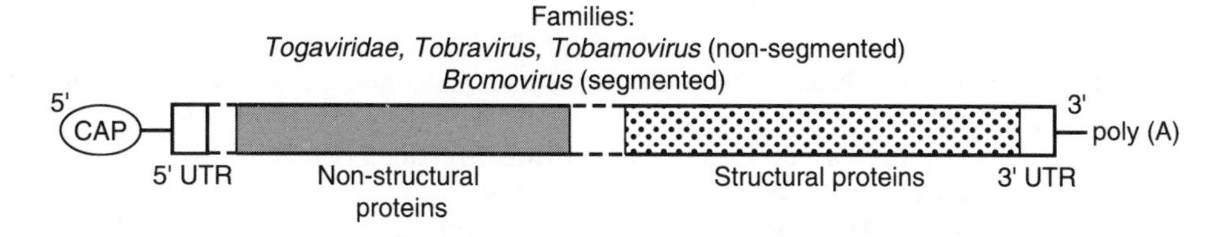

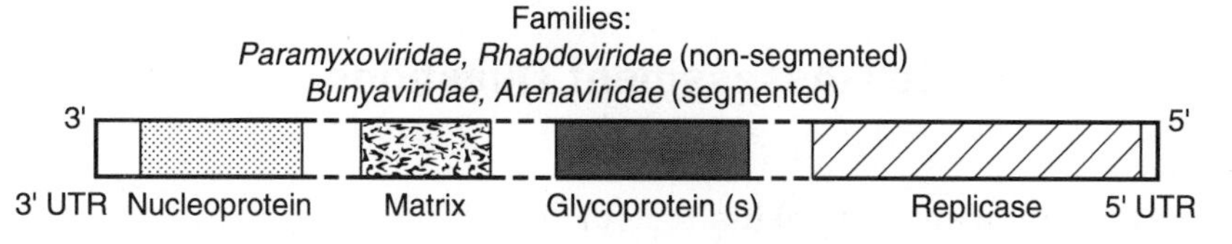

Figure 3.23 Similarities in the function and organization of the genomes of different virus families allow them to be grouped together as 'superfamilies'.

Summary

Molecular biology has put much emphasis on the structure and function of the virus genome. At first sight, this tends to emphasize the tremendous diversity of virus genomes. On closer examination, similarities and unifying themes become more apparent. Sequences and structures at the ends of virus genomes are in some ways functionally more significant than the unique coding regions within them. Common patterns of genetic organization seen in virus superfamilies suggest either that many viruses have evolved from common ancestors or that convergent evolution and exchange of genetic information between viruses has resulted in common solutions to common problems.

Further Reading

Baltimore, D. (1985). Retroviruses and retrotransposons: the role of reverse transcription in shaping the eukaryotic genome. *Cell* **40**: 481–2.
Belfort, M. (1989). Bacteriophage introns: parasites within parasites? *Trends Genet.* **5**: 209–213.
Conzelmann, K. (1996). Genetic manipulation of negative-strand RNA viruses. *J. Gen. Virol.* **77**: 381–89.
Dawson, W.O. and Lehto, K.M. (1990). Regulation of tobamovirus gene expression. *Adv. Vir. Res.* **38**: 307–342.
Holland, J.J. (Ed.). (1992). Genetic diversity of RNA Viruses. *Curr. Top. Microbiol. Immunol.*, Vol. **176**.
Morse, S.S. (Ed.). (1994). *The Evolutionary Biology of Viruses.* Raven Press, New York.
Saedler, H. and Gierl, A. (Eds). (1996). Transposable elements. *Curr. Top. Microbiol. Immunol.*, Vol. **204**.
Symonds, N., Toussaint, A., Van de Putte, P. and Howe, M. (Eds). (1987). *Phage Mu.* Cold Spring Harbor Laboratory Press, Cold Spring Harbor.
Zaccomer, B. *et al.* (1995). The remarkable variety of plant RNA virus genomes. *J. Gen. Virol.* **76**: 231–247.

Self-Assessment Questions

(3.1) Are the following statements true or false?
- (a) Virus genomes vary in size from 3500 nucleotides to approximately 235 kbp.
- (b) The genetic code of a virus must be distinct from that of the host organism.
- (c) Virus genomes consisting of (+)sense RNA are infectious when the purified vRNA is applied to cells in the absence of any virus proteins.
- (d) 'Reverse genetics' makes possible the manipulation of (−)sense RNA virus genomes.
- (e) Infection of cells caused by nucleic acid alone is referred to as transfection.

(3.2) Are the following statements true or false?
- (a) 'Superinfection' is where two viruses produce more virus particles than one.
- (b) The probability of recombination between two genetic markers is inversely proportional to the distance between them.
- (c) In viruses with segmented genomes, 'reassortment groups' are equivalent to the individual genome segments.
- (d) Populations of virus genomes consisting of mixtures of molecular variants are known as quasispecies.

(e) Site-specific molecular biological methods are commonly used to mutagenize virus genomes.

(3.3) Are the following statements true or false?
(a) Mutation rates in herpesvirus genomes may be as high as 10^{-3} to 10^{-4} per nucleotide incorporated.
(b) Mutation rates in retrovirus genomes may be as low as 10^{-8} to 10^{-11} per nucleotide incorporated.
(c) Every new HIV provirus formed contains, on average, at least one new mutation.
(d) Temperature sensitive (t.s.) mutants are very stable and rarely revert to the original phenotype.
(e) T.s. mutations usually result from mis-sense mutations in proteins.

(3.4) Are the following statements true or false?
(a) Deletion mutants are very useful for assigning structure–function relationships to virus genomes, because they are easily mapped by physical analysis.
(b) Deletion mutants can only revert to wild type by recombination, which only occurs at low frequencies.
(c) Complementation may be asymmetric or symmetrical.
(d) Phenotypic mixing is where individual progeny viruses from a mixed infection contain genes derived from both parental viruses.
(e) Most virus genomes are diploid.

(3.5) Herpesvirus genomes (true or false?):
(a) Vary in size from 5 to 25 kbp.
(b) Consist of linear, double-stranded DNA.
(c) In herpes simplex virus (HSV), each gene is expressed from its own promoter.
(d) The HSV genome is approximately 152 kbp.
(e) The HSV genome contains about 80 genes, densely packed and with overlapping reading frames.

(3.6) Are the following statements true or false?
(a) Parvovirus genomes consist of linear, non-segmented, single-stranded DNA.
(b) The ends of parvovirus genomes are palindromic sequences of about 115 nt required for replication.
(c) Polyomaviruses are about 5 kbp and contain six genes.
(d) In the virus particle, polyomavirus DNA assumes a supercoiled form and is associated with cellular histones.
(e) The polyomavirus genome contains overlapping reading frames on both strands of the DNA.

(3.7) (+)sense RNA virus genomes (true or false?):
(a) . . . can be up to 90 kbp in length.
(b) . . . are usually modified at the 5′ end.
(c) . . . are always polyadenylated at the 3′ end.

(d) . . . are all expressed as a polyprotein.
(e) . . . all require splicing for expression.

(3.8) (−)sense RNA virus genomes (true or false?):
(a) . . . are sometimes segmented.
(b) . . . cannot have an ambisense coding strategy.
(c) . . . are never capped at the 5′ end.
(d) . . . are never polyadenylated at the 3′ end.
(e) . . . are always circular.

(3.9) Retroviruses (true or false?):
(a) . . . are the only family of viruses to encode reverse transcriptase.
(b) . . . are the only RNA viruses whose genome is produced by cellular transcriptional machinery.
(c) . . . are the only (+)sense RNA viruses whose genome does not serve directly as mRNA immediately after infection.
(d) . . . have high mutation rates.
(e) . . . have a high rate of recombination.

(3.10) Are the following statements true or false?
(a) Segmentation of virus genomes reduces the probability of breakage due to shearing.
(b) Hepadnaviruses have large genomes of up to 55 kbp.
(c) The cauliflower mosaic virus (CaMV) genome consists of a gapped, circular double-stranded DNA molecule.
(d) The cauliflower mosaic virus (CaMV) genome is generated by reverse transcription.
(e) Virus superfamilies prove that many viruses had a common ancestor.

Answers to Self-Assessment Questions are given in Appendix 3.

Chapter 4

Replication

Overview of Virus Replication

The way in which viruses are classified has altered as our perception of them has changed:

(1) *By disease*: Many early civilizations, such as those of ancient Egypt and Greece, were well aware of the pathogenic effects of many different viruses. From these times we have several surprisingly accurate descriptions of diseases of humans, animals, and crops, although of course the nature of the agents responsible for these calamities was not realized at the time. Accurate though these descriptions were, a major problem with such a system of classification is that many diverse viruses cause similar symptoms; for example, respiratory infections with fever may be caused by many different viruses

(2) *By morphology*: As increasing numbers of viruses were isolated and techniques of analysis were improved, it became possible during the 1930s–1950s to classify viruses based on the structure of virus particles. Although this is an improvement on the above scheme, there are still problems in distinguishing between viruses that are morphologically similar but cause disparate clinical symptoms (e.g. the various picornaviruses). During this era, serology became an important aid in virus classification, and particle morphology continues to be an important strand in virus classification

(3) *Functional classification*: In recent years, more emphasis has been placed on the replication strategy of the virus. This is particularly true for the composition and structure of the virus genome and the constraints which this imposes on replication. This approach has accelerated since the development of molecular biology due to the central importance of the virus genome in this new technology. Molecular analysis of virus genomes permits rapid and unequivocal identification of individual virus strains, but is also predictive of the properties of a previously unknown or novel virus with a familiar genome structure. (The classification of viruses is described in Appendix 1.)

In a teleological sense (i.e. crediting an inanimate organism such as a virus with a conscious purpose), the sole objective of a virus is to replicate its genetic information. The nature of the virus genome is therefore pre-eminent in determining what steps are necessary to achieve this aim. In reality, there is a surprising amount of variation that can occur in these processes, even for viruses with seemingly similar genome structures. The reason for this lies in compartmentalization, both of eukaryotic cells into nuclear and cytoplasmic compartments, and of genetic information and biochemical capacity between the virus genome and that of the host cell.

The type of cell infected by the virus has a profound effect on the process of replication. For viruses of prokaryotes, replication reflects to some extent the relative simplicity of their host cells. For viruses with eukaryotic hosts, matters are frequently more complex. There are many examples of animal viruses undergoing different replicative cycles in different cell types. However, the coding capacity of the genome forces all viruses to choose a strategy for replication. This might be one involving heavy reliance on the host cell, in which case the virus genome can be very compact and need only encode the essential information for a few proteins, e.g. parvoviruses. Alternatively, large and complex virus genomes, such as those of poxviruses, encode most of the information necessary for replication, and the virus is only reliant on the cell for the provision of energy and the apparatus for macromolecular synthesis, such as ribosomes (see Chapter 1). Viruses with an RNA lifestyle, i.e. an RNA genome plus messenger RNAs, have no apparent need to enter the nucleus, although during the course of replication, many do. 'DNA viruses', as might be expected, mostly replicate in the nucleus where host cell DNA is replicated and where the biochemical apparatus necessary for this process is located. However, some viruses with DNA genomes (e.g. poxviruses) have evolved to contain sufficient biochemical capacity to be able to replicate in the cytoplasm, with minimal requirement for host cell functions. Most of this chapter will examine the process of virus replication and will look at some of the variations on the basic theme.

Investigation of Virus Replication

Bacteriophages have long been used by virologists as models to understand the biology of viruses. This is particularly true of virus replication. During the first half of this century, two particularly significant experiments which illustrated the fundamental nature of viruses were performed on bacteriophages. The first of these was accomplished by Ellis and Delbruck in 1939 and is usually referred to as the 'single-burst' experiment or 'one-step growth curve' (Figure 4.1). This was the first experiment to show the three essential phases of virus replication:

- Initiation of infection
- Replication and expression of the virus genome
- Release of mature virions from the infected cell.

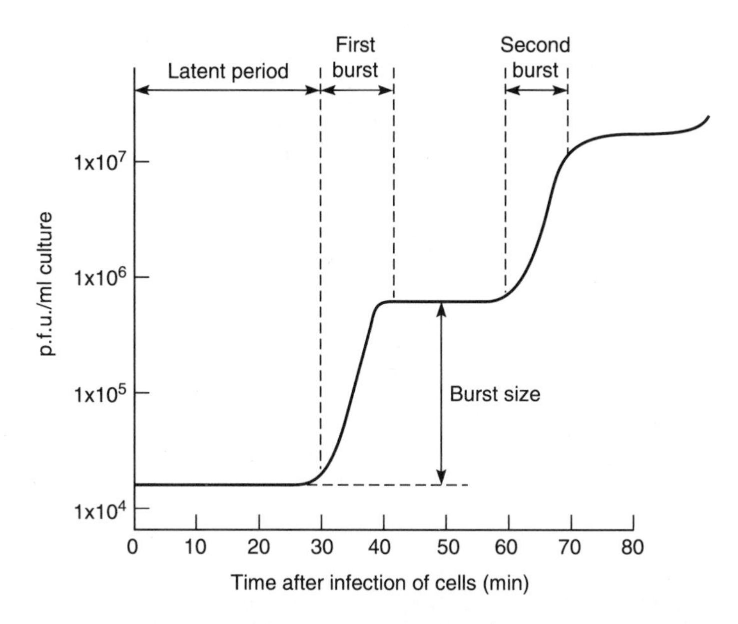

Figure 4.1 The 'one-step growth curve' or 'single-burst experiment'. First performed by Ellis and Delbruck in 1939, this classic experiment illustrates the true nature of virus replication. Details of the experiment are given in the text.

In this experiment, bacteriophages were added to a culture of rapidly growing bacteria and after a period of a few minutes, the culture was diluted, effectively preventing further interaction between the phage particles and the cells. This simple step is the key to the entire experiment, since it effectively synchronizes the infection of the cells and allows the subsequent phases of replication in a population of individual cells and virus particles to be viewed as if it were a single interaction (in much the same way that molecular cloning of nucleic acids allows analysis of populations of nucleic acid molecules as single species). Repeated samples of the culture were taken at short intervals and analysed for bacterial cells by plating onto agar plates and for phage particles by plating onto lawns of bacteria. As can be seen in Figure 4.1, there is a stepwise increase in the concentration of phage particles with time, each increase in phage concentration representing one replicative cycle of the virus. However, the data from this experiment can also by analysed in a different way, by plotting the number of **plaque-forming units** (p.f.u.) per bacterial cell against time (Figure 4.2). In this type of assay, a plaque forming unit can be either a single extracellular virus particle or an infected bacterial cell. These two can be distinguished by disruption of the bacteria with chloroform before plating, which releases any intracellular phage particles, giving the total virus count, i.e. intracellular plus extracellular particles.

Several additional features of virus replication are visible from the graph in Figure 4.2. Immediately after dilution of the culture, there is a phase of 10–15 min when no phage particles are detectable; this is known as the **eclipse period**. This represents a time when virus particles have broken down after penetrating

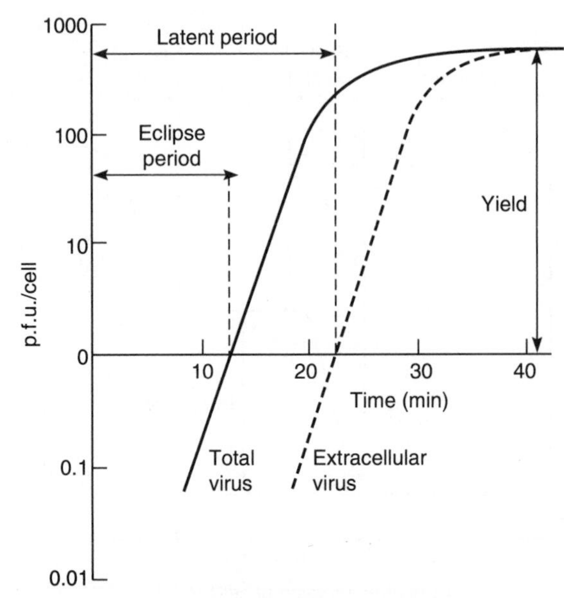

Figure 4.2 Analysis of data from a 'single-burst' experiment. Unlike Figure 4.1, which shows the total number of **plaque forming units (p.f.u.)** produced, here the data is plotted as p.f.u./bacterial cell, which reflects the events occurring in a 'typical' infected cell in the population. The phases of replication named on the graph are defined in the text.

cells, releasing their genomes as a prerequisite to replication. At this stage, they are no longer infectious and therefore cannot be detected by the plaque assay. The **latent period** is the time before the first new extracellular virus particles appear and is of the order of 20–25 min for most bacteriophages. About 40 min after the cells were infected, the curves for the total number of virus particles and for extracellular virus merge because the infected cells have lysed and released any intracellular phage particles by this time. The yield (i.e. number) of particles produced per infected cell can be calculated from the overall rise in phage titre.

Following the development of plaque assays for animal viruses in the 1950s, 'single burst' experiments have now been performed for many viruses of eukaryotes, with similar results (Figure 4.3). The major difference between these viruses and bacteriophages is the much longer time interval required for replication, which is measured in terms of hours and, in some cases, days, rather than minutes after infection. This difference reflects the much slower growth rate of eukaryotic cells, and in part, the complexity of viral replication in compartmentalized cells. Biochemical analysis of virus replication in eukaryotic cells has also been used to analyse the levels of viral and cellular protein and nucleic acid synthesis and to examine the intracellular events occurring during synchronous infections (Figure 4.4). The application of various metabolic inhibitors has also proved to be a valuable tool in such experiments. Examples of the use of such drugs will be given during the detailed consideration of the stages of virus replication later in this chapter.

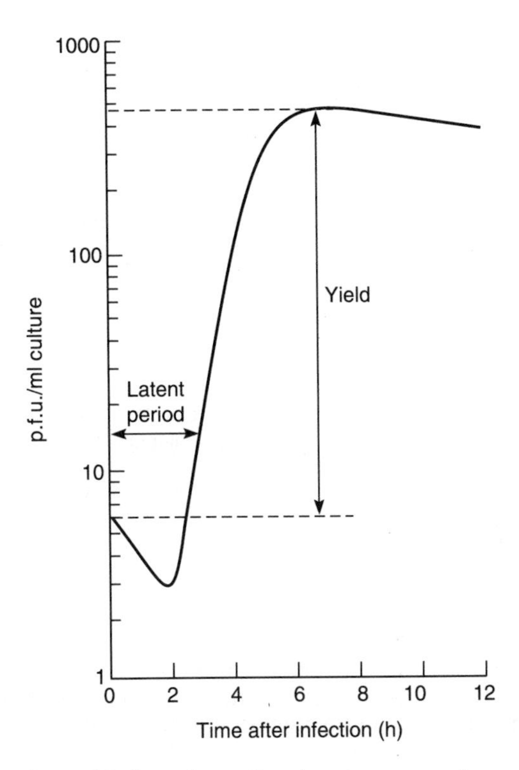

Figure 4.3 Replication of lytic eukaryotic viruses occurs in a similar way to that of bacteriophages. This figure shows a 'single-burst' type of experiment for a picornavirus (e.g. poliovirus). This type of data can only be produced from synchronous infections where a high **multiplicity of infection** is used.

The second key experiment on virus replication using bacteriophages was performed by Hershey and Chase in 1952. Bacteriophage T2 was propagated in *Escherichia coli* cells which had been 'labelled' with one of two radioisotopes, either ^{35}S, which is incorporated into sulphur-containing amino acids in proteins, or ^{32}P, which is incorporated into nucleic acids (which do not contain any sulphur) (Figure 4.5). Particles labelled in each of these ways were used to infect bacteria. After a short period to allow attachment to the cells, the mixture was homogenized briefly in a blender which did not destroy the bacterial cells but was sufficiently vigorous to knock the phage coats off the outside of the cells. Analysis of the radioactive content in the cell pellets and culture supernatant (containing the empty phage coats) showed that most of the radioactivity in the ^{35}S-labelled particles remained in the supernatant, while in the ^{32}P-labelled particles, most of the radiolabel had entered the cells. This experiment indicated that it was the DNA genome of the bacteriophage that entered the cells and initiated the infection.

Although this might seem obvious now, at the time this experiment settled a great controversy over whether a structurally simple polymer such as a nucleic acid, which was known to contain only four monomers, was complex enough to carry genetic information. (At the time, it was generally believed that proteins,

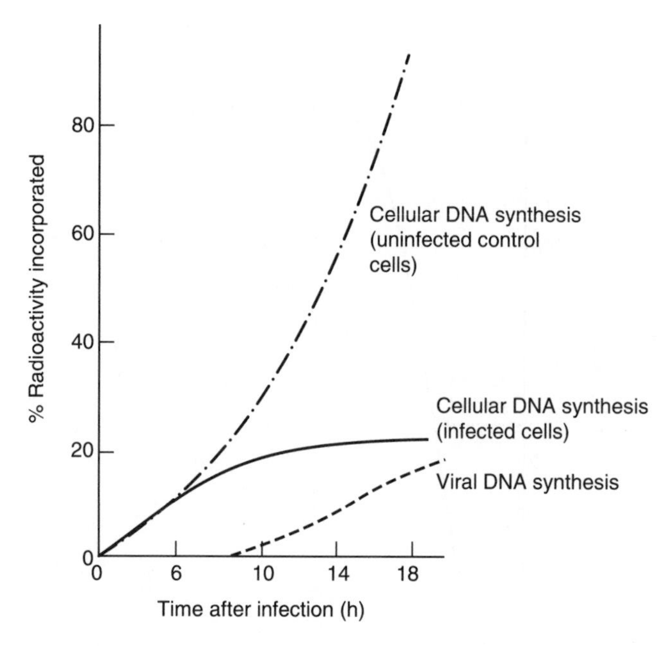

Figure 4.4 Biochemistry of virus infection. This graph shows the rate of cellular and viral DNA synthesis (based on the incorporation of radio-labelled nucleotides into high molecular weight material) in uninfected and herpesvirus-infected cells.

which consist of a much more complex mixture of more than 20 different amino acids, were the carriers of the genes and that DNA was probably a structural component of cells and viruses.) Together, these two experiments illustrate the process of virus replication. Virus particles enter susceptible cells and release their genomic nucleic acids. These are replicated and packaged into virus particles consisting of newly synthesized virus proteins, which are then released from the cell.

The Replication Cycle

Virus replication can be divided into eight stages, as shown in Figure 4.6. It should be emphasized that these are purely arbitrary divisions, used here for convenience in explaining the replication cycle of a non-existent 'typical' virus. For the sake of clarity, this chapter concentrates on viruses which infect animals. Viruses of bacteria, invertebrates and plants will be mentioned briefly, but the overall objective of this chapter is to illustrate similarities in the pattern of replication of different viruses. Regardless of their hosts, *all* viruses must undergo each of these stages in some form to successfully complete their replication cycles. Not all the steps described here are detectable as distinct stages for all viruses; often they 'blur' together and appear to occur almost simultaneously.

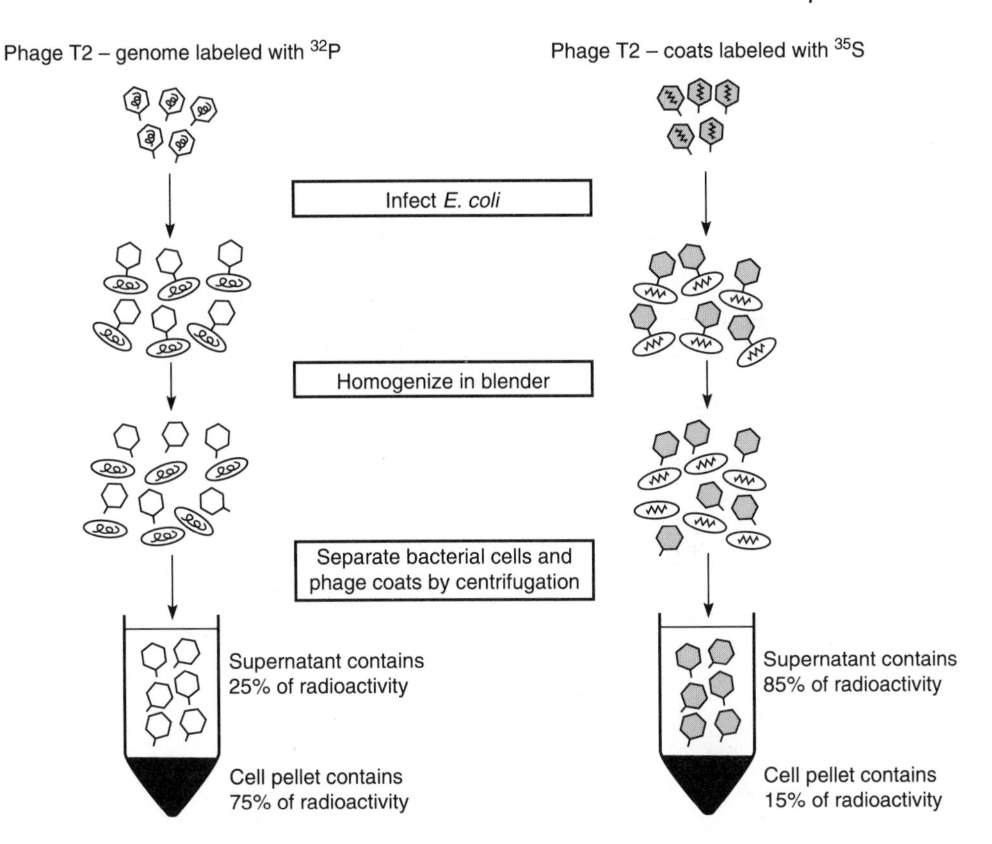

Figure 4.5 The Hershey–Chase experiment, first performed in 1952, demonstrated that virus genetic information was encoded by nucleic acids and not proteins. Details of the experiment are given in the text.

Some of the individual stages have been studied in great detail and a tremendous amount of information is known about them. Other stages are more mysterious and in some cases, virtually unstudied.

Attachment

Because the stages of viral replication described here are purely arbitrary, and because complete replication necessarily involves a cycle, it is possible to begin discussion of virus replication at any point. Arguably, it is most logical to consider the first interaction of a virus with a new host cell as the starting point of the cycle. Technically, virus attachment consists of specific binding of a **virus-attachment protein** (or 'antireceptor') to a cellular **receptor** molecule. Many

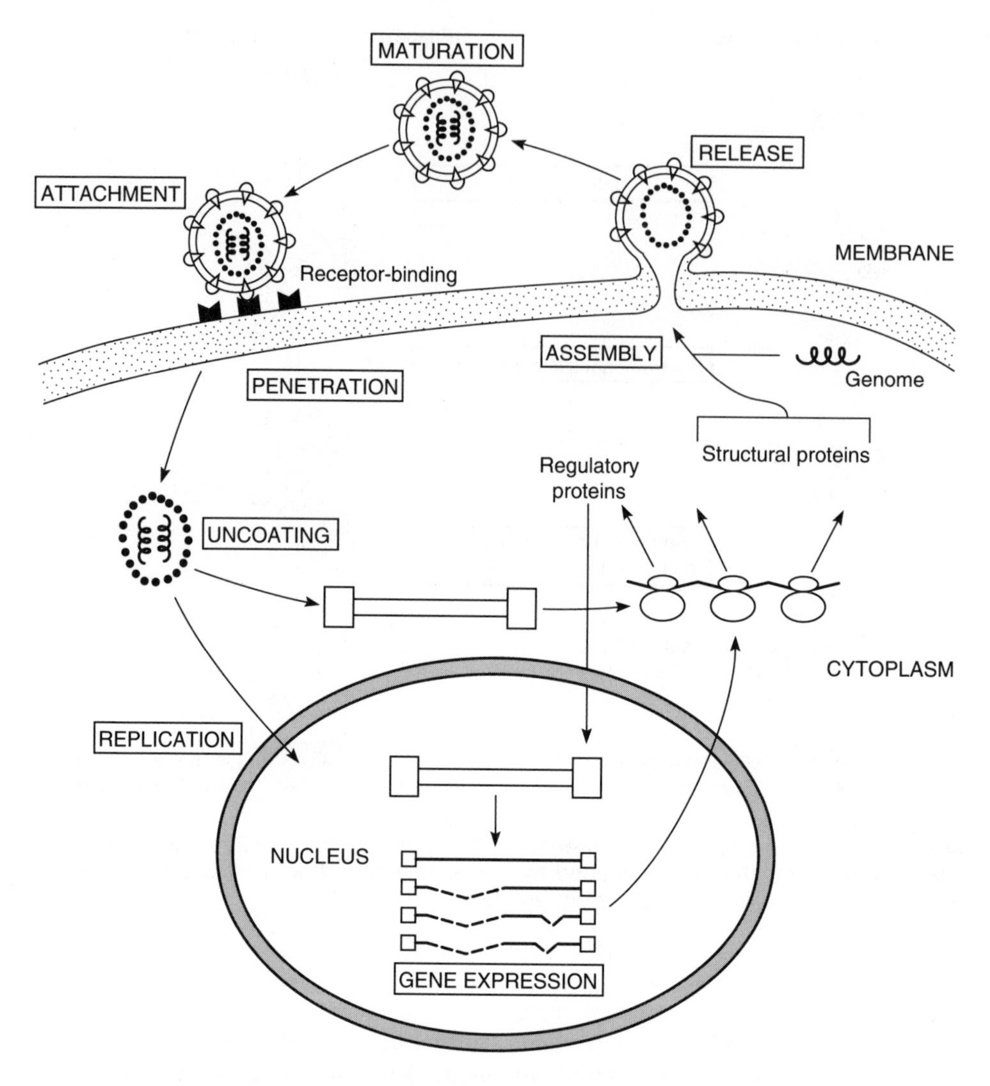

Figure 4.6 Generalized scheme for virus replication.

examples of virus receptors are now known (Table 4.1, Figure 4.7). The target receptor molecules on cell surfaces may be proteins (usually glycoproteins), or the carbohydrate residues present on glycoproteins or glycolipids. The former are usually specific receptors, in that a virus may use a particular protein as a receptor. Carbohydrate groups are usually less specific receptors because the same configuration of side-chains may occur on many different glycosylated membrane-bound molecules. Some complex viruses (e.g. poxviruses, herpesviruses) use more than one receptor and therefore have alternative routes of uptake into cells.

Plant viruses face special problems initiating an infection. The outer surfaces of plants are composed of protective layers of waxes and pectin, but more

Table 4.1 Known virus receptors

Virus (Family)	Receptor	Structure/Function
Amphotropic murine leukaemia virus (*Retroviridae*)	Phosphate transporter analogue	Transport protein
Bovine leukaemia virus (*Retroviridae*)	BLVRcp1	Unknown
Coxsackie virus A9, ECHO virus 22 (*Picornaviridae*)	Vibronectin	Integrin
EBV (*Herpesviridae*)	CR2 (CD 21)	Complement receptor
ECHO virus 1,8 (*Picornaviridae*)	VLA-2 (α-chain)	Integrin
ECHO virus 7 (*Picornaviridae*)	Decay-accelerating factor (CD55)	Involved in complement function
Ecotropic murine leukaemia virus (*Retroviridae*)	Cationic amino acid transporter	Transmembrane transport protein
EMCV (*Picornaviridae*)	Glycophorin A	Erythrocyte sialoglycoprotein
EMC-D (*Picornaviridae*)	Vascular cell adhesion molecule 1 (VCAM-1)	Immunoglobulin-like molecule
Feline immunodeficiency virus (*Lentiviridae*)	Leukocyte differentiation antigen (CD9)	Signalling receptor
Gibbon ape leukaemia virus (*Retroviridae*)	Sodium-dependent phosphate transporter	Transmembrane transport protein
Group A porcine rotavirus (*Rotaviridae*)	Sialic acid	Carbohydrate (glycoprotein)
HIV (*Lentiviridae*)	CD4 (first domain) plus β-chemokine receptors	Immunoglobulin-like molecule
HIV (*Lentiviridae*)	Galactosylceramide	Glycolipid
HHV-7 (*Herpesviridae*)	CD4	Immunoglobulin-like molecule
Human coronavirus 229E (*Coronaviridae*)	Aminopeptidase N	Metalloprotease
Human coronavirus OC43, bovine coronavirus (*Coronaviridae*)	Sialic acid	Carbohydrate (glycoprotein)
Influenza virus (*Orthmyxoviridae*)	Sialic acid	Carbohydrate (glycoprotein)
Lactate dehydrogenase elevating virus (*Arterivirus*)	Major histocompatibility complex I (MHC I)	Immunoglobulin-like molecule
Major group human rhinoviruses (90 serotypes), Coxsackie A viruses (*Picornaviridae*)	Inter-cell adhesion molecule 1 (ICAM-1) (first domain)	Immunoglobulin-like molecule
Measles virus (*Morbilliviridae*)	Membrane cofactor protein (CD46)	Regulator of complement activation
Measles virus (*Morbilliviridae*)	Moesin	Membrane organizing protein
Minor group human rhinoviruses (10 serotypes)	LDL receptor	Signalling receptor
Mouse cytomegalovirus (*Herpesviridae*)	Major histocompatibility complex I (MHC I)	Immunoglobulin-like molecule

Table 4.1 (Continued)

Virus (Family)	Receptor	Structure/Function
Mouse hepatitis virus (*Coronaviridae*)	Carcinoembryonic antigen(s) (CEA)	Immunoglobulin-like molecule
Parvovirus B19 (*Parvoviridae*)	Erythrocyte P antigen	Galactosylceramide
Polioviruses (*Picornaviridae*)	Poliovirus receptor (PVR) (first domain)	Immunoglobulin-like molecule
Rabies virus (*Rhabdoviridae*)	Acetylcholine receptor (α-1)	Signalling receptor
Reovirus (*Reoviridae*)	Glycophorin A	Erythrocyte sialoglycoprotein
Reoviruses (*Reoviridae*)	Sialic acid	Carbohydrate (glycoprotein)
Semliki Forest virus (*Togaviridae*)	Major histocompatibility complex I (MHC I)	Immunoglobulin-like molecule
Sendai virus (*Paramyxoviridae*)	GP-2	Sialoglycoprotein
Sindbis virus (*Togaviridae*)	High-affinity laminin receptor	Laminin receptor
Subgroup A avian leukosis and sarcoma viruses (*Retroviridae*)	LDL receptor	Signalling receptor
SV40 (*Papovaviridae*)	Major histocompatibility complex I (MHC I)	Immunoglobulin-like molecule
Visna virus (*Lentiviridae*)	Major histocompatibility complex II (MHC II)	Immunoglobulin-like molecule

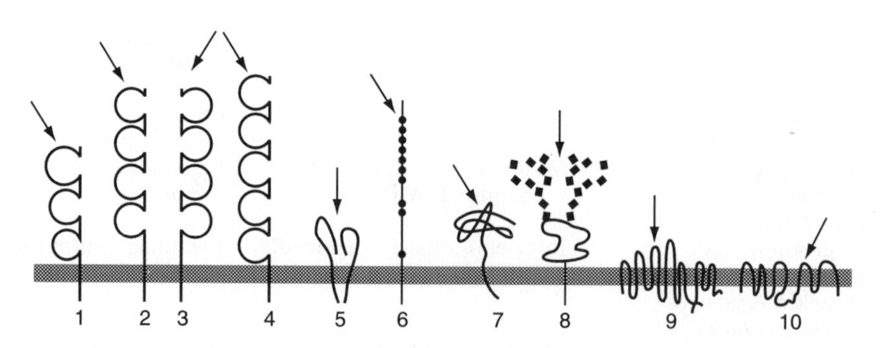

Figure 4.7 Schematic representation of known virus receptors. Arrows indicate virus attachment sites. (1) Poliovirus receptor (PVR); (2) CD4: HIV; (3) carcinoembryonic antigen(s): MHV (coronavirus); (4) ICAM-1: most rhinoviruses; (all immunoglobulin superfamily molecules). (5) VLA-2 integrin: ECHO viruses; (6) LDL receptor: some rhinoviruses; (7) aminopeptidase N: coronaviruses; (8) sialic acid (on glycoprotein): influenza, reoviruses, rotaviruses; (9) cationic amino acid transporter: murine leukaemia virus; (10) sodium-dependent phosphate transporter: gibbon ape leukaemia virus.

significantly, each cell is surrounded by a thick wall of cellulose overlying the cytoplasmic membrane. To date, no plant virus is known to use a specific cellular receptor of the type that animal and bacterial viruses use to attach to cells. Rather, plant viruses rely on a breach of the integrity of a cell wall to directly introduce a virus particle into a cell. This is achieved either by the vector associated with transmission of the virus or simply by mechanical damage to cells. After replication in an initial cell, the lack of receptors poses special problems for plant viruses in recruiting new cells to the infection. These are discussed in Chapter 6.

Some of the best understood examples of virus–receptor interactions are from the picornavirus family. Picornaviruses are unusual in that the virus–receptor interaction has been studied intensively from the viewpoints of both the structural features of the virus responsible for receptor binding and those of the receptor molecule itself. The major human rhinovirus (HRV) receptor molecule, ICAM-1 (intercellular adhesion molecule 1), is an adhesion molecule whose normal function is to bind cells to adjacent substrates. Structurally, ICAM-1 is similar to an immunoglobulin molecule, with constant (C) and variable (V) domains homologous to those of antibodies and is regarded as a member of the immunoglobulin superfamily of proteins (Figure 4.7). Similarly, the poliovirus receptor is an integral membrane protein which is also a member of this family, with one variable and two constant domains. Unlike ICAM-1, the normal function of the poliovirus receptor is not known, but a murine homologue has been identified which is structurally similar to the human molecule but does not function as a receptor for poliovirus. However, the existence of two conserved molecules offers an opportunity for studying the molecular interaction of poliovirus with its receptor.

Since the structure of a number of picornavirus capsids is known at a resolution of a few angstroms (Chapter 2), it has been possible in recent years to determine the features of the virus responsible for receptor binding. In human rhinoviruses (HRVs), there is a deep cleft known as the 'canyon' in the surface of each triangular face of the icosahedral capsid, which is formed by the flanking monomers, VP1, VP2, and VP3 (Figure 4.8). There is biochemical evidence from a class of inhibitory drugs which block attachment of HRV particles to cells that the interaction between ICAM-1 and the virus particle occurs on the floor of this shielded canyon (Figure 4.9). Unlike other areas of the virus surface, the amino acid residues forming the internal surfaces of the canyon are relatively invariant. They are protected from antigenic pressure because the antibody molecules are too large to fit into the cleft. This is important because radical changes in this region, although allowing the virus to escape an immune response, would disrupt receptor binding. In polioviruses, instead of a canyon, there is a trough which runs around each fivefold vertex of the capsid. The highly variant regions of the capsid to which antibodies bind are located on the 'peaks' on either side of this trough, which is too narrow to allow antibody binding to the residues at its base. The invariant residues at the base of the trough interact with the receptor.

Even within the picornavirus family there is variation. Although 90 serotypes of HRV use ICAM-1 as their receptor, some 10 serotypes use proteins related to

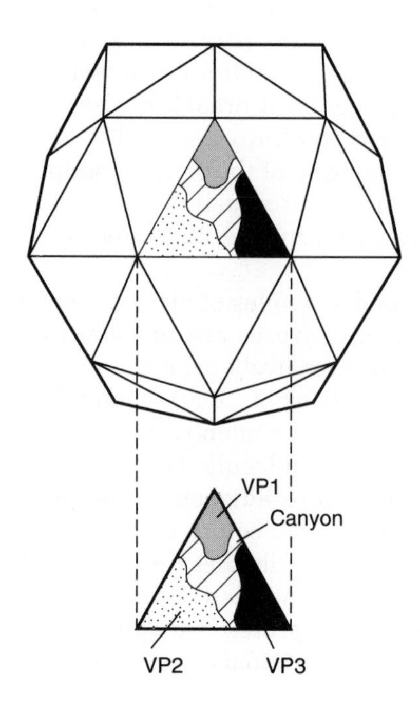

Figure 4.8 Rhinovirus particles have a deep surface cleft known as the 'canyon' between the three monomers (VP1, 2 and 3) making up each face of the particle.

the low density lipoprotein (LDL) receptor. EMCV has been reported to use the immunoglobulin molecule vascular cell adhesion factor (VCAM-1) or glycophorin A. Several picornaviruses use integrins as receptors: some ECHO viruses use VLA-2 or fibronectin; foot-and-mouth disease viruses (FMDV) have been reported to use an unidentified integrin-like molecule. Other ECHO viruses use decay-accelerating factor (DAF, CD55), a molecule involved in complement function. This list is given to illustrate that even within one structurally-related family of viruses, there is considerable variation in the receptor structures used.

Another well studied example of virus-receptor interaction is that of influenza virus. The haemagglutinin protein forms one of the two types of glycoprotein spikes on the surface of influenza virus particles (see Chapter 2), the other type being formed by the neuraminidase protein. Each haemagglutinin spike is composed of a trimer of three molecules, while the neuraminidase spike consists of a tetramer (Figure 4.10). The haemagglutinin spikes are responsible for binding the influenza virus receptor, which is sialic acid (*N*-acetyl neuraminic acid), a sugar group commonly found on a variety of glycosylated molecules. As a result, there is little cell-type specificity imposed by this receptor interaction and therefore influenza viruses bind to a wide variety of different cell types (e.g. causing haemagglutination of red blood cells) in addition to the cells in which productive infection occurs.

The neuraminidase molecule of influenza virus and paramyxoviruses illustrates another feature of this stage of virus replication. Attachment to cellular

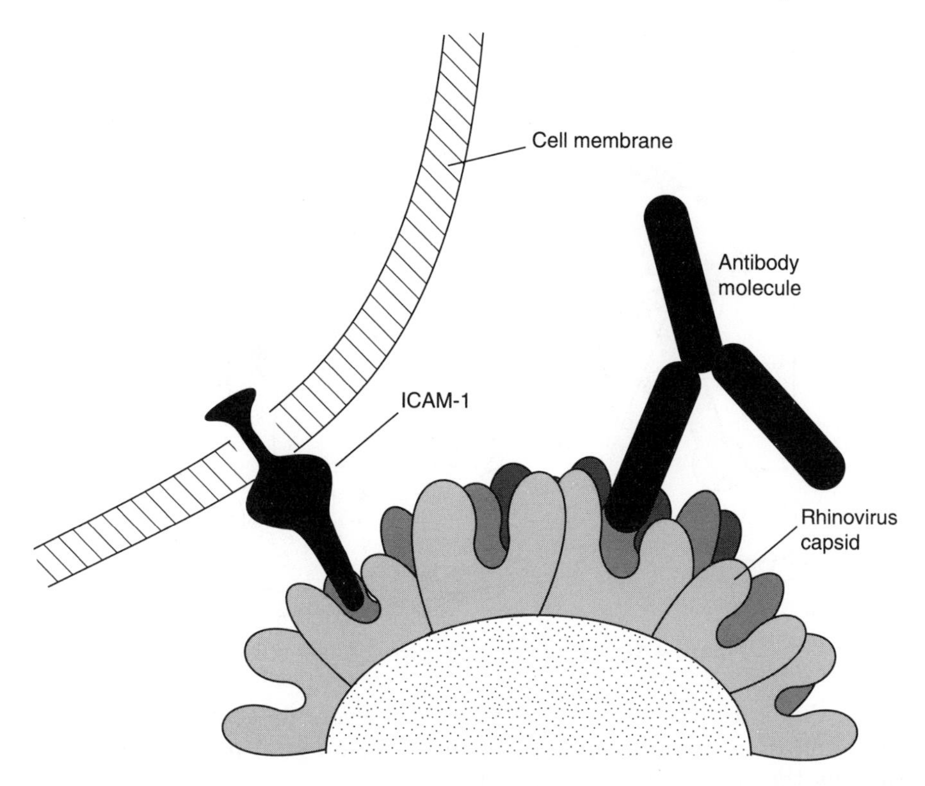

Figure 4.9 Receptor-binding by rhinoviruses. The cellular receptor (ICAM-1) is believed to interact with residues at the bottom of the 'canyon' on the surface of the virus particle (Figure 4.8). Antibody molecules are too large to penetrate the canyon, allowing the conserved receptor-binding site to avoid becoming a target for a neutralizing response.

receptors is in most cases a reversible process – if penetration of the cells does not ensue, the virus can elute from the cell surface. Some viruses have specific mechanisms for detachment and the neuraminidase protein is one of these. Neuraminidase is an esterase which cleaves sialic acid from sugar side-chains. This is particularly important for influenza. Because the receptor molecule is so widely distributed, the virus tends to bind inappropriately to a variety of cells and even cell debris. However, elution from the cell surface after receptor binding has occurred often leads to changes in the virus (e.g. loss or structural alteration of virus-attachment protein) which decrease or eliminate the possibility of subsequent attachment to other cells. Thus in the case of influenza, cleavage of sialic acid residues by neuraminidase leaves these groups bound to the active site of the haemagglutinin, preventing that particular molecule from binding to another receptor.

In many if not most cases, the expression (or absence) of receptors on the surface of cells largely determines the **tropism** of a virus, i.e. the type of host cell in which it is able to replicate. In a few cases, intracellular blocks at later stages of replication are responsible for determining the range of cell types in which a virus can carry out a productive infection, but this is not common.

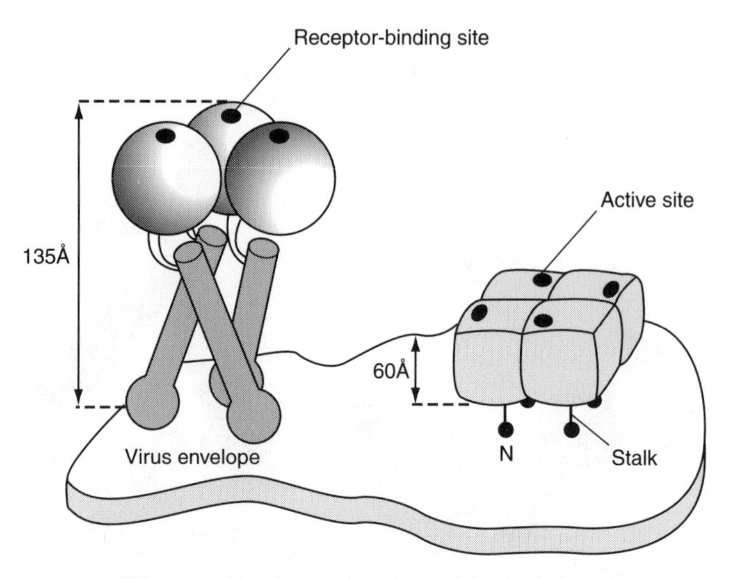

Figure 4.10 Influenza virus glycoprotein spikes.

Therefore, this initial stage of replication and the very first interaction between the virus and the host cell has a major influence on viral pathogenesis and in determining the course of a virus infection.

An example of the importance of receptor-determined tropism is seen in the replication of Epstein–Barr virus (EBV). EBV only infects human cells, but within the body can infect two cell types: epithelial cells, in which a **productive infection** occurs, and B-lymphocytes, which are usually non-productive. The infection of B-lymphocytes is known to occur via the CR2 (CD21) molecule on the surface of the cell. However, although they are infected by an unknown route *in vivo*, most epithelial cell lines lack CR2 and are therefore not infectable *in vitro*. This block can be overcome by **transfection** of epithelial cells with recombinant DNA constructions that express CR2; this renders the cell permissive for EBV infection.

In some cases, interactions with more than one protein are required for virus entry. These are not examples of alternative receptor use since neither protein alone is a functional receptor – both are required to act in concert. An example is the process by which adenoviruses enter cells. The initial event is the binding of the virus fibre protein to an unidentified cell surface protein related to nucleolin. However, this in itself is not sufficient for entry since a subsequent step is also required, namely the binding of the penton base of the virus to vitronectin. Thus vitronectin is not a receptor for adenovirus but is an accessory factor required for entry.

A similar observation has been made with HIV. The primary receptor for this virus is the helper T cell differentiation antigen, CD4. Transfection of human cells which do not normally express CD4 (such as epithelial cells) with recombinant CD4-expression constructs renders them permissive for HIV infection.

However, transfection of rodent cells with human CD4-expression vectors does not permit productive HIV infection. If HIV proviral DNA is inserted into rodent cells by transfection, virus is produced, indicating that there is no intracellular block to infection. Therefore, there must be one or more accessory factors in addition to CD4 which are required to form a functional HIV receptor. These have recently been identified as a family of proteins known as β-chemokine receptors. Several members of this family have been shown to play a role in the entry of HIV into cells, and their distribution may be the primary control for the tropism of HIV for different cell types (lymphocytes, macrophages, etc). Furthermore, there is evidence that in at least some cell types, HIV infection is not blocked by competing soluble CD4, indicating that in these cells a completely different receptor strategy may be being used. Several candidate molecules have been put forward to fill this role, e.g. galactosylceramide. However, if any or all of these do allow HIV to infect a range of CD4-negative cells, this process is much less efficient than the interaction of the virus with its major receptor complex.

In some cases, specific receptor binding can be side-stepped by non-specific or inappropriate interactions between virus particles and cells. It is possible that virus particles can be 'accidentally' taken up by cells via processes such as pinocytosis or phagocytosis (see below). However, in the absence of some form of physical interaction which holds the virus particle in close association with the cell surface, the frequency with which these accidental events happen is very low. On occasions, antibody-coated virus particles binding to Fc receptor molecules on the surface of monocytes and other blood cells can result in virus uptake. This phenomenon has been shown to occur in a number of cases, where antibody-dependent enhancement of virus uptake results in unexpected findings. For example, the presence of anti-viral antibodies can occasionally result in increased virus uptake by cells and increased pathogenicity rather than virus neutralization, as would normally be expected. It has been suggested that this mechanism may also be important in the uptake of HIV by macrophages and monocytes and that this might be a factor in the pathogenesis of AIDS, although this is not yet certain.

Penetration

Penetration of the target cell normally occurs a very short time after attachment of the virus to its receptor in the cell membrane. Unlike attachment, cell penetration is generally an energy-dependent process, i.e. the cell must be metabolically active for this to occur. Three main mechanisms are involved:

(1) *Translocation* of the entire virus particle across the cytoplasmic membrane of the cell (Figure 4.11). This process is relatively rare among viruses and is poorly understood. It must be mediated by proteins in the virus capsid and specific membrane receptors.

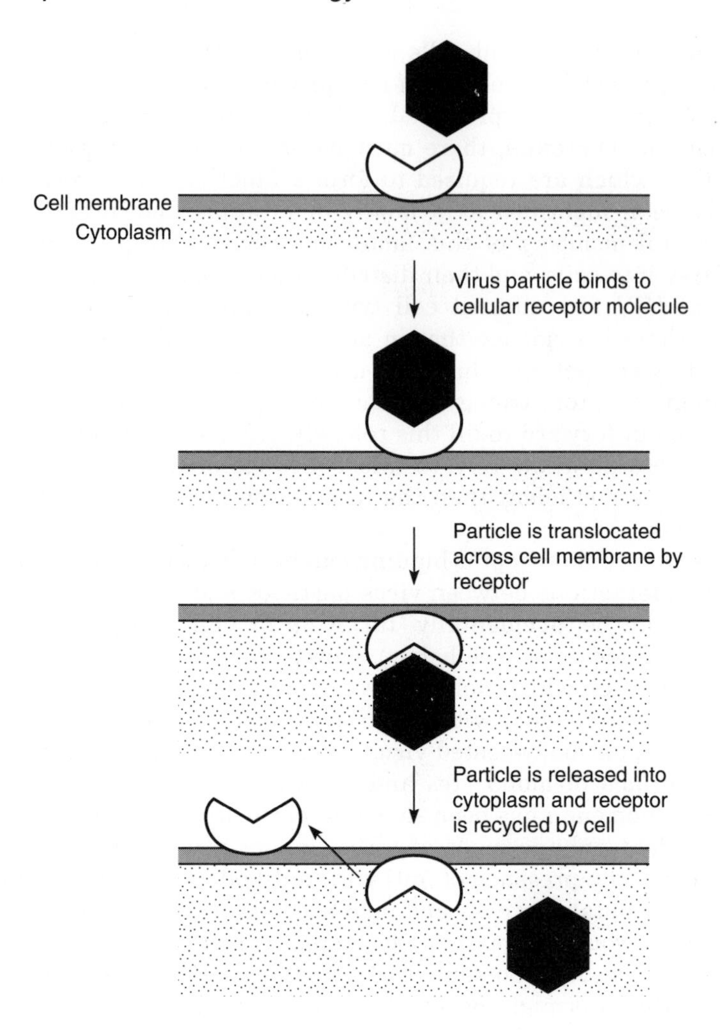

Figure 4.11 Translocation of entire virus particles across the cell membrane by cell surface receptors.

(2) *Endocytosis* of the virus into intracellular vacuoles (Figure 4.12). This is probably the most common mechanism of virus entry into cells. It does not require any specific virus proteins (other than those already utilized for receptor binding) but relies on the normal formation and internalization of coated pits at the cell membrane. Receptor-mediated endocytosis is an efficient process for taking up and concentrating extracellular macromolecules.

(3) *Fusion* of the virus envelope (only applicable to enveloped viruses) with the cell membrane, either directly at the cell surface or following endocytosis in a cytoplasmic vesicle (Figure 4.13). Fusion requires the presence of a specific **fusion protein** in the virus envelope, e.g. influenza haemagglutinin or retrovirus transmembrane (TM) glycoproteins. These proteins promote the

ENDOCYTOSIS

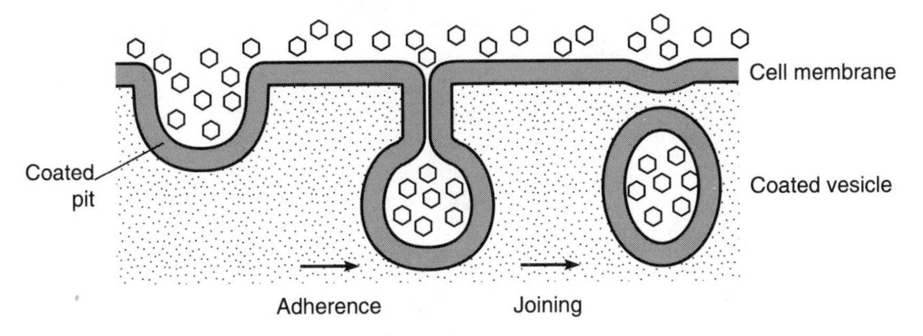

EXOCYTOSIS

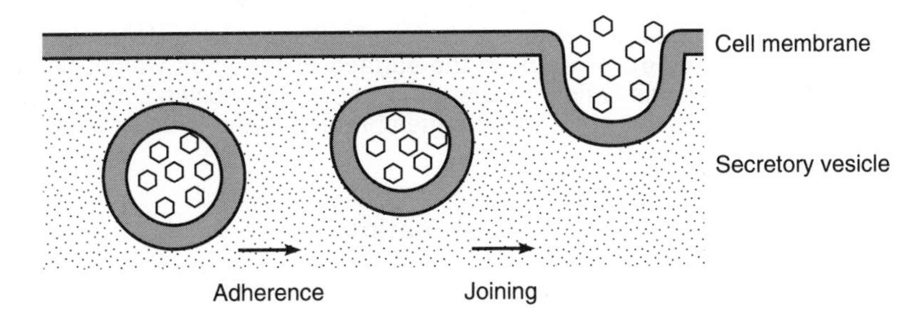

Figure 4.12 Endocytosis and exocytosis of virus particles.

joining of the cellular and virus membranes which results in the nucleocapsid being deposited directly in the cytoplasm. There are two types of virus-driven membrane fusion: one is pH-dependent and the other is pH-independent.

The process of endocytosis is almost universal in animal cells and requires further consideration (Figure 4.12). The formation of coated pits results in the engulfment of a membrane-bounded vesicle by the cytoplasm of the cell. The lifetime of these initial coated vesicles is very short. Within seconds, most fuse with endosomes, releasing their contents into these larger vesicles. At this point, any virus contained within these structures is still cut off from the cytoplasm by a lipid bilayer and therefore has not strictly entered the cell. Moreover, as endosomes fuse with lysosomes, the environment inside these vessels becomes progressively more hostile as they are acidified and the pH falls, while the concentration of degradative enzymes rises. This means that the virus must leave the vesicle and enter the cytoplasm before it is degraded. There are a number of mechanisms by which this occurs, including membrane fusion and rescue by transcytosis. The release of virus particles from endosomes and their passage into the cytoplasm is intimately connected with (and often impossible to separate from) the process of uncoating (see below).

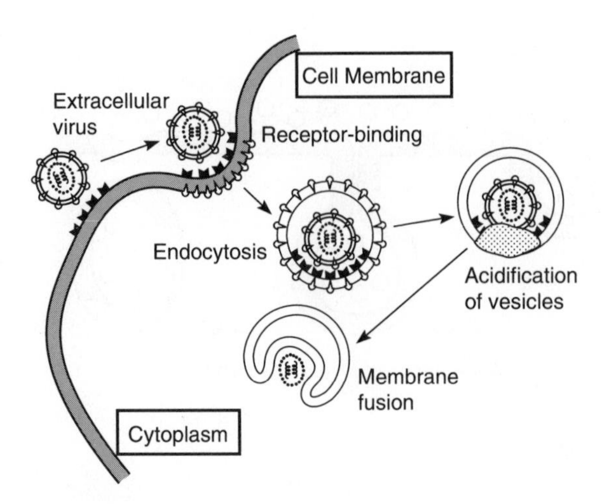

Figure 4.13 Virus-induced membrane fusion. This process is dependent on the presence of a specific **fusion protein** on the surface of the virus which, under particular circumstances (e.g. acidification of the virus-containing vesicle), becomes activated, inducing fusion of the vesicle membrane and the virus envelope.

Uncoating

Uncoating is a general term for the events which occur after penetration, in which the virus capsid is completely or partially removed and the virus genome exposed, usually in the form of a nucleoprotein complex. Unfortunately, uncoating is one of the stages of virus replication that has been least studied and is relatively poorly understood.

In one sense, the removal of a virus envelope which occurs during membrane fusion is part of the uncoating process. Fusion between virus envelopes and endosomal membranes is driven by viral fusion proteins. These are usually activated by the uncloaking of a previously hidden fusion domain as a result of conformational changes in the protein induced by the low pH inside the vesicle, although in some cases the fusion activity is triggered directly by receptor binding. The initial events in uncoating may occur inside endosomes, being triggered by the change in pH as the endosome is acidified, or directly in the cytoplasm. Ionophores such as monesin or nigericin or cations such as chloroquine and ammonium chloride can be used to block the acidification of these vesicles and to determine whether events are occurring following the acidification of endosomes (e.g. pH-dependent membrane fusion) or directly at the cell surface or in the cytoplasm (e.g. pH-independent membrane fusion). Endocytosis has some danger for viruses, since if they remain in the vesicle too long they will be irreversibly damaged by acidification or lysosomal enzymes. Some viruses can control this process, e.g. the influenza virus M2 protein is a membrane channel which allows entry of hydrogen ions into the nucleocapsid, facilitating uncoating. The M2 protein is multifunctional and also has a role in influenza virus maturation (below).

In picornaviruses, penetration of the cytoplasm by exit of virus from endosomes is tightly linked to uncoating (Figure 4.14). The acidic environment of the endosome causes a conformational change in the capsid which reveals hydrophobic domains not present on the surface of mature virus particles. The interaction of these hydrophobic patches with the endosomal membrane is believed to form pores through which the genome passes into the cytoplasm.

The product of uncoating depends on the structure of the virus nucleocapsid. In some cases, this might be relatively simple (e.g. picornaviruses have a small basic protein of approximately 23 amino acids (VPg) covalently attached to the 5′ end of the vRNA genome), or may be highly complex (e.g. retrovirus cores are highly ordered nucleoprotein complexes which contain, in addition to the diploid RNA genome, the reverse transcriptase enzyme responsible for converting the virus RNA genome into the DNA provirus). The structure and chemistry of the nucleocapsid determines the subsequent steps in replication. As discussed in Chapter 3, reverse transcription can only occur inside an ordered retrovirus core particle and cannot proceed with the components of the reaction free in solution. Herpesvirus, adenovirus, and papovavirus capsids undergo structural changes following penetration, but overall remain largely intact. These capsids contain sequences which are responsible for attachment to the cytoskeleton and this interaction allows the transport of the entire capsid to the nucleus. It is at the nuclear pores that uncoating occurs and the nucleocapsid passes into the nucleus. In reoviruses and poxviruses, complete uncoating does not occur, and many of the reactions of genome replication are catalysed by virus-encoded enzymes inside cytoplasmic particles which still resemble the mature virions.

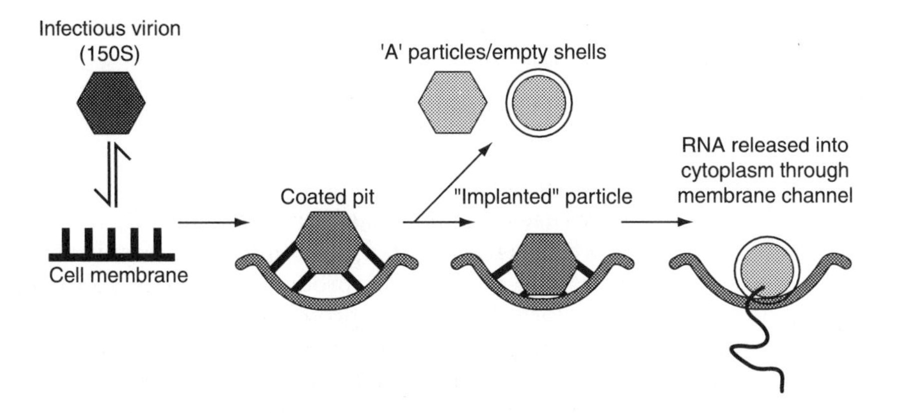

Figure 4.14 Cell penetration and uncoating of polioviruses.

Genome replication and gene expression

The replication strategy of any virus depends on the nature of its genetic material. In this respect, viruses can be divided into seven groups. Such a scheme was first proposed by David Baltimore in 1971. Originally, this classification included only six groups, but it has since been extended to include the scheme of genome replication used by the hepadnaviruses and caulimoviruses. For viruses with RNA genomes in particular, genome replication and the expression of genetic information are inextricably linked. Therefore, both of these criteria are taken into account in the scheme below. The control of gene expression determines the overall course of a virus infection (acute, chronic, persistent, or latent) and such is the emphasis placed on gene expression by molecular biologists that this subject is discussed in detail in Chapter 5. A schematic overview of the major events during replication of the different virus genomes is shown in Figure 4.15 and a complete list of all the families which constitute each class is given in Appendix 1.

- *Class I* Double-stranded DNA (Figure 4.15.a). This class can be subdivided into two further groups:
  - (a) Replication is exclusively nuclear. The replication of these viruses is relatively dependent on cellular factors
  - (b) Replication occurs in cytoplasm (*Poxviridae*). These viruses have evolved (or acquired) all the necessary factors for transcription and replication of their genomes and are therefore largely independent of the cellular machinery
- *Class II* Single-stranded DNA (Figure 4.15.b). Replication occurs in the nucleus, involving the formation of a double-stranded intermediate which serves as a template for the synthesis of single-stranded progeny DNA
- *Class III* Double-stranded RNA (Figure 4.15.c). These viruses have segmented genomes. Each segment is transcribed separately to produce individual monocistronic mRNAs
- *Class IV* Single-stranded (+)sense RNA. These can be subdivided into two groups:
  - (a) Viruses with polycistronic mRNA (Figure 4.15.dA). As with all the viruses in this class, the genome RNA forms the mRNA which is translated after infection, resulting in the synthesis of a polyprotein product, which is subsequently cleaved to form the mature proteins
  - (b) Viruses with complex transcription (Figure 4.15.dB). Two rounds of translation (e.g. Togaviruses) or subgenomic RNAs (e.g. Tobamoviruses) are necessary to produce the genomic RNA

- *Class V* Single-stranded (−)sense RNA. (Figure 4.15.e). As discussed in Chapters 3 and 5, the genomes of these viruses can be divided into two types:
  - (a) Segmented. The first step in replication is transcription of the (−)sense RNA genome by the virion RNA-dependent RNA polymerase to produce monocistronic mRNAs, which also serve as the template for subsequent genome replication. N.B. Some of these viruses also have an **ambisense** organization
  - (b) Non-segmented. Replication occurs as above, with monocistronic mRNAs for each of the virus genes produced by the virus transcriptase from the full-length virus genome (see Chapter 5)
- *Class VI* Single-stranded (+)sense RNA with DNA intermediate (Figure 4.15.f). Retrovirus genomes are (+)sense RNA but unique in that they are diploid, and that they do not serve directly as mRNA, but as a template for reverse transcription into DNA (see Chapter 3)
- *Class VII* Double-stranded DNA with RNA intermediate (Figure 4.15.g). This group of viruses also relies on reverse transcription, but unlike the retroviruses (class VI), this occurs inside the virus particle during maturation. On infection of a new cell, the first event to occur is repair of the gapped genome, followed by transcription (see Chapter 3).

Assembly

This process involves the collection of all the components necessary for the formation of the mature virion at a particular site in the cell. During assembly, the basic structure of the virus particle is formed. The site of assembly depends on the site of replication within the cell and on the mechanism by which the virus is eventually released from the cell and varies for different viruses. For example in picornaviruses, poxviruses and reoviruses assembly occurs in the cytoplasm; in adenoviruses, papovaviruses and parvoviruses it occurs in the nucleus.

As with the early stages of replication, it is not always possible to identify the assembly, maturation and release of virus particles as distinct and separate phases. The site of assembly has a profound influence on all these processes. In the majority of cases, cellular membranes are used to anchor virus proteins, and this initiates the process of assembly. In spite of considerable study, the control of virus assembly is generally not well understood. In general, it is thought that rising intracellular levels of virus proteins and genome molecules reach a critical concentration and this triggers the process. Many viruses

(a)

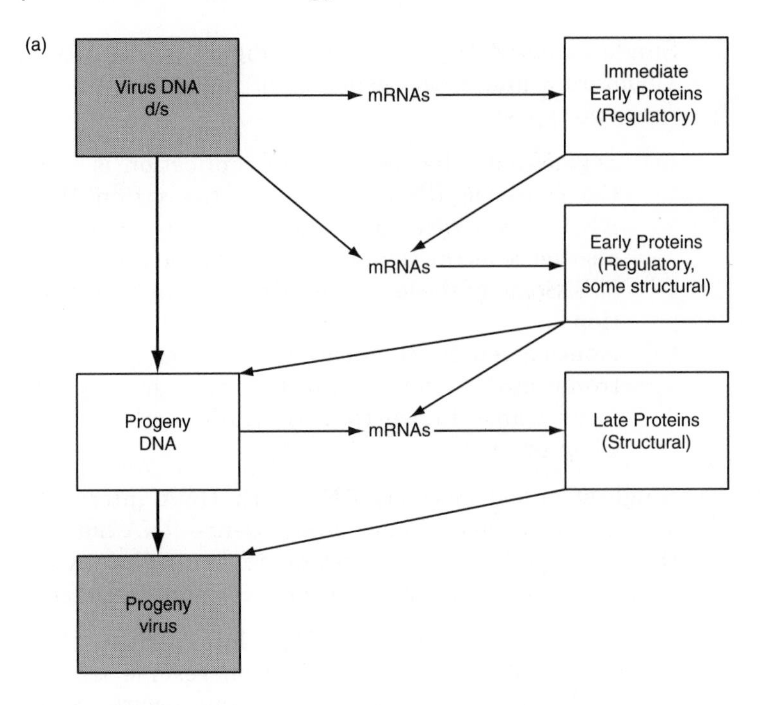
Virus DNA
d/s
mRNAs
Immediate
Early Proteins
(Regulatory)
mRNAs
Early Proteins
(Regulatory,
some structural)
Progeny
DNA
mRNAs
Late Proteins
(Structural)
Progeny
virus

(b)

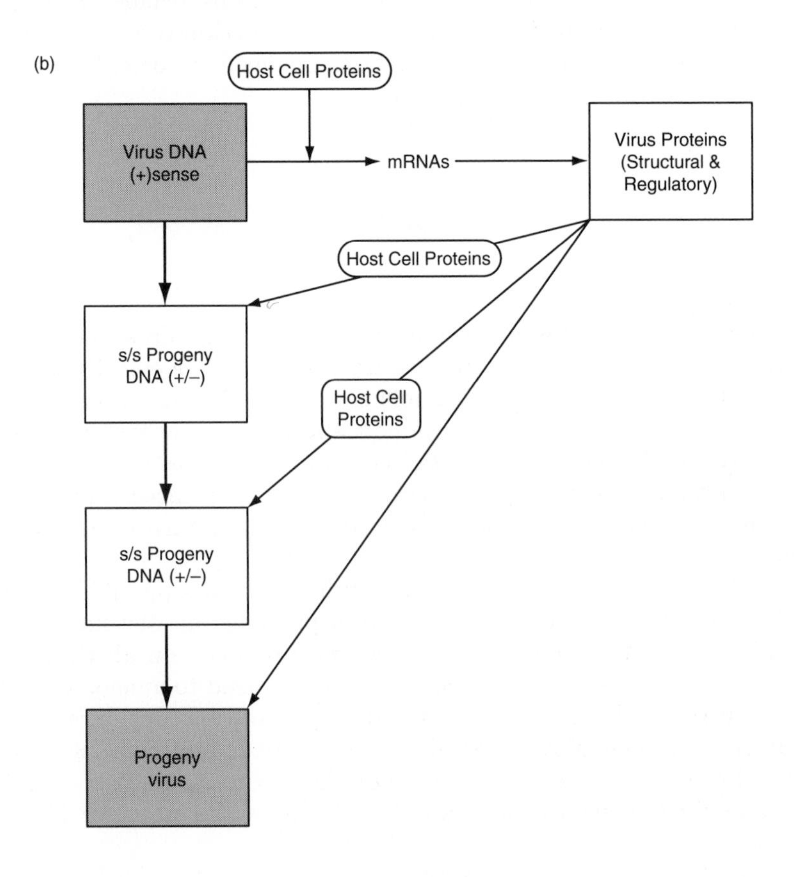
Host Cell Proteins
Virus DNA
(+)sense
mRNAs
Virus Proteins
(Structural &
Regulatory)
Host Cell Proteins
s/s Progeny
DNA (+/–)
Host Cell
Proteins
s/s Progeny
DNA (+/–)
Progeny
virus

(c)

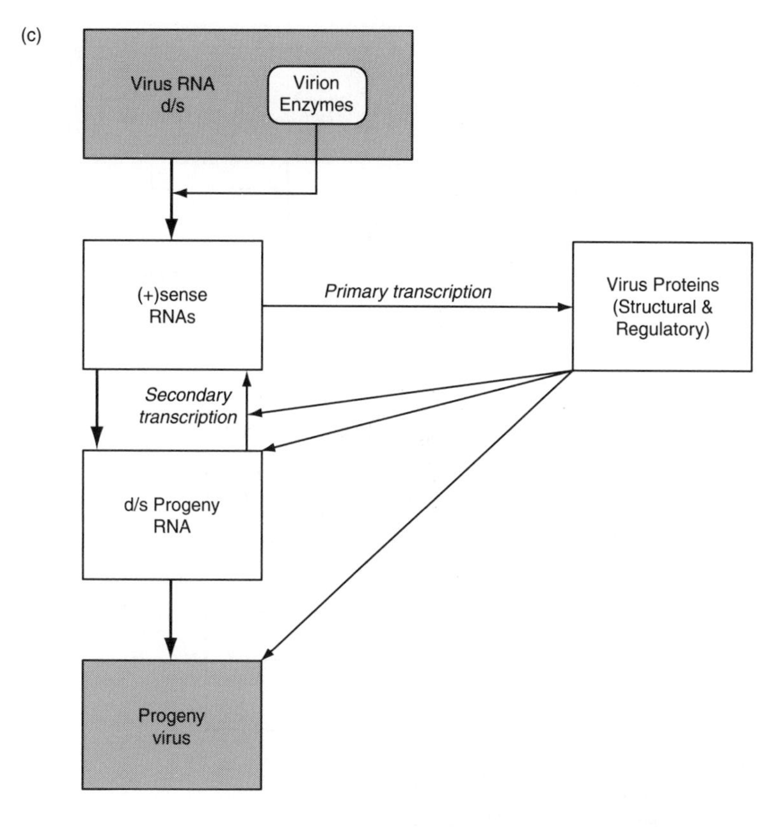
Virus RNA d/s
Virion Enzymes
(+)sense RNAs
Primary transcription
Virus Proteins (Structural & Regulatory)
Secondary transcription
d/s Progeny RNA
Progeny virus

(d)

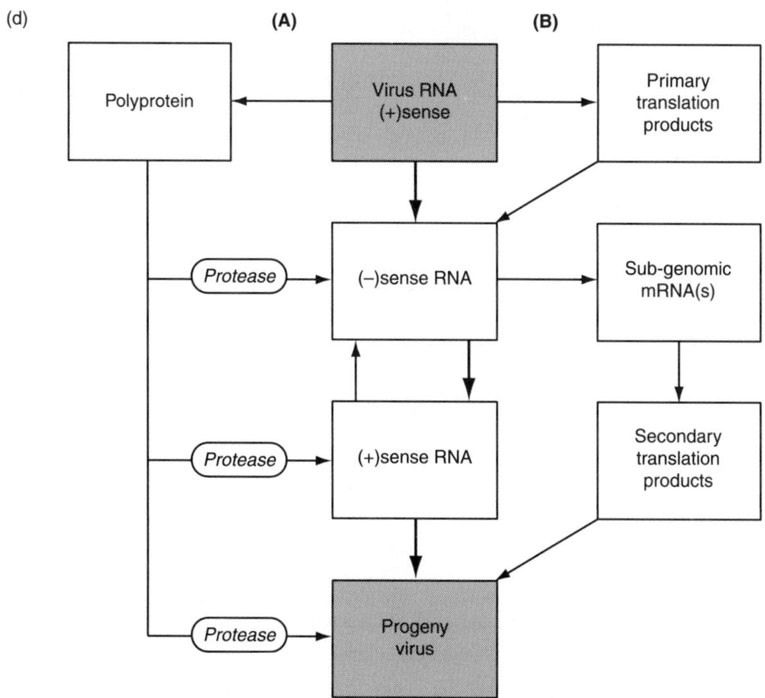
(A)
(B)
Polyprotein
Virus RNA (+)sense
Primary translation products
Protease
(–)sense RNA
Sub-genomic mRNA(s)
Protease
(+)sense RNA
Secondary translation products
Protease
Progeny virus

(e)

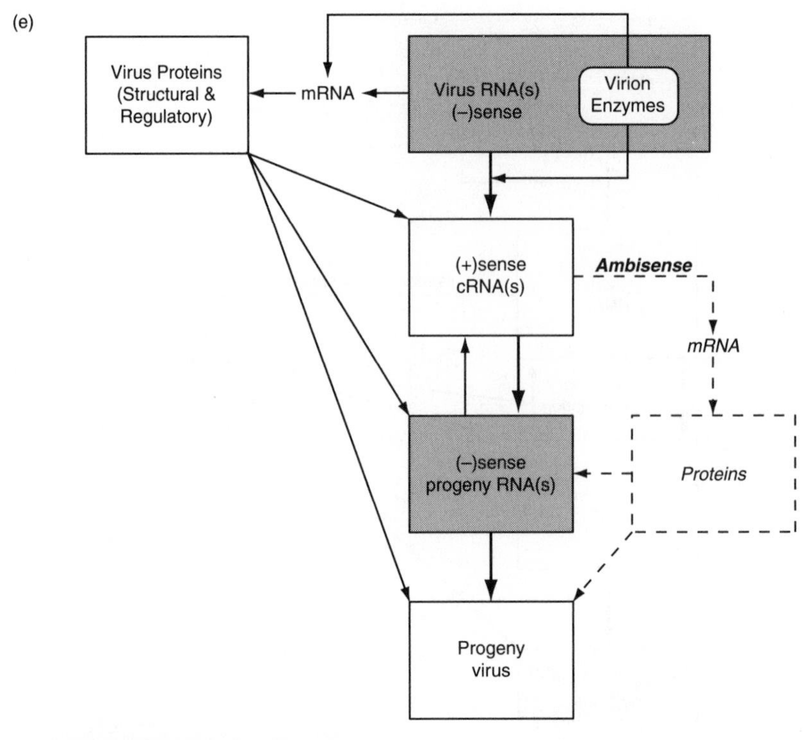
Virus Proteins
(Structural &
Regulatory)
mRNA
Virus RNA(s)
(–)sense
Virion
Enzymes
(+)sense
cRNA(s)
Ambisense
mRNA
(–)sense
progeny RNA(s)
Proteins
Progeny
virus

(f)

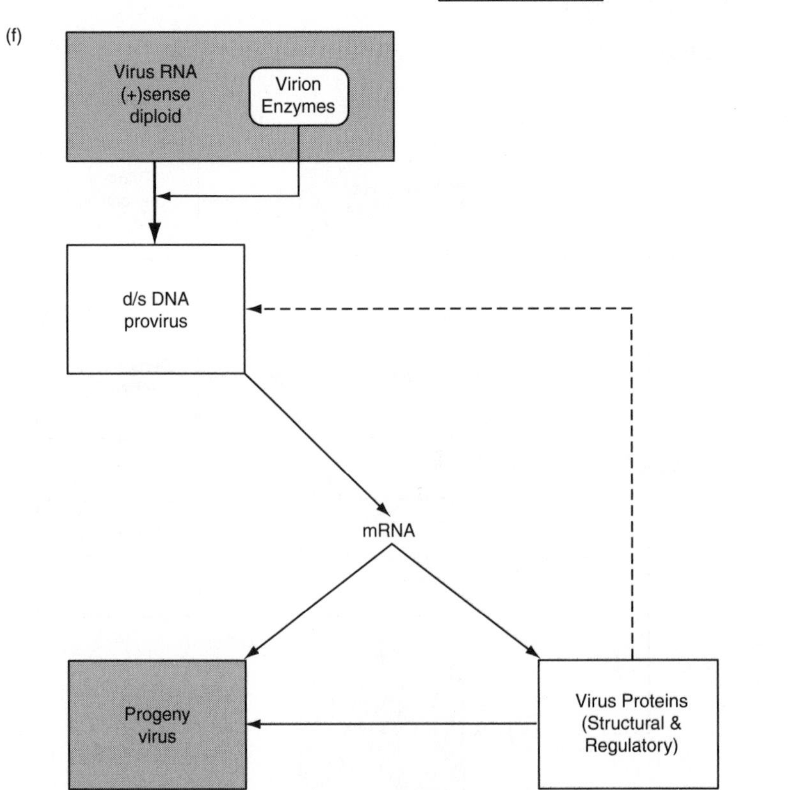
Virus RNA
(+)sense
diploid
Virion
Enzymes
d/s DNA
provirus
mRNA
Progeny
virus
Virus Proteins
(Structural &
Regulatory)

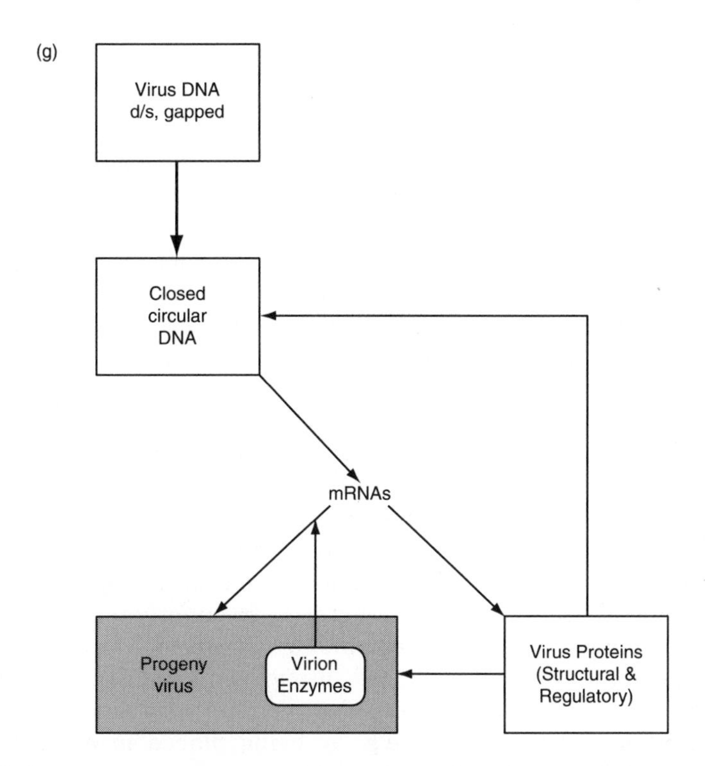

Figure 4.15 Schematic representation of the pattern of replication of virus genomes. (a) Class I; (b) Class II; (c) Class III; (d) Class IV; (e) Class V; (f) Class VI; (g) Class VII. Details of the events which occur for genomes of each type are given in the text.

achieve high levels of newly synthesized structural components by concentrating these into subcellular compartments, visible in light microscopes, which are known as **inclusion bodies**. These are a common feature of the late stages of infection of cells by many different viruses. The size and location of inclusion bodies in infected cells is often highly characteristic of particular viruses, e.g. rabies virus infection results in large perinuclear 'Negri bodies,' first observed by Adelchi Negri in 1903. Alternatively, local concentrations of viral structural components can be boosted by lateral interactions between membrane-associated proteins. This mechanism is particularly important in enveloped viruses which are released from the cell by **budding** (see below).

As discussed in Chapter 2, the formation of virus particles may be a relatively simple process which is driven only by interactions between the subunits of the capsid and controlled by the rules of symmetry. In other cases, assembly is a highly complex, multi-step process involving not only virus structural proteins, but in addition virus-encoded and cellular scaffolding proteins which act as templates to guide the assembly of virions. The encapsidation of the virus genome may occur either early in the assembly of the particle (e.g. many viruses with helical symmetry are nucleated on the genome), or at a late stage, when the genome is stuffed into an almost completed protein shell.

Maturation

This is the stage of the life-cycle at which the virus becomes infectious. Maturation usually involves structural changes in the virus particle which may result from specific cleavages of capsid proteins to form the mature products or conformational changes in proteins during assembly. Such events frequently lead to substantial structural changes in the capsid which may be detectable by criteria such as differences in the antigenicity of incomplete and mature virus particles, which in some cases (e.g. picornaviruses) alters radically. Alternatively, internal structural alterations, for example the condensation of nucleoproteins with the virus genome, often results in such changes.

Virus proteases are frequently involved in maturation, although cellular enzymes or a mixture of virus and cellular enzymes are used in some cases. Clearly, there is a danger in relying on cellular proteolytic enzymes in that their relative lack of substrate specificity could easily completely degrade the capsid proteins. However, virus-encoded proteases are usually highly specific for particular amino acid sequences and structures, frequently only cutting one particular peptide bond in a large and complex virus capsid. Moreover, they are often further controlled by being packaged into virus particles during assembly and are only activated when brought into close contact with their target sequence by the conformation of the capsid, e.g. by being placed in a local hydrophobic environment or by changes of pH or metal ion concentration inside the capsid. Retrovirus proteases are good examples of enzymes involved in maturation which are under this tight control. The retrovirus core particle is composed of proteins from the *gag* gene and the protease is packaged into the core before its release from the cell on budding. At some stage of the budding process (the exact timing varies for different retroviruses) the protease cleaves the gag protein precursors into the mature products – the capsid, nucleocapsid, and matrix proteins of the mature virus particle (Figure 4.16).

Not all protease cleavage events involved in maturation are this tightly regulated. Native influenza virus haemagglutinin undergoes post-translational modification (glycosylation in the golgi apparatus) and at this stage exhibits receptor-binding activity. However, the protein must be cleaved into two fragments (HA_1 and HA_2) to be able to promote membrane fusion during infection. Cellular trypsin-like enzymes are responsible for this process, which occurs in secretory vesicles as the virus buds into them prior to release at the cell surface. Amantadine and rimantadine are two drugs which are active against influenza A viruses (Chapter 6). The action of these closely related agents is complex and incompletely understood, but they are believed to block cellular membrane ion channels. The target for both drugs is the influenza matrix protein (M_2), but resistance to the drug may also map to the haemagglutinin gene. The replication of some strains of influenza virus is inhibited at the cell penetration stage and that of others at maturation. The biphasic action of these drugs results from the inability of drug-treated cells to lower the pH of the endosomal compartment (a function normally controlled by the M_2 gene product), and hence to cleave

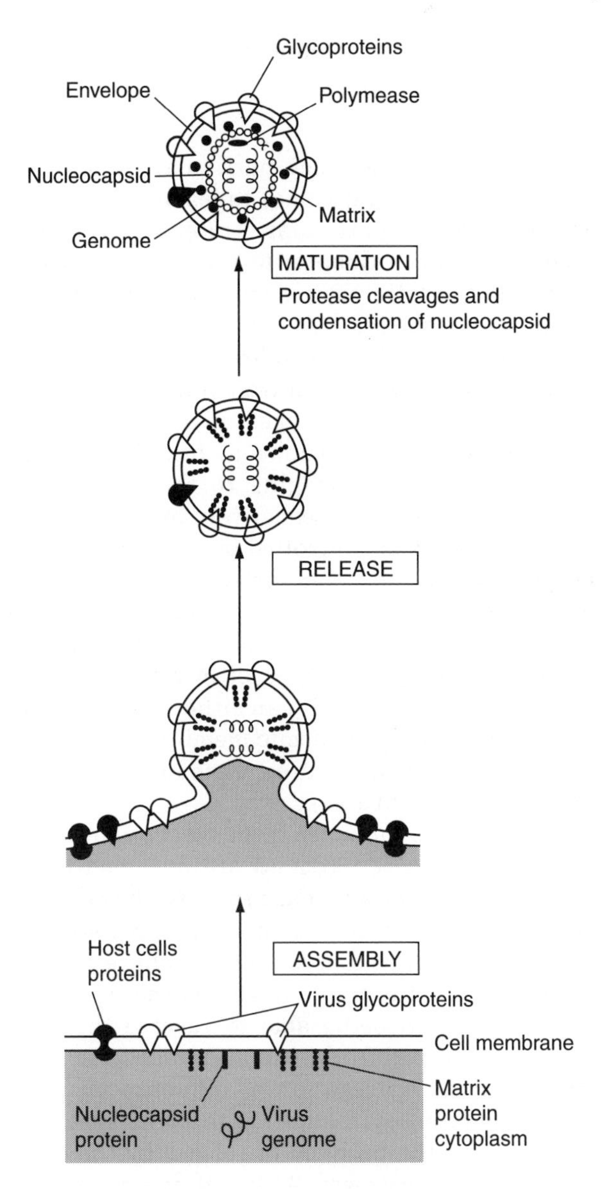

Figure 4.16 Virus release by budding. This is the process by which enveloped virus particles acquire their membranes and associated proteins.

haemagglutinin during maturation. Similarly, retrovirus envelope glycoproteins require cleavage into the surface (SU) and transmembrane (TM) proteins for activity. This process is also carried out by cellular enzymes, but is in general poorly understood.

As already stated, for some viruses, assembly and maturation occur inside the cell and are inseparable, whereas for others, maturation events may occur only after release of the virus particle from the cell. In all cases, the process of maturation prepares the particle for the infection of subsequent cells.

Release

As described earlier, plant viruses face particular difficulties imposed by the structure of plant cell walls when it comes to leaving cells and infecting others. In response, they have evolved particular strategies to overcome this problem which are discussed in detail in Chapter 6. All other viruses escape the cell by one of two mechanisms. For lytic viruses (such as most non-enveloped viruses), release is a simple process – the infected cell breaks open and releases the virus. Enveloped viruses acquire their lipid membrane as the virus buds out of the cell through the cell membrane or into an intracellular vesicle prior to subsequent release. Virion envelope proteins are picked up during this process as the virus particle is extruded. This process is known as **budding**. Release of virus particles in this way may be highly damaging to the cell (e.g. paramyxoviruses, rhabdoviruses and togaviruses) or alternatively may not be (e.g. retroviruses), but in either case, the process is controlled by the virus – the physical interaction of the capsid proteins on the inner surface of the cell membrane forces the particle out through the membrane (Figure 4.16). As mentioned earlier, assembly, maturation, and release are usually simultaneous processes for viruses which are released by budding. The type of membrane from which the virus buds depends on the virus concerned. It can either be the cell surface membrane (retroviruses; togaviruses), the cytoplasmic membrane (orthomyxoviruses; paramyxoviruses; bunyaviruses; coronaviruses; rhabdoviruses; hepadnaviruses), or the nuclear membrane (herpesviruses).

The release of mature virus particles from susceptible host cells by budding presents a problem in that these particles are designed to enter, rather than leave, cells. How do these particles manage to leave the cell surface? The details are not known but there are clues as to how the process is achieved. Certain viral envelope proteins are involved in the release phase of replication as well as in the initiating steps. A good example of this is the neuraminidase protein of influenza virus. In addition to being able to reverse the attachment of virus particles to cells via haemagglutinin, neuraminidase is also believed to be important in preventing the aggregation of influenza virus particles and may well have a role in virus release. In addition to using specific proteins, viruses which bud have also solved the problem of release by the careful timing of the assembly–maturation–release pathway. Although it may not be possible to separate these stages by means of biochemical analysis, this does not mean that careful spatial separation of these processes has not evolved as a means to solve this problem. Similarly, although we may not understand all the subtleties of the many conformation changes which occur in virus capsids and envelopes during these late stages of replication, virus replication clearly works, despite our deficiencies!

Summary

In general terms, virus replication involves three broad stages which are carried out by all types of virus: the initiation of infection, replication and expression of the genome, and finally, the release of mature virions from the infected cell. At a detailed level, there are many differences between the replication processes of different viruses. These are imposed by the biology of the host cell and the nature of the virus genome. Nevertheless, it is possible to derive an overview of virus replication with common stages which, in one form or another, are followed by all viruses.

Further Reading

Clapham, P.R. *et al.* (1989). Soluble CD4 blocks the infectivity of diverse strains of HIV and SIV for T-cells and monocytes but not for brain and muscle cells. *Nature* **337**: 368–70.

Ellis, E.L. and Delbruck, M. (1939). The growth of bacteriophage. *J. Gen. Physiol.* **22**: 365–84.

Guo P. (Ed.). Virus Assembly. *Sem. Virol.* **5**: No.1 (1994).

Hershey, A.D. and Chase, M. (1952). Independent functions of viral protein and nucleic acid in growth of bacteriophage. *J. Gen. Physiol.* **26**: 36–56.

Li, Q.X. *et al.* (1992). Epstein–Barr virus infection and replication in a human epithelial cell line. *Nature* **356**: 347–50.

Weiss, R.A. and Clapham, P. (1996). Hot fusion of HIV. *Nature* **381**: 647–648.

Wimmer, E. (Ed.). (1994). *Cellular Receptors for Animal Viruses.* Cold Spring Harbor Laboratory Press, Cold Spring Harbor.

Self-Assessment Questions

(4.1) Virus replication involves the following distinct phases (true or false?):
- (a) Initiation of infection.
- (b) Replication of the virus genome.
- (c) Expression of the virus genome.
- (d) Cleavage of the virus genome.
- (e) Release of mature virions from the infected cell.

(4.2) Are the following statements true or false?
- (a) The 'single burst' experiment was performed by Ellis and Delbruck in 1989.
- (b) The 'single burst' experiment demonstrates that new virus particles result from the intracellular assembly of preformed components and are released in a batch-like process.

(c) The 'single burst' experiment demonstrates the essential phases of virus replication.
(d) The 'Hershey–Chase' experiment demonstrates the essential phases of virus replication.
(e) The 'Hershey–Chase' experiment demonstrates that nucleic acid is the genetic material.

(4.3) Are the following statements true or false?
(a) Viruses use proteins exclusively as their receptors.
(b) Viruses only have one type of receptor molecule.
(c) Plant viruses all use proteins as receptors.
(d) Sialic acid is the human rhinovirus receptor.
(e) The expression of receptors on the surface of cells is the major factor which determines the tropism of a virus.

(4.4) The following are mechanisms by which virus particles penetrate cells (true or false?):
(a) Translocation.
(b) Transfection.
(c) Exocytosis.
(d) Endocytosis.
(e) Fusion.

(4.5) Are the following statements true or false?
(a) Viral penetration of cells is an energy-dependent process.
(b) Influenza haemagglutinin is a virus fusion protein.
(c) Uncoating can occur before or after penetration of the host cell.
(d) During uncoating, all proteins are removed from the virus genome.
(e) Fusion proteins are responsible for nuclear localization of virus genomes.

(4.6) Are the following statements true or false?
(a) There are six classes of virus genome structure.
(b) Class I viruses have circular double-stranded DNA genomes.
(c) Class II viruses have (−)sense single-stranded DNA genomes.
(d) Class VI viruses (retroviruses) have (−)sense single-stranded RNA genomes.
(e) Class VII viruses have (+)sense single-stranded RNA genomes.

(4.7) Are the following statements true or false?
(a) Adenoviruses encode their own DNA polymerase but are dependent on cellular factors for genome replication.
(b) Poxviruses encode their own DNA polymerase but are dependent on cellular factors for genome replication.
(c) Reovirus genomes are transcribed by cellular RNA polymerase.
(d) RNA virus genomes always replicate in the cytoplasm of the host cell.
(e) Retrovirus genomes are transcribed by cellular RNA polymerase.

(4.8) The following viruses replicate their genomes in the cytoplasm of infected cells (true or false?):
 (a) Picornaviruses, e.g. poliovirus.
 (b) Poxviruses, e.g. vaccinia virus.
 (c) Rhabdoviruses, e.g. rabies virus.
 (d) Paramyxoviruses, e.g. respiratory syncytial virus.
 (e) Hepadnaviruses, e.g. hepatitis B virus.

(4.9) The following viruses replicate their genomes in the nucleus of infected cells (true or false?):
 (a) Herpesviruses, e.g. Epstein–Barr virus.
 (b) Parvoviruses, e.g. B19.
 (c) Orthomyxoviruses, e.g. influenza virus.
 (d) Rotaviruses, e.g. group A rotaviruses.
 (e) Coronaviruses, e.g. OC43.

(4.10) Are the following statements true or false?
 (a) Maturation usually involves structural changes in the virus particle.
 (b) Maturation frequently involves protease cleavage of virus proteins.
 (c) Enveloped viruses acquire their lipid membranes by budding through cellular membranes.
 (d) Virus envelope proteins are acquired during the process of budding.
 (e) Virus envelope proteins are involved in release from as well as attachment to host cells.

Answers to Self-Assessment Questions are given in Appendix 3.

Chapter 5

Expression

Expression of Genetic Information

The most critical interaction between a virus and its host cell is the requirement by the virus for the cellular apparatus of nucleic acid and protein synthesis. The course of virus replication is determined by tight control of gene expression. There are fundamental differences in the control of these processes in prokaryotic and eukaryotic cells. These differences inevitably affect the viruses that utilize them as hosts. In addition, the relative simplicity and compact size of virus genomes (compared with those of even the simplest cell) imposes further constraints. Cells have evolved varied and complex mechanisms for controlling gene expression by utilizing their extensive genetic capacity. Conversely, viruses have had to achieve highly specific quantitative, temporal, and spatial control of expression with much more limited genetic resources. Aspects of this problem have already been discussed in Chapters 3 and 4.

Viruses have counteracted their genetic limitations by the evolution of a range of solutions to these problems. These mechanisms include:

- Powerful positive and negative signals which promote or repress gene expression
- Highly compressed genomes in which overlapping reading frames are commonplace
- Control signals which are frequently nested within other genes
- Several strategies designed to create multiple polypeptides from a single messenger RNA.

Gene expression involves regulatory loops mediated by signals which act either in ***cis*** (affecting the activity of contiguous genetic regions) or in ***trans*** (giving rise to diffusible products which act on regulatory sites whether or not these are contiguous with the site from which they are produced). For example, transcription **promoters** are *cis*-acting sequences that are located adjacent to the genes whose transcription they control, while proteins such as 'transcription factors' which bind to specific sequences present on any stretch of nucleic acid present in the cell are examples of *trans*-acting factors. The (perceived) relative simplicity of virus genomes and the elegance of their control mechanisms are models that form the basis of current understanding of genetic regulation.

This chapter assumes familiarity with the mechanisms involved in cellular control of gene expression. However, to illustrate the details of virus gene expression, a very brief summary of some pertinent aspects is given below.

Control of Prokaryote Gene Expression

Bacterial cells are second only to viruses in the specificity and economy of their genetic control mechanisms and possibly foremost in terms of the intensity with which these mechanisms have been studied. Genetic control is exercised both at the level of transcription and at subsequent (post-transcriptional) stages of gene expression.

The initiation of transcription is regulated primarily in a negative fashion by the synthesis of *trans*-acting repressor proteins, which bind to operator sequences upstream of protein coding sequences. Collections of metabolically related genes are grouped together and co-ordinately controlled as 'operons'. Transcription of these operons typically produces a **polycistronic** mRNA that encodes several different proteins. During subsequent stages of expression, transcription is also regulated by a number of mechanisms which act, in Mark Ptashne's famous phrase, as 'genetic switches,' turning on or off the transcription of different genes. Such mechanisms include antitermination, which is controlled by *trans*-acting factors that promote the synthesis of longer transcripts encoding additional genetic information, and by various modifications of RNA polymerase. Bacterial σ (sigma) factors are apoproteins which affect the specificity of the RNA polymerase holoenzyme for different promoters. Several bacteriophages (e.g. phage SP01 of *Bacillus subtilis*) encode proteins which function as alternative σ factors, sequestering RNA polymerase and altering the rate at which phage genes are transcribed. Phage T4 of *Escherichia coli* encodes an enzyme which carries out a covalent modification (ADP-ribosylation) of the host cell RNA polymerase. This is believed to eliminate the requirement of the polymerase holoenzyme for σ factor and to achieve an effect similar to the production of modified σ factors by other bacteriophages.

At a post-transcriptional level, gene expression in bacteria is also regulated by control of translation. The best known viral examples of this phenomenon come from study of bacteriophages of the family *Leviviridae*, such as R17, MS2, and Qβ. In these phages, the secondary structure of the single-stranded RNA phage genome not only regulates the quantities of different phage proteins that are translated, but also operates temporal control of a switch in the ratios between the different proteins produced in infected cells.

Control of Expression in Bacteriophage λ

The genome of phage λ has probably been studied in more detail than any other and illustrates several of the mechanisms described above, including the action of repressor proteins in regulating lysogeny versus lytic replication and anti-termination of transcription by phage-encoded *trans*-acting factors. Such has been the impact of these discoveries that no discussion of the control of virus gene expression could be considered complete without detailed examination of this phage.

Phage λ was discovered at the Pasteur Institute by André Lwoff (who later helped to initiate the idea of operons in bacteria) when he observed that some strains of *E. coli,* when irradiated with ultraviolet light, stopped growing and subsequently lysed, releasing a crop of bacteriophage particles. Together with Francois Jacob and Jaques Monod, he subsequently showed that all the cells of these bacterial strains carried the bacteriophage in a dormant form, known as a **prophage**, and that the phage could be made to alternate between the **lysogenic** (non-productive) and **lytic** (productive) growth cycles. After many years of study, our understanding of λ has been refined into a picture that represents one of the best understood and most elegant genetic control systems yet to be investigated.

A simplified genetic map of λ is shown in Figure 5.1. For regulation of the growth cycle of the phage, the structural genes encoding the head and tail components of the virus capsid can be ignored. The pertinent components involved in genetic control are as follows:

- P_L is the promoter responsible for transcription of the left-hand side of the λ genome, including *N* and *cIII*
- O_L is a short non-coding region of the phage genome (approximately 50 bp) which lies between the *cI* and *N* genes next to P_L
- P_R is the promoter responsible for transcription of the right-hand side of the λ genome, including *cro*, *cII*, and the genes encoding the structural proteins

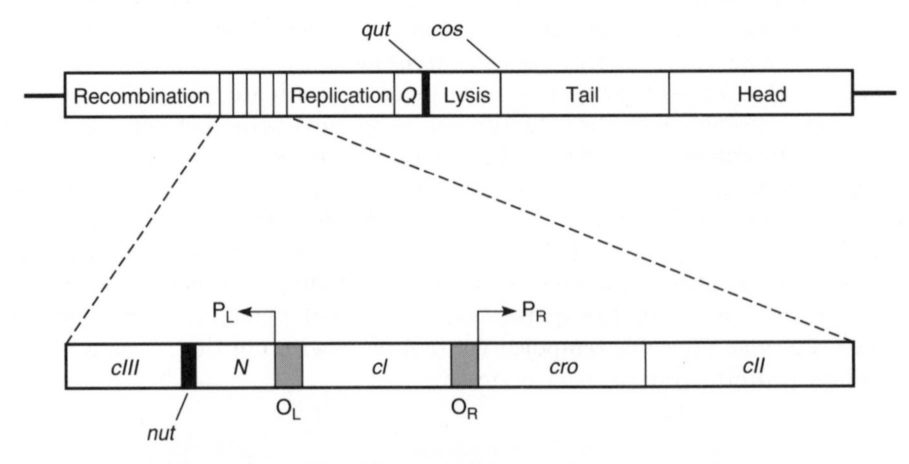

Figure 5.1 Simplified genetic map of bacteriophage λ.

- O_R is a short non-coding region of the phage genome (approximately 50 bp) which lies between the *cI* and *cro* genes next to PR
- *cI* is transcribed from its own promoter and encodes a repressor protein of 236 amino acids which binds to O_R, preventing transcription of *cro* but allowing transcription of *cI*, and to O_L, preventing transcription of *N* and the other genes in the left-hand end of the genome
- *cII* and *cIII* encode activator proteins which bind to the genome, enhancing the transcription of the *cI* gene
- *cro* encodes a 66 amino acid protein which binds to O_R, blocking binding of the repressor to this site
- *N* encodes an antiterminator protein which acts as an alternative rho factor for host cell RNA polymerase, modifying its activity and permitting extensive transcription from P_L and P_R
- *Q* is an antiterminator similar to *N*, but only permits extended transcription from P_R.

In a newly infected cell, *N* and *cro* are transcribed from P_L and P_R respectively (Figure 5.2). The N protein allows RNA polymerase to transcribe a number of phage genes, including those responsible for DNA recombination and integration of the prophage, as well as *cII* and *cIII*. The N protein acts as a positive transcription regulator. In the absence of the N protein, the RNA polymerase holoenzyme stops at certain sequences located at the end of the *N* and *Q* genes, known as the nut and qut sites, respectively. However, RNA polymerase–N protein complexes are able to overcome this restriction and permit full transcription from P_L and P_R. The RNA polymerase–Q protein complex results in extended transcription from P_R only. As levels of the cII and cIII proteins in the cell build up, transcription of the *cI* repressor gene from its own promoter is turned on.

It is at this point that the critical event occurs which determines the outcome of the infection. The cII protein is constantly degraded by proteases present in the cell. If levels of cII remain below a critical level, transcription from P_R and P_L continues and the phage undergoes a productive replication cycle which culminates in lysis of the cell and the release of phage particles. This is the sequence of events that occurs in the vast majority of infected cells. However, in a few rare instances, the concentration of cII protein builds up, transcription of *cI* is enhanced and intracellular levels of the cI repressor protein rise. The repressor binds to O_R and O_L, which prevents transcription of all phage genes (and in particular *cro*: see below) except itself. The level of cI protein is maintained automatically by a negative feedback mechanism, since at high concentrations the repressor also binds to the left-hand end of O_R and prevents transcription of *cI* (Figure 5.3). This autoregulation of cI synthesis keeps the cell in a stable state of lysogeny.

If this is the case, how do such cells ever leave this state and enter a productive, lytic replication cycle? Physiological stress and particularly ultraviolet irradiation of cells results in the induction of a host cell protein, RecA. This protein, whose normal function is to induce the expression of cellular genes which permit the cell to adapt to and survive in altered environmental

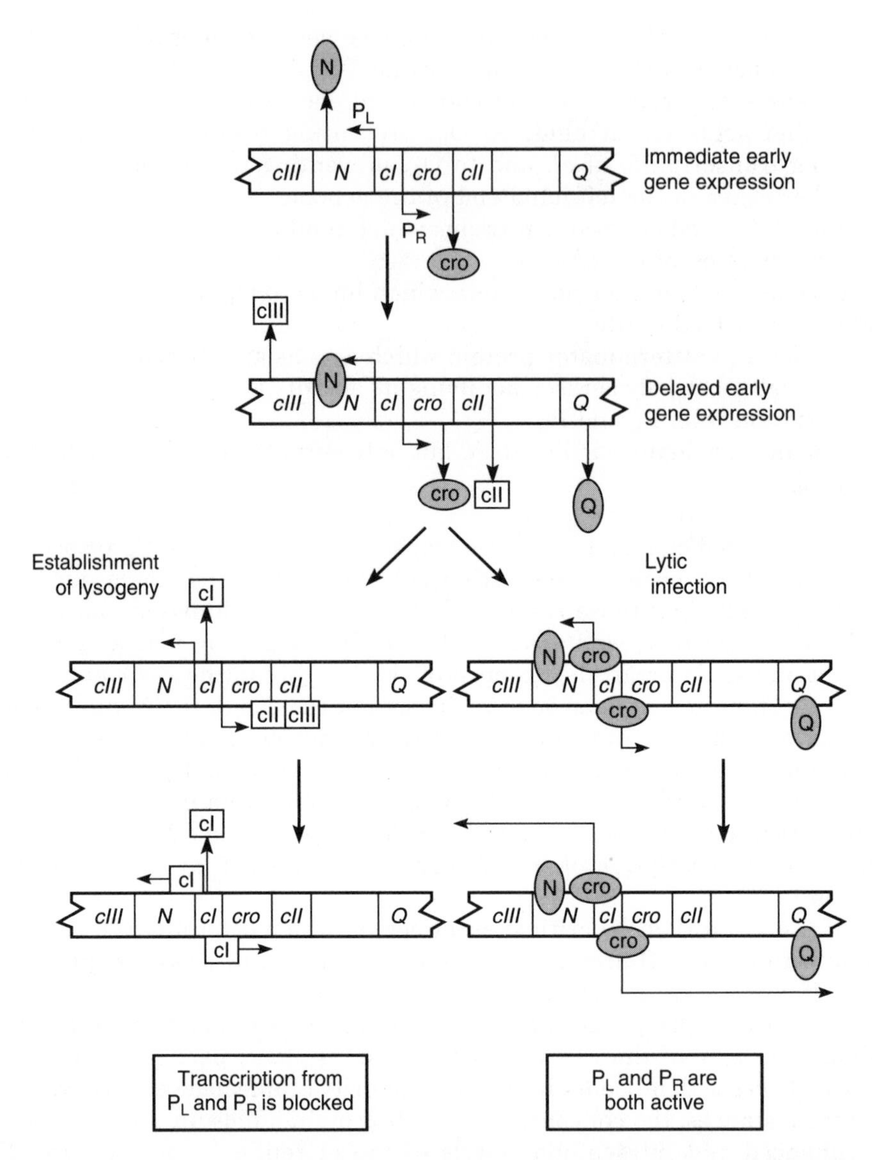

Figure 5.2 Control of expression of the bacteriophage λ genome. A detailed description of the events which occur in a newly infected cell and during lytic infection or lysogeny is given in the text.

conditions, cleaves the cI repressor protein. In itself, this would not be sufficient to prevent the cell re-entering the lysogenic state. However, when repressor protein is not bound to O_R, *cro* is transcribed from P_R. This protein also binds to O_R but unlike cI, which preferentially binds the right-hand end of O_R, the cro protein binds preferentially to the left-hand end of O_R, preventing the transcription of *cI* and enhancing its own transcription in a positive feedback loop. Thus, the phage is locked into a lytic cycle and cannot return to the lysogenic state.

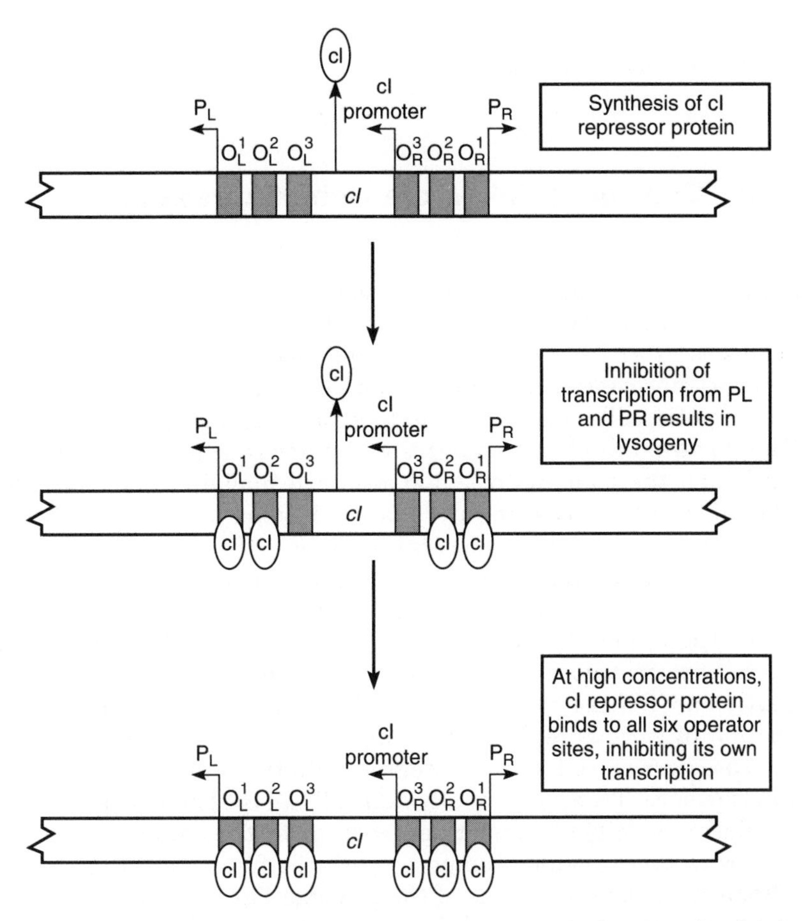

Figure 5.3 Control of lysogeny in bacteriophage λ. See text for details.

This description is a highly simplified version of the genetic control of expression in phage λ. A great deal of detail is known about the molecular mechanisms by which the above systems work, but since this could easily fill a whole book on its own there is insufficient space to recount all of it here. The molecular details of the λ system are not only of isolated interest, but have shaped our understanding of genetic regulation in prokaryotic and also in eukaryotic cells. Determination of the structures of the proteins involved in the above scheme has identified the fundamental principles by which many proteins from unrelated organisms can recognize and bind to specific sequences in DNA molecules. The concepts of proteins with independent DNA-binding and dimerization domains, protein cooperativity in DNA binding and DNA looping allowing proteins bound at distant sites to interact with one another have all arisen from the study of λ. It is important to read the further reading given at the end of this chapter to understand fully the nuances of gene expression in this complex bacteriophage, but also to remember the design and operation of this system when reading the examples of the regulation of gene expression described in the rest of this chapter.

Control of Eukaryote Gene Expression

Control of gene expression in eukaryotic cells is much more complex than in prokaryotic cells, and involves a 'multilayered' approach in which diverse control mechanisms exert their effects at multiple levels. The first level of control occurs prior to transcription and depends on the local configuration of the DNA. DNA in eukaryotic cells has an elaborate structure, forming complicated and dynamic but far from random complexes with numerous proteins to form **chromatin**. Although the contents of eukaryotic cell nuclei appear amorphous in electron micrographs (at least in interphase), they are actually highly ordered. Chromatin interacts with the structural backbone of the nucleus, the nuclear matrix, and these interactions are thought to be important in controlling gene expression. Locally, nucleosome configuration and DNA conformation, particularly the formation of left-handed helical 'Z-DNA', are also important. DNAse I digestion of chromatin does not give an even, uniform digestion pattern, but reveals particular 'DNAse hypersensitive' sites believed to indicate differences in the function of various regions of the chromatin. It is possible, for example, that retroviruses are more likely to integrate into the host cell genome at these sites than elsewhere. Transcriptionally active DNA is also hypomethylated, i.e. there is a relative paucity of nucleotides modified by the covalent attachment of methyl groups in these regions compared with the frequency of methylation in transcriptionally quiescent regions of the genome. The methylation of Moloney murine leukaemia virus sequences in pre-implantation mouse embryos has been shown to suppress the transcription of the provirus genome.

The second level of control rests in the process of transcription itself, which again is much more complex than in prokaryotes. There are three forms of RNA polymerase in eukaryotic cells which can be distinguished by their relative sensitivities to the drug α-amanitin and which show specificities for different classes of genes (Table 5.1). The rate at which transcription is initiated is a key control point in eukaryotic gene expression. Initiation is influenced dramatically by sequences upstream of the transcription start site, which function by

Table 5.1 Forms of RNA polymerase in eukaryotic cells

RNA polymerase	Sensitivity to α-amanatin	Cellular genes transcribed	Virus genes transcribed
I	Unaffected	Ribosomal RNAs	–
II	Highly sensitive	Most single-copy genes	Most DNA virus genomes
III	Moderately sensitive	5S rRNA, tRNAs	Adenovirus VA RNAs

Table 5.2 Eukaryotic transcription factors

Transcription factor	Molecular weight	Recognition site
SP1	85 000	5′-GGGCGG-3′
NF1 (‘CTF’)	52 000–66 000	5′-GCCAAT-3′
ATF (‘CREB’)	43 000	5′-(T/G)(TA)CGTCA-3′
AP2	48 000	5′-CCCCAGGG-3′
NFκB	50 000–65 000	5′-GGGACTTTCC-3′

acting as recognition sites for families of highly specific DNA-binding proteins, known colloquially as ‘transcription factors’ (Table 5.2). Immediately upstream of the transcription start site is a relatively short region known as the **promoter**. It is at this site that transcription complexes, consisting of RNA polymerase plus accessory proteins, bind to the DNA and transcription begins. However, the sequences further upstream from the promoter influence the efficiency with which transcription complexes form. The rate of initiation depends on the combination of transcription factors bound to these transcription **enhancers**. The properties of these enhancer sequences are remarkable in that they can be inverted and/or moved around relative to the position of the transcription start site without losing their activity and can exert their influence even from a distance of several kilobases away. This emphasizes the flexibility of DNA, which allows proteins bound at distant sites to interact with one another, as also shown by the protein–protein interactions seen in regulation of phage λ gene expression (above). Transcription of eukaryotic genes results in the production of **monocistronic** mRNAs, each of which is transcribed from its own individual promoter.

At the next stage, gene expression is influenced by the structure of the mRNA produced. The stability of eukaryotic mRNAs varies considerably, some having comparatively long half-lives in the cell (e.g. many hours). The half-lives of others, typically those which encode regulatory proteins, may be very short (e.g. a few minutes). The stability of eukaryotic mRNAs depends on the speed with which they are degraded. This is determined by factors such as its terminal sequences, which consist of a methylated cap structure at the 5′ end and polyadenylic acid at the 3′ end, as well as on the overall secondary structure of the message. However, gene expression is also regulated by differential **splicing** of heterogeneous (heavy) nuclear RNA (**hnRNA**) precursors in the nucleus, which can alter the genetic ‘meaning’ of different mRNAs transcribed from the same gene. In eukaryotic cells, control is also exercised during export of RNA from the nucleus to the cytoplasm.

Finally, the process of translation offers further opportunities for control of expression. The efficiency with which different mRNAs are translated varies greatly. These differences result largely from the efficiency with which ribosomes bind to different mRNAs, recognize AUG translation initiation codons in different sequence contexts and the speed at which different sequences are converted into protein. Certain sequences act as translation enhancers, performing a function analogous to that of transcription enhancers.

The point of giving this extensive list of eukaryotic gene expression mechanisms is that they are all utilized by viruses to control gene expression. Examples of each type are given in the sections below. If this seems remarkable, it must be remembered that the control of gene expression in eukaryotic cells was unravelled largely by utilizing viruses as model systems. Therefore, finding examples of these mechanisms in viruses is really only a self-fulfilling prophecy.

Genome Coding Strategies

In Chapter 4, genome structure was one element of an arbitrary classification used to divide virus genomes into seven groups. The other part of such a scheme is the way in which the genetic information of each class of virus genomes is expressed. The replication and expression of virus genomes are inseparably linked, and this is particularly true in the case of RNA viruses. Here, the seven classes of virus genome described in Chapter 4 and Appendix 1 will be reviewed again, this time examining the way in which the genetic information of each class is expressed.

Class I: Double-stranded DNA

It was stated in Chapter 4 that this class of virus genomes can be subdivided into two further groups: those in which genome replication is exclusively nuclear (e.g. *Adenoviridae, Papovaviridae, Herpesviridae*) and those in which replication occurs in the cytoplasm (*Poxviridae*). In one sense, all of these viruses can be considered as similar, since their genomes all resemble double-stranded cellular DNA, they are essentially transcribed by the same mechanisms as cellular genes. However, there are profound differences between them relating to the degree to which each family is reliant on the host cell machinery.

Papovaviruses

Papovaviruses are heavily dependent on cellular machinery both for replication and gene expression. Polyoma viruses encode *trans*-acting factors (T-antigens) which stimulate transcription (and genome replication). Papillomaviruses in particular are dependent on the cell for replication, which only occurs in terminally differentiated keratinocytes and not in other cell types, although they do encode several *trans*-regulatory proteins (Chapter 7).

Adenoviruses

Adenoviruses are also heavily dependent on the cellular apparatus for transcription, but they possess various mechanisms that specifically regulate virus gene expression. These include *trans*-acting transcriptional activators such as the E1A protein and post-transcriptional regulation of expression, which is achieved by alternative splicing of mRNAs and the virus-encoded VA RNAs (Chapter 7). Adenovirus infection of cells is divided into two stages, early and late, the latter phase commencing at the time when genome replication occurs. However, in adenoviruses, these phases are less distinct than in herpesviruses (below).

Herpesviruses

These viruses are less reliant on cellular enzymes than the above groups. They encode many enzymes involved in DNA metabolism (e.g. thymidine kinase) and several *trans*-acting factors that regulate the temporal expression of virus genes, controlling the phases of infection. Transcription of the large, complex genome is sequentially regulated in a cascade fashion (Figure 5.4). At least 50 virus-encoded proteins are produced after transcription of the genome by host cell RNA polymerase II. Three distinct classes of mRNAs are made:

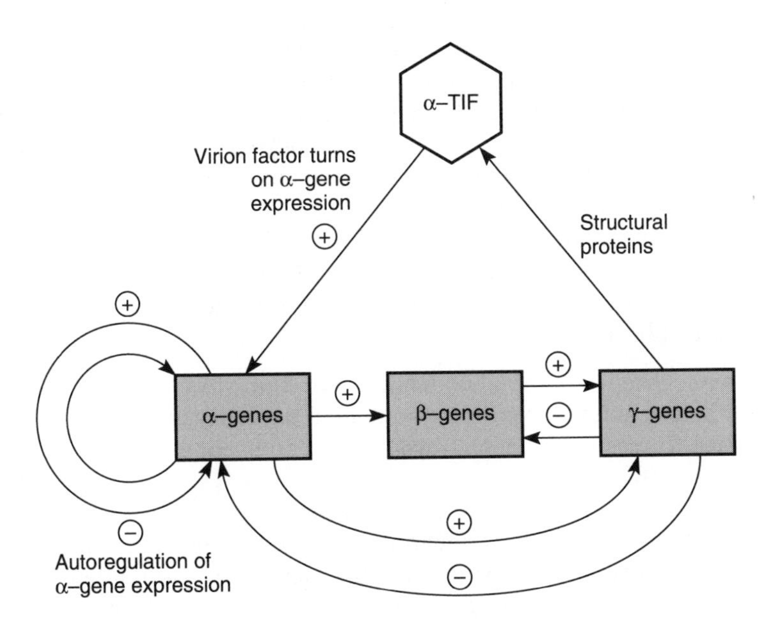

Figure 5.4 Control of expression of the herpesvirus genome. Herpesvirus particles contain a protein called α-TIF (α-gene Transcription Initiation Factor) which turns on α-gene expression in newly-infected cells, beginning a cascade of closely-regulated events which control the expression of the entire complement of the 70-odd genes in the virus genome.

- α: Immediate-early (IE) mRNAs encode five *trans*-acting regulators of virus transcription
- β: (Delayed) early mRNAs encode further non-structural regulatory proteins and some minor structural proteins
- γ: Late mRNAs encode the major structural proteins.

Gene expression in herpesviruses is tightly and co-ordinately regulated, as indicated by the following observations (see Figure 5.4):

- If translation is blocked shortly after infection (e.g. by treating cells with cycloheximide), early mRNAs immediately accumulate in the nucleus but no further virus mRNAs are transcribed
- Synthesis of the early gene product turns off the immediate-early products and initiates genome replication
- Some of the late structural proteins (γ1) are produced independently of genome replication, others (γ2) are only produced after replication.

Both the immediate-early and early proteins are required to initiate genome replication. A virus-encoded DNA-dependent DNA polymerase and a DNA-binding protein are involved in genome replication, together with a number of enzymes (e.g. thymidine kinase) which alter cellular biochemistry. The production of all of these proteins is closely controlled.

Poxviruses

Genome replication and gene expression in poxviruses are almost independent of cellular mechanisms (except for the requirement for host cell ribosomes). Poxvirus genomes encode numerous enzymes involved in DNA metabolism, viral gene transcription, and post-transcriptional modification of mRNAs. Many of these enzymes are packaged within the virus particle (which contains >100 proteins), enabling transcription and replication of the genome to occur in the cytoplasm (rather than in the nucleus, like all the families described above) almost totally under the control of the virus. Gene expression is carried out by viral enzymes associated with the core of the particle and is divided into two rather indistinct phases:

- *Early genes*: These comprise about 50% of the poxvirus genome and are expressed before genome replication inside a partially uncoated core particle (see Chapter 2), resulting in the production of 5′ capped, 3′ polyadenylated but unspliced mRNAs
- *Late genes*: These are expressed after genome replication in the cytoplasm, but their expression is also dependent on virus-encoded rather than on cellular transcription proteins (which are located in the nucleus). Like herpesviruses, late gene promoters are dependent on prior DNA replication for activity.

More detailed consideration of some of the mechanisms mentioned above is given later in this chapter (see Transcriptional Control of Expression and Post-Transcriptional Control of Expression below).

Class II: Single-stranded DNA

Both the autonomous and the helper virus-dependent parvoviruses are highly reliant on external assistance for gene expression and genome replication. This is presumably because the very small size of their genomes does not permit them to encode the necessary biochemical apparatus. Thus, they appear to have evolved an extreme form of parasitism, utilizing the normal functions present in the nucleus of their host cells for both expression and replication (Figure 5.5). The members of the replication-defective *Dependovirus* genus of the parvovirus family are entirely dependent on adenovirus or herpesvirus superinfection for the provision of further helper functions essential for their replication. The adenovirus genes required as helpers are the early, transcriptional regulatory genes such as E1A rather than late structural genes, but it has been shown that treatment of cells with u.v., cycloheximide, or some carcinogens can replace the requirement for helper viruses. Therefore, the help required appears to be for a

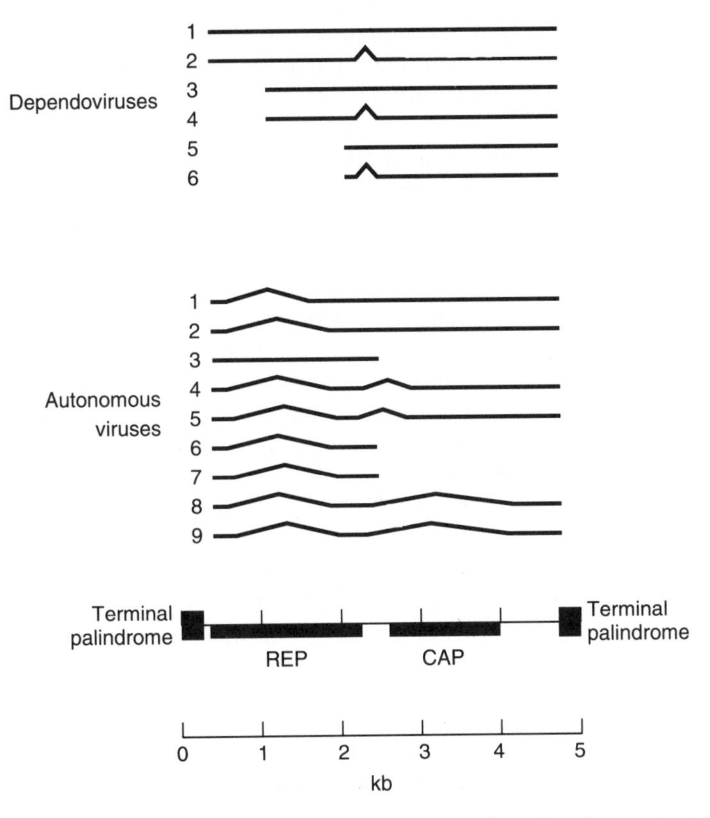

Figure 5.5 Transcription of parvovirus genomes is heavily dependent on host cell factors and results in the synthesis of a series of spliced, sub-genomic mRNAs which encode two proteins, rep, which is involved in genome replication and cap, the capsid protein (see text).

modification of the cellular environment (probably affecting transcription of the defective parvovirus genome) rather than for a specific virus protein.

The geminiviruses of plants also fall into this class of genome structures (Figure 3.15). The expression of their genomes is quite different from that of parvoviruses, but nevertheless still relies heavily on host cell functions. There are open reading frames in both orientations in the virus DNA, which means that both (+) and (−)sense strands are transcribed during infection. The mechanisms involved in control of gene expression have not been fully investigated, but at least some geminiviruses (subgroup I) may use splicing.

Class III: Double-stranded RNA

All viruses with RNA genomes differ fundamentally from their host cells, which of course possess double-stranded DNA genomes. Therefore, although each virus must be biochemically 'compatible' with its host cell, there are fundamental differences in the mechanisms of virus gene expression from those of the host cell. Reoviruses have multipartite genomes (see Chapter 3) and replicate in the cytoplasm of the host cell. Characteristically for viruses with segmented RNA genomes, a separate monocistronic mRNA is produced from each segment (Figure 5.6).

Early in infection, transcription of the d/s RNA genome segments by virus-specific transcriptase activity occurs inside partially uncoated sub-viral particles. At least seven enzymatic activities are present in reovirus particles to carry out this process, although these are not necessarily all separate peptides (Table 5.3, Figure 2.12). This primary transcription results in capped transcripts that are not polyadenylated and which leave the virus core to be translated in the cytoplasm. The various genome segments are transcribed/translated at different frequencies, which is perhaps the main advantage of a segmented genome. RNA is transcribed conservatively, i.e. only (−)sense strands are used, resulting in synthesis of (+)sense mRNAs, which are capped inside the core (all this occurs without *de novo* protein synthesis). Secondary transcription occurs later in infection inside new particles produced in infected cells and results in uncapped, non-polyadenylated transcripts. The genome is replicated in a conservative fashion (c.f. semi-conservative DNA replication). An excess of (+)sense strands are produced which serve as late mRNAs and as template for (−)sense strand synthesis (i.e. each (−) strand leads to many (+) strands, not one-for-one as in semi-conservative replication).

Class IV: Single-stranded (+)sense RNA

This type of genome occurs in many animal viruses and plant viruses (Appendix 1). In terms of both the number of different families and the number of individual

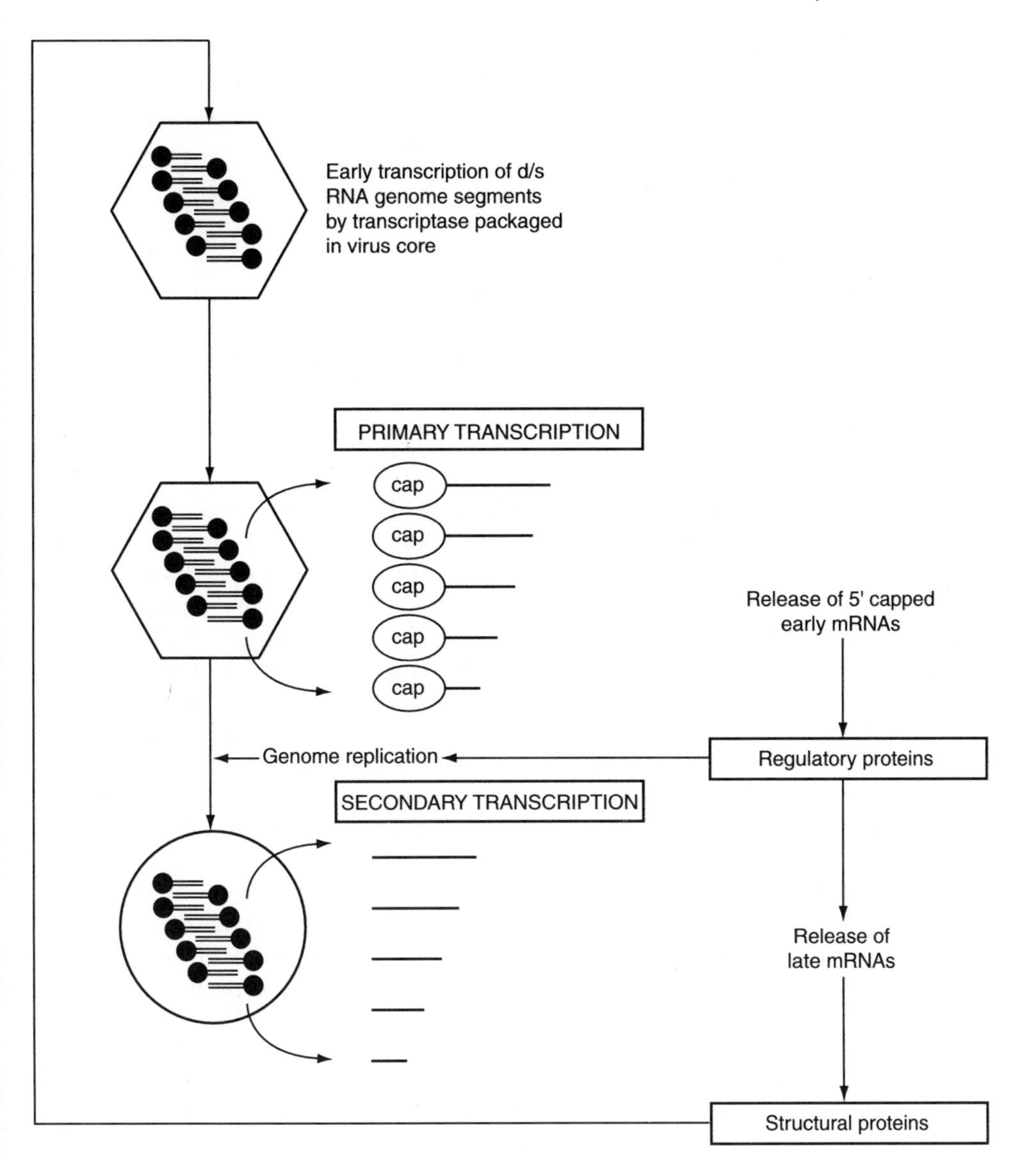

Figure 5.6 Expression of reovirus genomes is initiated by a transcriptase enzyme packaged inside every virus particle. Subsequent events occur in a tightly-regulated pattern, the expression of late mRNAs encoding the structural proteins being dependent on prior genome replication.

viruses, this is the largest single class of virus genome. Fundamentally, these virus genomes act as messenger RNAs and are themselves translated immediately after infection of the host cell (see Chapter 3). Not surprisingly with so many representatives, this class of genomes displays a very diverse range of strategies for controlling gene expression and genome replication. However, in

Table 5.3 Enzymes in reovirus particles

Activity	Virus protein	Encoded by genome segment
d/s RNA-dependent RNA polymerase	λ3	L
RNA triphosphatase	???	???
Guanyltransferase	λ2	L
Methyltransferase 1	λ2?	L?
Methyltransferase 2	λ2?	L?
Helicase	???	???

very broad terms, the viruses in this class can be subdivided into two groups as follows:

(1) Production of a polyprotein encompassing the whole of the viral genetic information, which is subsequently cleaved by proteases to produce precursor and mature polypeptides. These cleavages can be a subtle way of regulating the expression of genetic information. Alternative cleavages result in the production of various proteins with distinct properties from a single precursor, e.g. in picornaviruses and potyviruses (Figure 5.7). Certain plant viruses with multipartite genomes utilize a very similar strategy for controlling gene expression, although a separate polyprotein is produced from each of the genome segments. The best studied example of this are the comoviruses, whose genome organization is very similar to that of the picornaviruses and may represent another member of this 'superfamily' (Figure 5.7).
(2) Production of subgenomic mRNAs, resulting from two or more rounds of translation of the genome. This strategy is used to achieve temporal separation of what are essentially 'early' and 'late' phases of replication, in which non-structural proteins, including a viral replicase, are produced during the 'early' phase followed by structural proteins in the 'late' phase (Figure 5.8). The proteins produced in each of these phases may result from proteolytic processing of a polyprotein precursor, although this encompasses only part of the virus genome rather than the entire genome, as above. Proteolytic processing offers further opportunities for regulation of the ratio of different polypeptides produced in each phase of replication (e.g. in togaviruses and tymoviruses). In addition to proteolysis, some viruses employ another strategy to produce alternative polypeptides from a subgenomic mRNA, either by read-through of a 'leaky' translation stop codon, e.g. tobamoviruses such as TMV (Figure 3.12), or by deliberate ribosomal frameshifting at a particular site (see Post-Transcriptional Control of Expression below).

All the viruses in this class have evolved mechanisms that allow them to regulate their gene expression both in terms of the ratios of different virus-encoded proteins and the stage of the replication cycle when they are produced. Compared with the two classes of DNA virus genomes described above, these

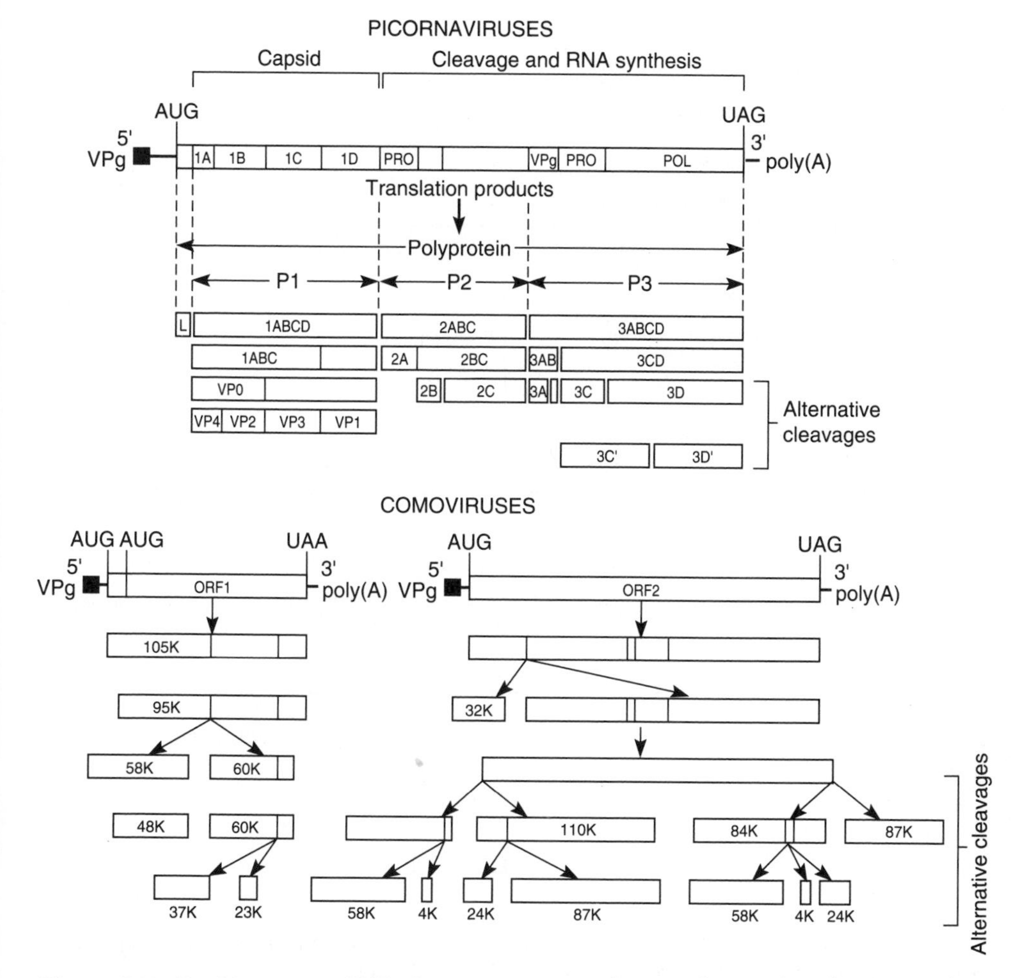

Figure 5.7 Positive-sense RNA virus genomes are frequently translated to form a long polyprotein, which is subsequently cleaved by a highly specific virus-encoded protease to form the mature polypeptides.

mechanisms operate largely independently of those of the host cell. The power and flexibility of these strategies is reflected very clearly in the overall success of the viruses in this class, as determined by the number of different representatives known and the number of different hosts they infect.

Class V: Single-stranded (−)sense RNA

As discussed in Chapter 3, the genomes of these viruses may be either segmented or non-segmented. The first step in the replication of segmented orthomyxovirus genomes is transcription of the (−)sense vRNA by the virion-associated RNA-dependent RNA polymerase to produce (predominantly) monocistronic mRNAs,

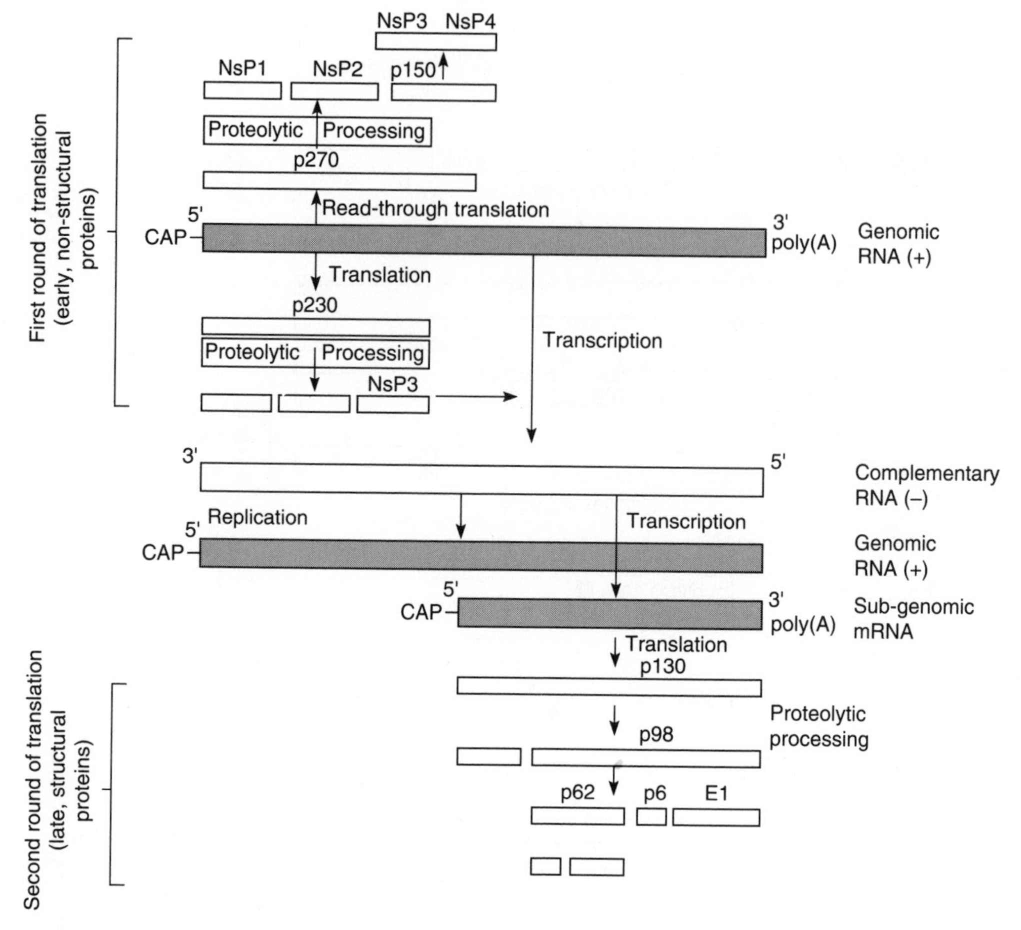

Figure 5.8 Some positive-sense RNA virus genomes (e.g. togaviruses) are expressed by two separate rounds of translation, involving the production of a subgenomic mRNA at a later stage of replication.

which also serve as the template for subsequent genome replication (Figure 5.9). As with all (−)sense RNA viruses, packaging of this virus-specific transcriptase/replicase within the virus nucleocapsid is essential because no host cell contains any enzyme capable of decoding and copying the RNA genome.

In the other families that have non-segmented genomes, monocistronic mRNAs are also produced. Here, however, these messages must be produced from a single, long (−)sense RNA molecule. Exactly how this is achieved is not clear. It is possible that a single, genome-length transcript is cleaved after transcription to form the separate mRNAs, but it is more likely that these are produced individually by a stop and start mechanism of transcription regulated by the conserved intergenic sequences present between each of the virus genes (see Chapter 3). Splicing mechanisms cannot be used because these viruses replicate in the cytoplasm.

On the surface, such a scheme of gene expression might appear to offer few opportunities for regulation of the relative amounts of different virus proteins. If

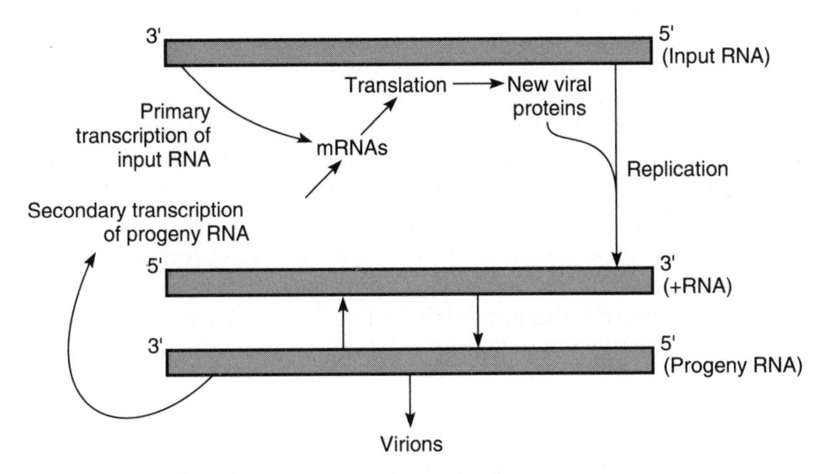

Figure 5.9 Generalized scheme for the expression of negative-sense RNA virus genomes.

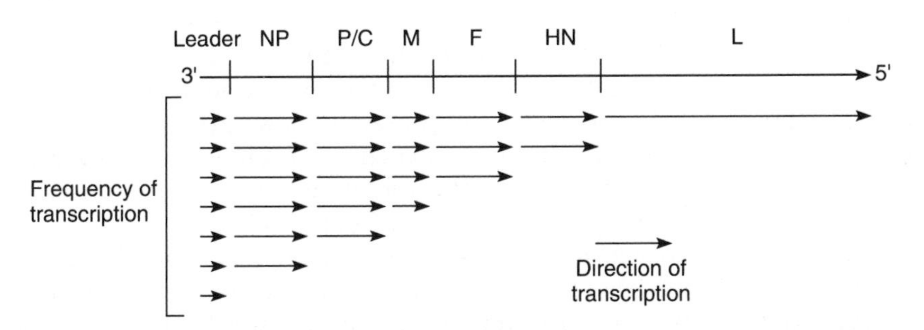

Figure 5.10 Paramyxovirus genomes exhibit 'transcriptional polarity'. Transcripts of genes at the 3′ end of the virus genome are more abundant than those of genes at the 5′end of the genome, permitting regulation of the relative amounts of structural (3′ genes) and non-structural (5′ genes) proteins produced.

this were true, it would be a major disadvantage, since all viruses require far more copies of the structural proteins (e.g. nucleocapsid protein) than of the non-structural proteins (e.g. polymerase) for each virion produced. In practice, the ratio of different proteins is regulated both during transcription and afterwards. In paramyxoviruses for example, there is a clear polarity of transcription from the 3′ end of the virus genome to the 5′ end, which results in the synthesis of far more mRNAs for the structural proteins encoded in the 3′ end of the genome than for the non-structural proteins located at the 5′ end (Figure 5.10). Similarly, the advantage of producing monocistronic mRNAs is that the translational efficiency of each message can be varied with respect to the others (see Post-Transcriptional Control of Expression below).

Class VI: Single-stranded (+)sense RNA with DNA intermediate

The retroviruses are perhaps the ultimate case of reliance on host cell transcription machinery. The RNA genome does not serve as mRNA, but as a template for reverse transcription into DNA (see Chapter 3). Once integrated into the host cell genome, the DNA provirus is under the control of the host cell and is transcribed exactly as are other cellular genes. Some retroviruses, however, have evolved a number of transcriptional and post-transcriptional mechanisms that allow them to control the expression of their genetic information and these are discussed in detail in the relevant sections later in this chapter.

Class VII: Double-stranded DNA with RNA intermediate

Expression of the genomes of these viruses is complex and relatively poorly understood. The hepadnaviruses contain a number of overlapping reading frames clearly designed to squeeze as much coding information as possible into a compact genome. The X gene encodes a transcriptional *trans*-activator believed to be analagous to the HTLV tax protein (below). At least two mRNAs are produced from independent promoters, each of which encodes several proteins and the larger of which is also the template for reverse transcription during the formation of the virus particle (see Chapter 3).

Expression of caulimovirus genomes is similarly complex, although there are similarities with hepadnaviruses in that two major transcripts are produced, 35S and 19S. Each of these encodes several polypeptides and the 35S transcript is the template for reverse transcription during the formation of the virus genome.

Transcriptional Control of Expression

Having reviewed the general strategies used by different groups of viruses to regulate gene expression, the rest of this chapter will concentrate on more detailed explanations of specific examples from some of the viruses mentioned above, beginning with control of transcription in SV40, a member of the *Papovaviridae* family.

Few other genomes, viral or cellular, have been studied in such detail as that of SV40, which has been a paradigm for the study of eukaryotic transcription mechanisms (and particularly, DNA replication: see Chapter 6) for many years. In this sense, SV40 provides a eukaryotic parallel with the bacteriophage λ genome. *In vitro* systems exist for both transcription and replication of the SV40 genome and it is believed that all the viral and cellular DNA-binding

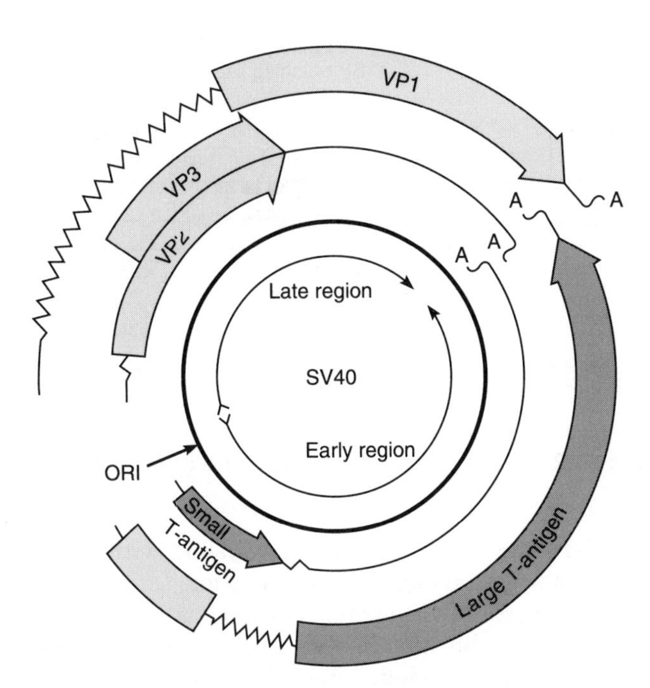

Figure 5.11 Organization and protein-coding potential of the SV40 genome.

proteins involved in both of these processes are known. The SV40 genome encodes two T-antigens ('tumour antigens') known as large T-antigen and small T-antigen after the sizes of the proteins (Figure 5.11). Replication of the double-stranded DNA genome of SV40 occurs in the nucleus of the host cell. Transcription of the genome is carried out by host cell RNA polymerase II and large T-antigen plays a vital role in regulating transcription of the virus genome. Small T-antigen is not essential for virus replication, but allows virus DNA to accumulate in the nucleus. Both proteins contain 'nuclear localization signals' which results in their accumulation in the nucleus, where they migrate after being synthesized in the cytoplasm.

Soon after infection of permissive cells, early mRNAs are expressed from the early promoter, which contains a strong transcription enhancer element (the 72 bp sequence repeats), allowing it to be active in newly infected cells (Figure 5.12). The early proteins synthesized are the two T-antigens. As the concentration of large T-antigen builds up in the nucleus, transcription of the early genes is repressed by direct binding of the protein to the origin region of the virus genome, preventing transcription from the early promoter and causing the switch to the late phase of infection. As already mentioned, large T-antigen is also required for replication of the genome, which is considered further in Chapter 7. After DNA replication has occurred, transcription of the late genes occurs from the late promoter and results in the synthesis of the structural proteins, VP1, VP2 and VP3. Therefore, the role of the SV40 T-antigen in controlling the transcription of the genome is comparable to that of a 'switch' and

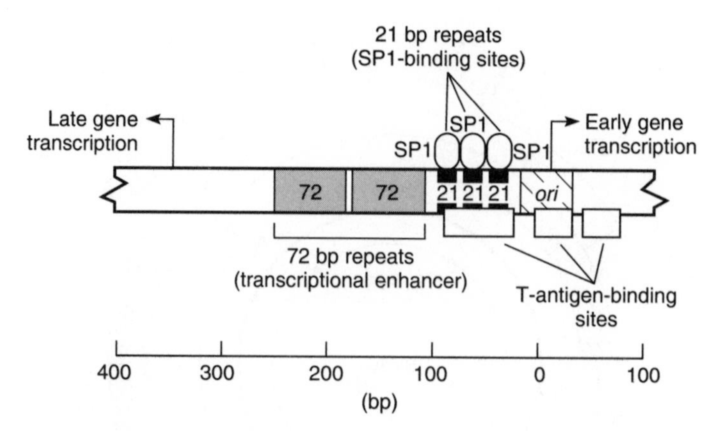

Figure 5.12 Control of transcription of the SV40 genome.

readers should compare the functioning of this system with the description of bacteriophage λ gene expression control given earlier.

Another area where control of viral transcription has received much attention over the last few years are the human retroviruses, human T-cell leukaemia virus (HTLV) and human immunodeficiency virus (HIV). Integrated DNA proviruses are formed by reverse transcription of the RNA retrovirus genome, as described in Chapter 3. The presence of numerous binding sites for cellular transcription factors in the long terminal repeats (LTRs) of these viruses have been analysed by 'DNAse I footprinting' and 'gel-shift' assays (Figure 5.13). Together, the 'distal' elements (such as NFκB and SP1 binding-sites) and 'proximal' elements (such as the TATA box) make up a functional transcription promoter in the U3 region of the LTR (see Chapter 3). However, the basal activity of these promoters on their own is relatively weak, and results in only limited transcription of the provirus genome by RNA polymerase II. Both HTLV and HIV encode proteins which are *trans*-acting positive regulators of transcription, the tax protein of HTLV and the HIV tat protein (Figure 5.14). These proteins act to increase transcription from the viral LTR by a factor of at least 50–100 times that of the basal rate from the 'unaided' promoter. Unlike T-antigen and the early promoter of SV40, neither the tax nor the tat protein (which have no structural similarity to one another) bind directly to their respective LTRs. Recent work has illustrated how these proteins act.

HTLV tax forms protein–protein interactions with a considerable number of different cellular transcription factors, e.g. p105, an inactive precursor of transcription factor NF-κB. NF-κB plays a central role in controlling transcription and immune activation of lymphoid cells. Antisense inhibition of NF-κB has been shown to ablate tax-transformed tumours transplanted into mice, indicating the importance of this protein for tax function. The promiscuous nature of tax *trans*-activation is also explained by a report that tax enhances dimerization of 'bZIP' proteins that bind to DNA via a basic domain–leucine zipper region. Such proteins include a number of factors known to be involved in transcription of the HTLV genome. Dimerization of these proteins is required

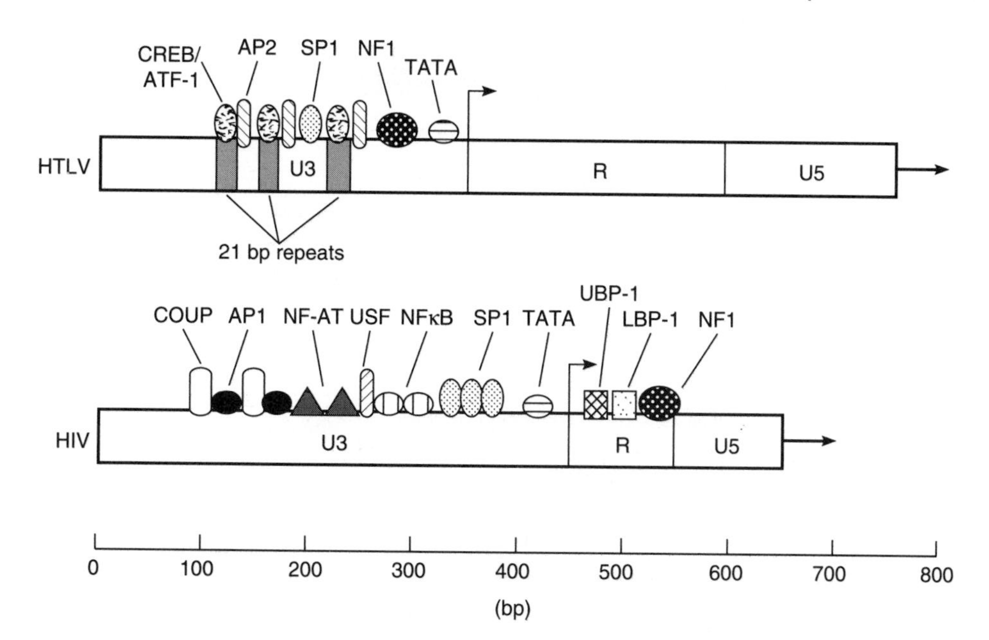

Figure 5.13 Cellular transcription factors which interact with retrovirus LTRs. These cellular DNA-binding proteins are involved in regulating both the basal and *trans*-activated levels of transcription from the promoter in the U3 region of the LTR.

for DNA binding and thus for transcriptional activation, and tax stimulates this process, even in the absence of DNA.

The HIV tat protein binds to a stem-loop structure at the 5′ end of mRNAs transcribed from the LTR, known as the 'TAR' (Trans-Activation Response) element. Several cellular proteins also bind to TAR, although the way in which they interact with tat has not been clearly defined. At present, there is no consensus as to precisely how tat increases steady-state levels of HIV mRNAs in infected cells, but there are several possibilities:

- Tat and TAR might function as antiterminators of transcription in a way analogous to the N protein/nut site in phage λ. This would result in increased transcription elongation efficiency
- Tat might act, directly or indirectly, by modifying the transcription initiation complex which is assembled at the promoter site in the viral LTR. This would result in increased transcription initiation efficiency.

The HTLV tax and HIV tat proteins are positive regulators of the basal promoter in the provirus LTR and are under the control of the virus, since synthesis of these proteins is dependent on the promoters which they themselves activate (Figure 5.15). On its own, this would be an unsustainable system because it would result in unregulated positive feedback, which might be acceptable in a lytic replication cycle, but would not be appropriate for a retrovirus integrated into the genome of the host cell. Therefore, these viruses each encode an additional protein, the rex and rev proteins from HTLV and HIV, respectively, which

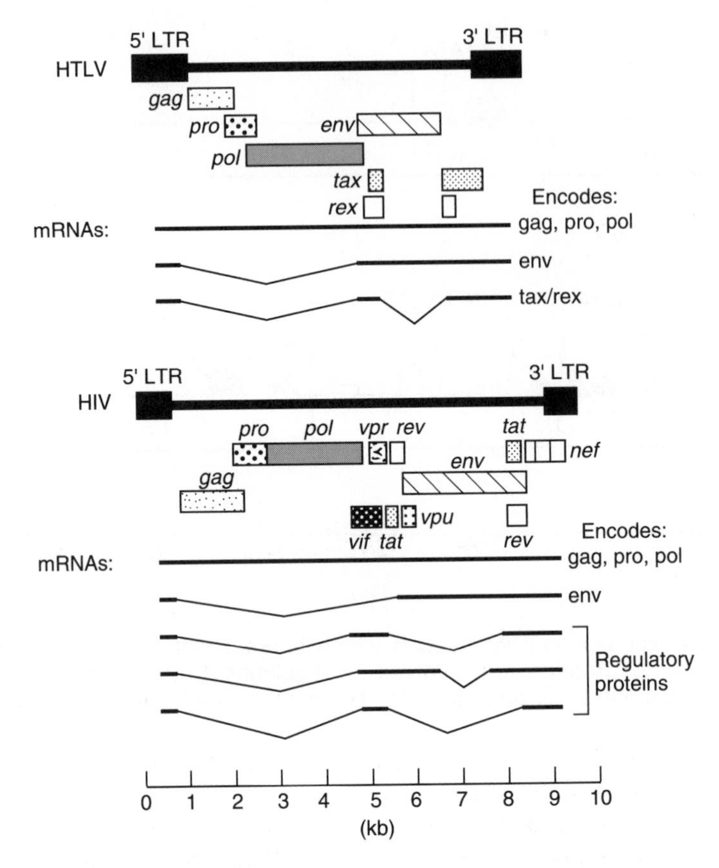

Figure 5.14 Expression of the HTLV and HIV genomes.

further regulate gene expression at a post-transcriptional level (see Post-Transcriptional Control of Expression).

Control of transcription is a critical step in viral replication and in all cases, is closely regulated. Even some of the simplest virus genomes, such as SV40, encode proteins that regulate their transcription. Many virus genomes encode *trans*-acting factors that modify and/or direct the cellular transcription apparatus. Examples of this include HTLV and HIV as described above, but also the X protein of hepadnaviruses, rep protein of parvoviruses, E1A protein of adenoviruses (see below) and the immediate early proteins of herpesviruses. The expression of RNA virus genomes is similarly tightly controlled, but this process is carried out by virus-encoded transcriptases and has been less intensively studied and is generally much less well understood than transcription of DNA genomes.

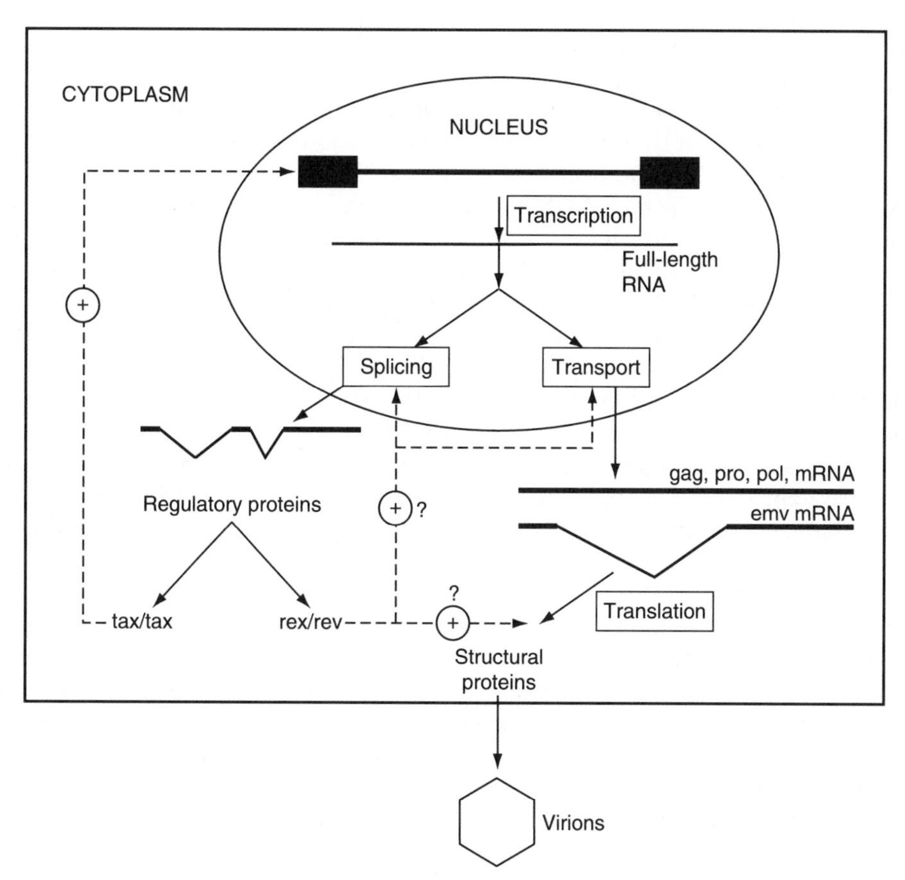

Figure 5.15 *Trans*-acting regulation of HTLV and HIV gene expression by virus-encoded proteins. The tax and tat proteins (from HTLV and HIV respectively) act at a transcriptional level and stimulate the expression of all virus genes. The rex (HTLV) and rev (HIV) proteins act post-transcriptionally and regulate the balance of expression between virion proteins and regulatory proteins.

Post-Transcriptional Control of Expression

In addition to control of the process of transcription itself, the expression of viral genetic information is also governed at a number of additional stages between the formation of the primary RNA transcript and the completion of the finished polypeptide. There are many generalized, subtle controls, such as the differential stability of various mRNAs, which are undoubtedly employed by viruses to regulate the flow of genetic information from their genomes into proteins. This section however will describe only a few well-researched specific examples of post-transcriptional regulation.

Many DNA viruses that replicate in the nucleus encode mRNAs which must be spliced by cellular mechanisms to remove intervening sequences (introns) before being translated. This type of modification applies only to viruses which replicate in the nucleus (and not, for example, to poxviruses) since it requires the processing of mRNAs by nuclear apparatus before they are transported into the cytoplasm for translation. However, several virus families have taken advantage of this capacity of their host cells to compress more genetic information into their genomes. Some of the best examples of such a reliance on splicing are the parvoviruses, transcription of which results in multiple spliced, polyadenylated transcripts in the cytoplasm of infected cells, enabling them to produce multiple proteins from their 5 kb genomes (Figure 5.5), and similarly, papovaviruses such as SV40 (Figure 5.11). In contrast, the large genetic capacity of herpesviruses makes it possible for these viruses to produce mostly unspliced monocistronic mRNAs, each of which is expressed from its own promoter, thereby rendering unnecessary extensive splicing to produce the required repertoire of proteins.

One of the best-studied examples of the splicing of viral mRNAs is the expression of the adenovirus genome (Figure 5.16). Several 'families' of adenovirus genes are expressed via differential splicing of precursor **hnRNA** transcripts. This is particularly true for the early genes that encode *trans*-acting regulatory proteins expressed immediately after infection. The first proteins to be expressed, E1A and E1B, are encoded by a transcriptional unit on the r-strand at the extreme left-hand end of the adenovirus genome (Figure 5.16). These proteins are primarily transcriptional *trans*-regulatory proteins comparable to the tax and tat proteins described above, but are also involved in transformation of adenovirus-infected cells (see Chapter 6). Five polyadenylated, spliced mRNAs are produced (13S, 12S, 11S, 10S, 9S) which encode five related E1A polypeptides (containing 289, 243, 217, 171, and 55 amino acids respectively) (Figure 5.17). All of these proteins are translated from the same reading frame and have the same

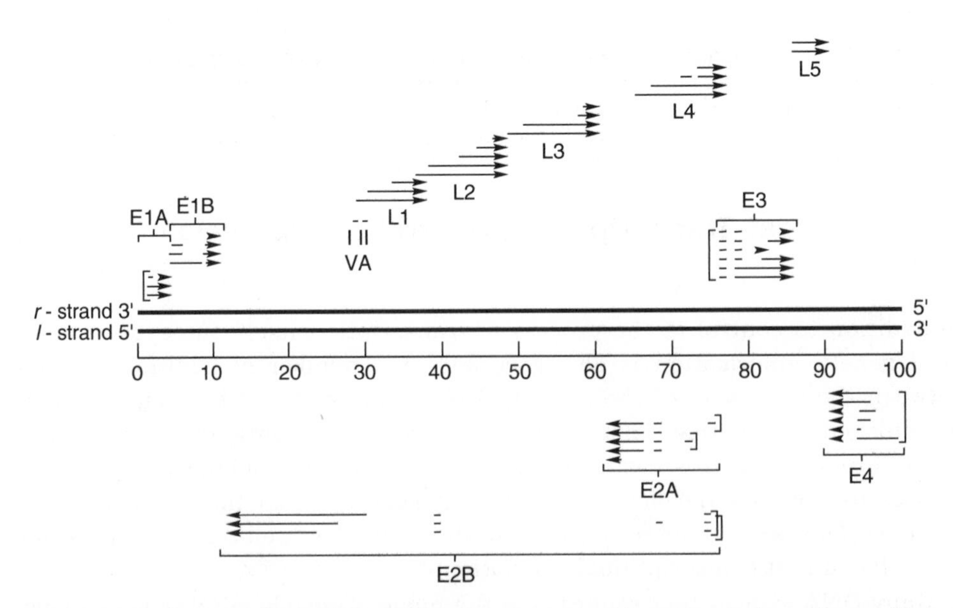

Figure 5.16 Transcription of the adenovirus genome. The arrows show the position of exons in the virus genome which are joined by splicing to produce families of virus proteins.

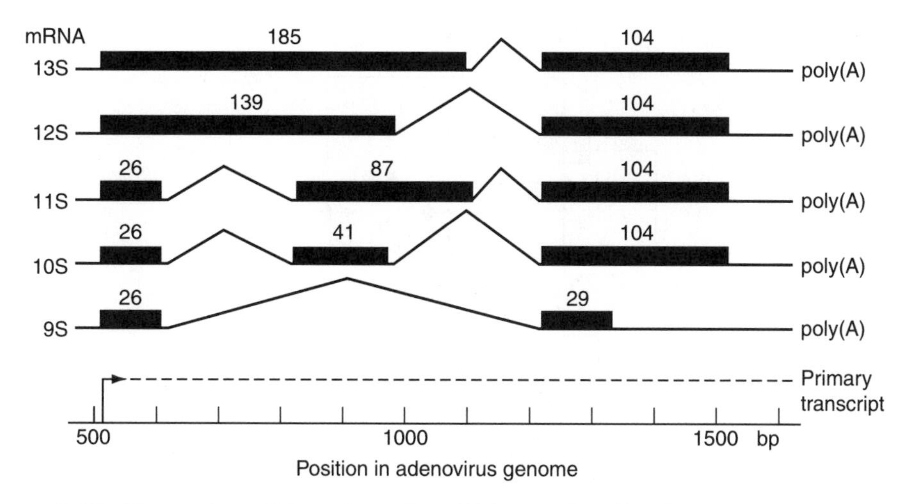

Figure 5.17 Expression of the adenovirus E1A proteins. Numbers shown above each box are the number of amino acids encoded by each exon.

amino and carboxy termini. The differences between them are a consequence of differential splicing of the E1A transcriptional unit and result in major differences in their functions. The 289 and 243 amino acid peptides are transcriptional activators. Although these proteins activate transcription from all the early adenovirus promoters, it has been discovered that they also seem to be 'promiscuous', activating most RNA polymerase II-responsive promoters that contain a TATA box. There are no obvious common sequences present in all of these promoters and there is no evidence that the E1A proteins bind directly to DNA. Like the tax and tat proteins, E1A functions indirectly, either by forming complexes with cellular transcription factors or by binding to promoters via protein–protein interactions with an intermediate DNA-binding protein such as the TATA-binding promoter (TBP), AP1 or ATF/CREB. The 217, 171, and 55 amino acid proteins are not active as transcriptional *trans*-activators, but are involved in cell transformation and possibly in regulation of expression.

Synthesis of E1A starts a cascade of transcriptional activation by turning on transcription of the other adenovirus early genes, E1B, E2, E3, and E4 (Figure 5.17). After the virus genome has been replicated, this cascade eventually results in transcription of the late genes encoding the structural proteins. The transcription of the E1A itself is a balanced, self-regulating system. The immediate early genes of DNA viruses typically have strong enhancer elements upstream of their promoters. This is because in a newly infected cell, there are no virus proteins present and the enhancer is required to 'kick-start' expression of the virus genome. The immediate early proteins synthesized are transcriptional activators that turn on expression of other virus genes, and E1A functions in exactly this way. However, although E1A *trans*-activates its own promoter, the protein represses the function of the upstream enhancer element and therefore at high concentrations, also down-regulates its own expression (Figure 5.18).

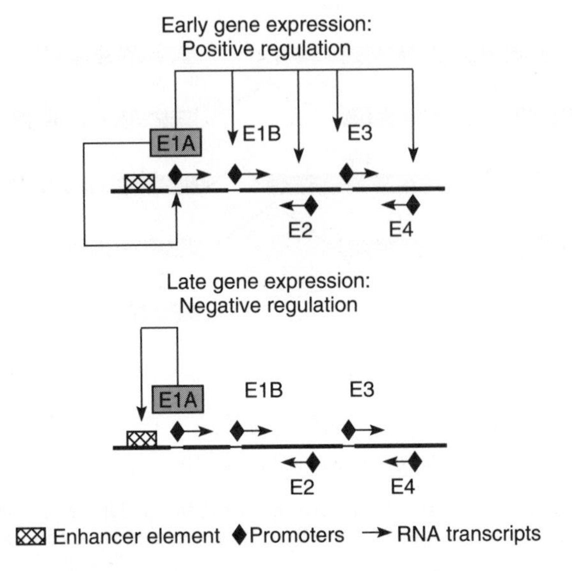

Figure 5.18 Regulation of adenovirus gene expression.

The next stage at which expression can be regulated is during export of mRNA from the nucleus and preferential translation in the cytoplasm. Again, the best studied example of this phenomenon comes from the adenovirus family. The VA genes encode two small RNAs transcribed from the r-strand of the genome by RNA polymerase III (whose normal function is to transcribe similar small RNAs such as 5S ribosomal RNA and tRNAs) during the late phase of virus replication (Figure 5.16). Both VA RNA_I and VA RNA_{II} have a high degree of secondary structure and neither molecule encodes any polypeptide – in these two respects they are similar to tRNAs. The way in which these two RNAs act is not completely understood, but their net effect is to boost the synthesis of adenovirus late proteins. The VA RNAs may be associated with the small nuclear ribonucleoprotein complexes (snRNPs) which are part of the nuclear splicing apparatus, but there is also evidence that VA RNA_I selectively promotes the translation of adenovirus mRNAs in the following way. Virus infection of cells stimulates the production of interferons (see Chapter 6). One of the actions of interferons is to activate a cellular protein kinase known as PKR that inhibits the initiation of translation. VA RNA_I binds to this kinase, preventing its activity and therefore relieves the inhibition on translation. The effects of interferons on the cell are generalized (discussed in Chapter 6), and result in inhibition of the translation of both cellular and viral mRNAs. VA RNA_I may be able to promote selectively the translation of adenovirus mRNAs at the expense of cellular mRNAs whose translation remains inhibited.

The HTLV rex and HIV rev proteins mentioned earlier also act to promote the selective translation of specific virus mRNAs. These proteins regulate the differential expression of the virus genome but do not, as far as is known, substantially alter the expression of cellular mRNAs. Both of these proteins appear to function in a similar way, and although not related to one another in terms of their amino acid sequences, the HTLV rex protein has been shown to be able to substitute functionally for the HIV rev protein. Certain negative-regulatory sequences in the HIV and HTLV genomes appear to cause the retention of virus mRNAs in the nucleus of the infected cell. These sequences are located in the intron regions that are removed from spliced mRNAs encoding the tax/tat and rex/rev proteins (Figure 5.14). Therefore, these proteins are expressed immediately after infection. Tax and tat stimulate enhanced transcription from the virus LTR (Figure 5.15). However, unspliced or singly spliced mRNAs encoding the *gag*, *pol*, and *env* gene products are only expressed when sufficient rex/rev protein is present in the cell. Both proteins bind to a region of secondary structure formed by a particular sequence in the mRNA. Whether, after binding, they act in the nucleus by controlling the transport of mRNA from the nucleus to the cytoplasm, or alternatively, in the cytoplasm, promoting translation of mRNA by enhancing the formation or activity of polyribosomes, is not yet clear.

The efficiency with which different mRNAs are translated varies considerably. This is determined by a number of factors, including the stability and secondary structure of the RNA, but the main one appears to be the particular nucleotide sequence surrounding the AUG translation initiation codon which is recognized by ribosomes. The most favourable sequence for initiation is GCC(A/G)CCAUGGG, although there can be considerable variation within this sequence. A number of viruses use variations of this sequence to regulate the amounts of protein synthesized from a single mRNA. Examples of this are the tax and rex proteins of HTLV, which are encoded by overlapping reading frames in the same doubly spliced 2.1 kb mRNA (Figure 5.14). The AUG initiation codon for the rex protein is upstream of that for tax, but provides a less favourable context for initiation of translation than the sequence surrounding the tax AUG codon. This is known as the 'leaky scanning' mechanisms because it is believed that the ribosomes scan along the mRNA before initiating translation. Therefore, the relative abundance of rex protein in HTLV-infected cells is considerably less than that of the tax protein, even though both are encoded by the same mRNA.

Picornavirus genomes illustrate an alternative mechanism for controlling the initiation of translation. Although these genomes are genetically economical (i.e. have discarded most *cis*-acting control elements and express their entire coding capacity as a single polyprotein), they have retained very long non-coding regions at their 5′ ends (5′ NCR), comprising approximately 10% of the entire genome. These sequences are involved in the replication and possibly packaging of the virus genome. Translation of most cellular mRNAs is initiated when ribosomes recognize the 5′ end of the mRNA and scan along the nucleotide sequence until they reach an AUG initiation codon. Picornavirus genomes are not translated in this way. The 5′ end of the RNA is not capped and thus is not recognized by ribosomes in the same way as other mRNAs, but is modified by the addition of the VPg protein (see Chapters 3 and 6). There are also multiple AUG codons in the 5′ NCR upstream of the start of the polyprotein coding sequences which are not recognized by ribosomes. In picornavirus-infected cells, a virus protease cleaves the 220 kDa 'cap-binding complex' (CBC) involved in binding the m7G cap structure at the 5′ end of the mRNA during initiation of translation. Translation of artificially mutated picornavirus mRNAs *in vitro* and the construction of bicistronic picornavirus genomes bearing additional 5′ NCR signals in the middle of the polyprotein have resulted in the concept of the ribosome 'landing pad', or **internal ribosomal entry site (IRES)**. Rather than scanning along the RNA from the 5′ end, ribosomes bind to the RNA via the IRES and begin translation internally. This is a precise method for controlling the translation of virus proteins. Very few cellular mRNAs utilize this mechanism but it has been shown to be used by a variety of viruses, including picornaviruses, coronaviruses and flaviviruses.

Many viruses belonging to different families compress their genetic information by encoding different polypeptides in overlapping reading frames. The problem with this strategy lies in decoding the information. If each polypeptide is expressed from a monocistronic mRNA transcribed from its own promoter, the additional *cis*-acting sequences required to control and co-ordinate expression might cancel out any genetic advantage gained. More importantly, there is the problem of co-ordinately regulating the transcription and translation of multiple different messages. Therefore, it is highly desirable to express several polypeptides from a single RNA transcript, and the examples described above illustrate several mechanisms by which this can be achieved, namely, differential splicing and control of RNA export from the nucleus or initiation of translation.

There is, however, an additional mechanism known as ribosomal frameshifting used by several groups of viruses to achieve the same end. The best studied examples of this phenomenon come from retrovirus genomes, but a number of

Table 5.4 Ribosomal frameshift signals in viral RNAs

Family/Group	Virus	Gene overlap
Astroviridae	Human astrovirus serotype-1 (HAst-1)	*orf1a/orf1b*
Coronaviridae	Infectious bronchitis virus (IBV)	*orf1a/orf1b*
	Mouse hepatitis virus (MHV)	*orf1a/orf1b*
	Human coronavirus (HCV)	*orf1a/orf1b*
	Transmissible gastroenteritis virus (TGEV)	*orf1a/orf1b*
	Torovirus (Berne virus (BEV))	*orf1a/orf1b*
	Arterivirus (Equine arteritis virus (EAV))	*orf1a/orf1b*
Dianthoviridae	Red clover necrotic mosaic virus (RCNMV)	*p27/p57*
Luteoviridae	Barley yellow dwarf virus (BYDV)	*39K/60K*
	Beet western yellows virus (BWYV)	*orf2/orf3*
	Potato leaf roll virus (PLRV)	*orf2a/orf2b*
Podoviridae	Bacteriophage T7	*10A/10B*
Retroviridae	Human immunodeficiency virus	*gag/pol*
	Feline immunodeficiency virus (FIV)	*gag/pol*
	Rous sarcoma virus (RSV)	*gag/pol*
	Mouse mammary tumour virus (MMTV)	*gag/pol*
	Simian retrovirus type I (SRV-I)	*gag/pol*
	HTLV/BLV	*gag/pro, pro/pol*
Siphoviridae	Bacteriophage λ	*gpG/T*
Totiviridae	*Giardia lamblia* virus (GLV)	*orf1/orf2*
	Saccharomyces cerevisiae dsRNA virus L-A	*gag/pol*
	S. cerevisiae dsRNA virus L1	*gag/pol*

virus groups use a similar mechanism (Table 5.4). Such frameshifting is also known to occur during the expression of certain bacterial transposons and one bacterial gene (the *dnaX* gene of *E. coli*) but to date, has not been observed in eukaryotic cells other than during virus infection. Retrovirus genomes are transcribed to produce at least two 5′ capped, 3′ polyadenylated mRNAs. Spliced mRNAs encode the envelope proteins and, in more complex retroviruses such as HTLV and HIV, additional proteins such as tax/tat and rex/rev proteins (Figure 5.14). A long, unspliced transcript encodes the *gag*, *pro*, and *pol* genes and also forms the genomic RNA packaged into virions. The problem faced by retroviruses is how to express three different proteins from one long transcript. The arrangement of the three genes varies in different viruses. In some cases (e.g. HTLV), they occupy three different reading frames, while in others (e.g. HIV), the protease (*pro*) gene forms an extension at the 5′ end of the *pol* gene (Figure 5.19). In the latter case, the protease and polymerase (i.e. reverse transcriptase) are expressed as a polyprotein which is autocatalytically cleaved into the mature proteins in a process that is similar to the cleavage of picornavirus polyproteins.

At the boundary between each of the three genes there is a particular sequence which usually consists of a tract of reiterated nucleotides, such as

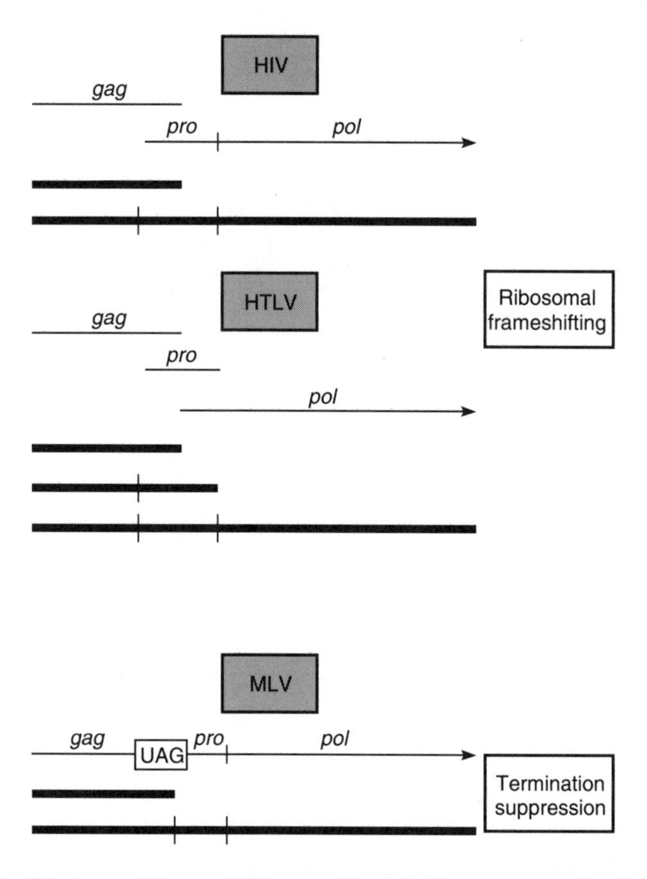

Figure 5.19 Ribosomal frameshifting and termination suppression in retroviruses.

UUUAAAC (Figure 5.20). Remarkably, this sequence is rarely found in protein-coding sequences and therefore appears to be specifically used for this type of regulation. Most ribosomes encountering this sequence will translate it without difficulty and continue on along the transcript until a translation stop codon is reached. However, a proportion of the ribosomes which attempt to translate this sequence will slip back by one nucleotide before continuing to translate the message, but now in a different (i.e. −1) reading frame. Because of this, the UUUAAAC sequence has been termed the 'slippery sequence', and the result of this −1 frameshifting is the translation of a polyprotein containing alternative

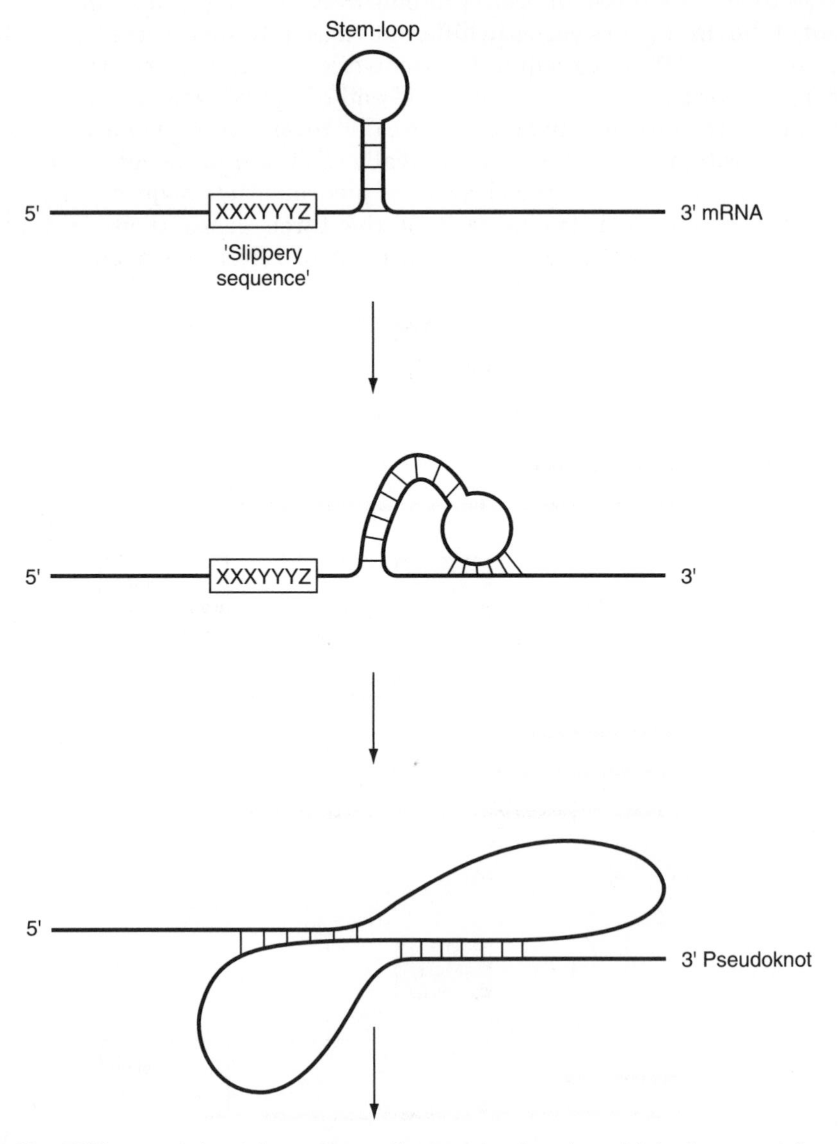

Figure 5.20 RNA pseudoknot formation – the mechanism by which ribosomal frame-shifting occurs.

information from a different reading frame. This mechanism also allows the virus to control the ratios of the proteins produced. Because only a proportion of ribosomes undergo frameshifting at each slippery sequence, there is a gradient of translation from the reading frames at the 5′ end of the mRNA to those at the 3′ end.

However, the slippery sequence alone only results in a low frequency of frameshifting, which appears to be inadequate to produce the amount of protease and reverse transcriptase protein required by the virus. Therefore, there are additional sequences that further regulate this system and increase the frequency of frameshift events. A short distance downstream of the slippery sequence there is an inverted repeat which allows the formation of a stem-loop structure in the mRNA (Figure 5.20). A little further on, there is an additional sequence complementary to the nucleotides in the loop which allows base-pairing between these two regions of the RNA. The net result of this combination of sequences is the formation of what is known as an RNA ‘pseudoknot’. This secondary structure in the mRNA causes ribosomes translating the message to pause at the position of the slippery sequence upstream, and this slowing or pausing of the ribosome during translation increases the frequency at which frameshifting occurs, thus boosting the relative amounts of the proteins encoded by the downstream reading frames. It is easy to imagine how this system can be fine-tuned by subtle mutations that alter the stability of the pseudoknot structure and thus the relative expression of the different genes.

The final method of translational control to be considered is termination suppression. This is a mechanism similar in many respects to frameshifting and permits multiple polypeptides to be expressed from individual reading frames in a single mRNA. In some retroviruses, such as murine leukaemia virus (MLV), the *pro* gene is separated from the *gag* gene by a UAG termination codon rather than a ‘slippery sequence’ and pseudoknot (Figure 5.19). In the majority of cases, translation of MLV mRNA terminates at this sequence, giving rise to the gag proteins. However, in a few instances, the UAG stop codon is suppressed and translation continues, producing a gag–pro–pol polyprotein, which subsequently cleaves itself to produce the mature proteins. The overall effect of this system is much the same as ribosomal frameshifting, with the relative ratios of gag and pro/pol proteins being controlled by the frequency with which ribosomes traverse or terminate at the UAG stop codon.

Summary

Control of gene expression is a vital element of virus replication. Co-ordinate expression of groups of virus genes results in successive phases of gene expression. Typically, immediate early genes encode ‘activator’ proteins, early genes encode further regulatory proteins and late genes encode virus structural proteins. Viruses make use of the biochemical apparatus of their host cells to express their genetic information as proteins and, consequently, utilize the appropriate biochemical language recognized by the cell. Thus viruses of prokaryotes produce polycistronic mRNAs while viruses with eukaryotic hosts produce mainly monocistronic mRNAs. Some viruses of eukaryotes do produce polycistronic mRNA to assist with the co-ordinate regulation of multiple genes. In addition, viruses rely on specific *cis-* and *trans*-acting mechanisms to manipulate the biology of their host cells and to enhance and co-ordinate the expression of their own genetic information.

Further Reading

Alberts, B., Bray, D., Lewis, J., Raff, M,. Roberts, K. and Watson, J.D. (Eds). (1994). *Molecular Biology of the Cell* (3rd edition). Garland, New York.

Boulanger, P.A. and Blair, G.E. (1991). Expression and interactions of human adenovirus oncoproteins. *Biochem. J.* **275**: 281–99.

Brierley, I. (1995). Ribosomal frameshifting on viral RNAs. *J. Gen. Virol.* **76**: 1885–1982.

Cann, A.J .and Chen, I.S.Y. (1996). Human T-cell leukemia virus types I and II. In *Virology*, Fields, B.N., Knipe D.M., Howley, P.M. (Eds), pp. 1849–1880 (3rd edition). Raven Press, New York.

Ehrenfeld, E. (Ed). (1993). Translational regulation in virus-infected cells. *Sem. Virol.* **4**: 199–268.

Gaynor, R. (1992). Cellular transcription factors involved in the regulation of HIV-1 gene expression. *AIDS* **6**: 347–63.

Kozak, M. (1991). Structural features in eukaryotic messenger RNAs that modulate the initiation of translation. *J. Biol. Chem.* **266**: 19867–70.

Peterlin, B.M. (Ed.). (1993). Transcriptional regulation of viruses. *Sem. Virol.* **4**: 1–80.

Ptashne, M. (1986). *A Genetic Switch.* Cell Press/Blackwell Scientific, Oxford.

Ptashne, M. (1986). Gene regulation by proteins acting nearby and at a distance. *Nature* **322**: 697–701.

Sarnow, P. (Ed.). (1995). Cap-Independent Translation. *Curr. Top. Microbiol. Immunol.* Vol. **203**.

Self-Assessment Questions

(5.1) Are the following statements true or false?

- (a) The course of virus replication is independent of virus gene expression.
- (b) Transcription promoters are *cis*-acting sequences located adjacent to the genes whose transcription they control.
- (c) Transcription factors are *trans*-acting proteins which bind to specific sequences present anywhere in the cell.
- (d) Transcription of bacterial operons typically produces polycistronic mRNAs.

(e) Gene expression in bacteria is regulated exclusively at the level of transcription.

(5.2) Viruses have counteracted their genetic limitations by evolving of a range of solutions to these problems, including (true or false?):
(a) Powerful positive *cis*- and *trans*-acting signals which promote gene expression.
(b) Powerful *cis*-acting negative signals which repress gene expression.
(c) Highly compressed genomes with overlapping reading frames common.
(d) Control signals which are frequently nested within other genes.
(e) Several strategies designed to create multiple polypeptides from a single mRNA.

(5.3) Bacteriophage λ (true or false?):
(a) Can alternate between the lysogenic and lytic growth cycles.
(b) The λ N protein acts as a negative transcription regulator.
(c) In the absence of the N protein, RNA polymerase stops at the nut site.
(d) The cI repressor protein binds to O_R but not O_L.
(e) Lysogeny is maintained by autoregulation of cI transcription.

(5.4) Are the following statements true or false?
(a) Transcription of eukaryotic cellular genes results in monocistronic mRNAs, each transcribed from its own promoter.
(b) Post-transcriptional control of gene expression in eukaryotes is achieved by splicing of hnRNA and differential export of mRNA from the nucleus to the cytoplasm.
(c) Eukaryotic mRNAs are translated with variable efficiency owing to the speed at which ribosomes are degraded.
(d) The stability of eukaryotic mRNAs depends on the speed at which they are degraded.
(e) Translation enhancers perform a function analogous to that of transcription enhancers.

(5.5) Are the following statements true or false?
(a) Papovaviruses are heavily dependent on cellular machinery for gene expression.
(b) Herpesviruses make three classes of mRNA: early, late and structural.
(c) Geminivirus genomes are replicated and expressed in the nucleus of infected cells because they are heavily reliant on host cell functions.
(d) Reovirus transcription occurs inside virus core particles.
(e) Segmented virus genomes are usually transcribed to produce monocistronic mRNAs.

(5.6) Viruses with (+)sense RNA genomes (true or false?):
(a) Are heavily dependent on cellular mechanisms for control of gene expression.
(b) Are all translated to produce a long polyprotein.

(c) Are all transcribed to produce subgenomic RNAs.
(d) Are all non-segmented.
(e) Can only produce an equal ratio of all the polypeptides they encode.

(5.7) Are the following statements true or false?
(a) Viruses with non-segmented (−)sense genomes produce monocistronic mRNAs.
(b) Viruses with (−)sense genomes all contain a virus-specific transcriptase/replicase within the virus nucleocapsid.
(c) Viruses with non-segmented (−)sense genomes can only produce an equal ratio of all the polypeptides they encode.
(d) Retroviruses are the only RNA viruses whose genomes are transcribed by cellular RNA polymerase.
(e) Hepadnavirus and caulimovirus genomes contain a number of overlapping reading frames.

(5.8) Are the following statements true or false?
(a) SV40 large T-antigen binds to the origin of replication in the virus genome.
(b) SV40 large T-antigen represses transcription of the virus early genes.
(c) SV40 large T-antigen promotes export of virus mRNAs from the nucleus to the cytoplasm.
(d) The HTLV tax protein binds to the TAR sequence in the provirus genome.
(e) The HIV tat protein binds to cellular transcription factors, not virus mRNAs.

(5.9) Are the following statements true or false?
(a) With the exception of retroviruses, no RNA virus genome undergoes splicing.
(b) The HTLV rex protein promotes the export of virus structural protein mRNAs from the nucleus.
(c) The abbreviation 'IRES' stands for 'Initial Replication Exit Site'.
(d) All (+)sense RNA viruses contain an IRES element.
(e) All eukaryotic mRNAs contain an IRES element.

(5.10) Are the following statements true or false?
(a) Ribosomal frameshifting allows viruses to produce several different proteins from a polycistronic mRNA.
(b) Ribosomal frameshifting allows viruses to produce varying amounts of different proteins from a polycistronic mRNA.
(c) RNA pseudoknots are produced by *trans*-acting proteins such as the HIV tat protein.
(d) RNA pseudoknots increase the frequency of ribosomal frameshifting.
(e) Suppression of termination codons allows some viruses to produce multiple proteins from a polycistronic mRNA.

Answers to Self-Assessment Questions are given in Appendix 3.

Chapter 6

Infection

Virus infection of higher organisms is the cumulative result of all the processes of replication and gene expression described in earlier chapters. Together, these determine the overall course of each infection. Virus infections range in complexity and duration from a very brief, superficial interaction between the virus and its host to infections which may encompass the entire life of the host organism, from before birth to its eventual death and in which many different tissues and organs are infected. One of the most common misconceptions is that virus infection inevitably results in disease. In reality, the converse is true – only a small minority of virus infections give rise to any disease symptoms.

This chapter provides an overview of the numerous patterns of virus infection and forms an introduction to the consideration of viral pathogenesis in Chapter 7. Most of this chapter is concerned with the infection of eukaryotes by viruses. Unlike previous and subsequent chapters, this chapter deals primarily with the interaction of viruses with intact organisms rather than with the molecular biologist's usual concern about the interaction between a virus and the cell.

Virus Infections of Plants

The overall course of virus replication is determined by a dynamic interaction between the virus and its host organism. Clearly, there are major differences between virus infections of plants and those of vertebrates. In economic terms, viruses are only of importance if it is likely that they will spread to crops during their commercial lifetime, which of course varies greatly between very short extremes in horticultural production and very long extremes in forestry. Some estimates put total worldwide damage due to plant viruses as high as US$ 6×10^{10} per year. The mechanism by which plant viruses are transmitted between hosts is therefore of great importance. There are a number of routes by which plant viruses may be transmitted:

- *Seeds*: These may transmit virus infection either by external contamination of the seed with virus particles, or by infection of the living tissues of the embryo. Transmission by this route leads to early outbreaks of disease in new crops which are usually initially focal in distribution, but may subsequently be transmitted to the remainder of the crop by other mechanisms (below)
- *Vegetative propagation/grafting*: These techniques are cheap and easy methods of plant propagation, but provide the ideal opportunity for viruses to spread to new plants
- *Vectors*: Many different groups of living organisms can act as vectors and spread viruses from one plant to another:
  Bacteria (e.g. *Agrobacterium tumefaciens* – the Ti plasmid of this organism has been used experimentally to transmit virus genomes between plants)
  Fungi
  Nematodes
  Arthropods: Insects – aphids, leafhoppers, planthoppers, beetles, thrips, etc.
  Arachnids – mites
- *Mechanical*: Mechanical transmission of viruses is the most widely used method for experimental infection of plants and is usually achieved by rubbing virus-containing preparations into the leaves, which in most plant species are particularly susceptible to infection. However, this is also an important natural method of transmission. Virus particles may contaminate soil for long periods and may be transmitted to the leaves of new host plants as wind-blown dust or as rain-splashed mud.

The problems plant viruses face in initiating infections of host cells have already been described (Chapter 4), as has the fact that no known plant virus employs a specific cellular receptor of the types that animal and bacterial viruses use to attach to cells. Transmission of plant viruses by insects is therefore of particular agricultural importance. Extensive areas of monoculture and the inappropriate use of pesticides which kill natural predators can result in massive population booms of insects such as aphids. Plant viruses rely on a mechanical breach of the integrity of a cell wall to directly introduce a virus particle into a cell. This is achieved either by the vector associated with transmission of the virus or simply by mechanical damage to cells. Transfer by insect vectors is a particularly efficient means of virus transmission. In some instances, viruses are transmitted mechanically from one plant to the next by the vector and the insect is merely a means of distribution, through flying or being carried on the wind for long distances (sometimes hundreds of miles). Insects which bite or suck plant tissues are, of course, the ideal means of transmitting viruses to new hosts. This is known as **non-propagative transmission**. However, in other cases (e.g. many plant rhabdoviruses) the virus may also infect and multiply in the tissues of the insect (**propagative transmission**) as well as those of host plants. In these cases, the vector serves as a means not only of distributing the virus, but also of amplifying the infection.

Initially, most plant viruses multiply at the site of infection, giving rise to localized symptoms such as necrotic spots on the leaves. The virus may subsequently, be distributed to all parts of the plant either by direct cell-to-cell spread or by the vascular system, resulting in a systemic infection involving the whole plant. However, the problem these viruses face in reinfection and recruitment of new cells is the same as they face initially – how to cross the barrier of the plant cell wall. Plant cell walls necessarily contain channels called plasmodesmata which allow plant cells to communicate with each other and to pass metabolites between them. However, these channels are too small to allow the passage of virus particles or genomic nucleic acids. Many (if not most) plant viruses have evolved specialized **movement proteins** which modify the plasmodesmata. One of the best known examples of this is the 30 k protein of tobacco mosaic virus (TMV). This protein is expressed from a subgenomic mRNA (Figure 3.12) and its function is to modify plasmodesmata causing genomic RNA coated with 30 k protein to be transported from the infected cell to neighbouring cells (Figure 6.1). Other viruses, such as cowpea mosaic virus (CPMV, Comovirus family) have a similar strategy but employ a different molecular mechanism. In CPMV, the 58/48 k proteins form tubular structures allowing the passage of intact virus particles to pass from one cell to another (Figure 6.1).

Typically, virus infections of plants might result in effects such as growth retardation, distortion, mosaic patterning on the leaves, yellowing, wilting, etc. These macroscopic symptoms result from:

- Necrosis of cells, caused by direct damage due to virus replication
- Hypoplasia, i.e. localized retarded growth frequently leading to mosaicism (the appearance of thinner, yellow areas on the leaves)
- Hyperplasia, which is excessive cell division or the growth of abnormally large cells, resulting in the production of swollen or distorted areas of the plant.

Plants might be seen as sitting targets for virus infection – unlike animals, they cannot run away. However, plants exhibit a range of responses to virus infections

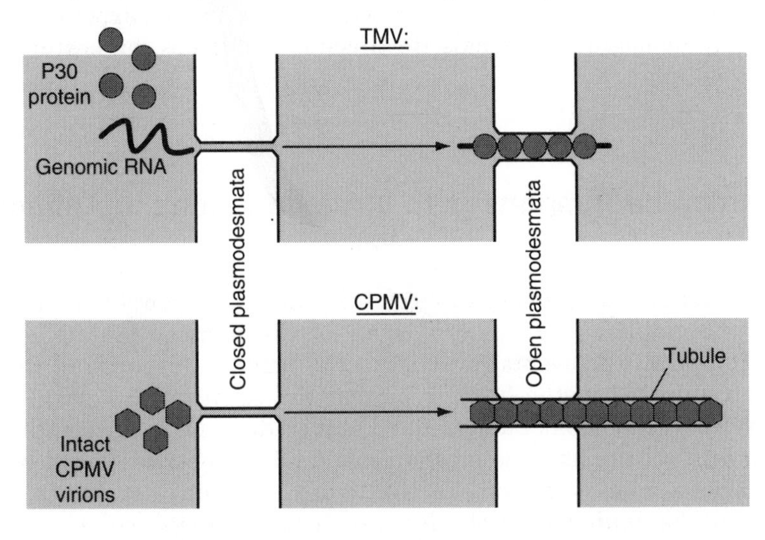

Figure 6.1 Plant movement proteins.

designed to minimize their effects. Initially, infection results in a 'hypersensitive response', manifested as the synthesis of a range of new proteins, the 'PR proteins'. Although this system is poorly understood, at least some of these proteins have been characterized and have been shown to be proteases, which presumably destroy virus proteins, limiting the spread of the infection. There is some similarity here between this response and the production of interferons by animals (see below). In addition, systemic resistance to virus infection is a naturally occurring phenomenon in some strains of plant. This is clearly a highly desirable characteristic and is highly prized by plant breeders, who try to spread this attribute to economically valuable crop strains. There are probably many different mechanisms involved in systemic resistance, but in general terms there is a tendency towards increased local necrosis as substances such as proteases and peroxidases are produced by the plant to destroy the virus and to prevent its spread and subsequent systemic disease. An example of this is the *N* gene, which when present in plants causes TMV to produce a localized, necrotic infection rather than the systemic mosaic symptoms normally seen.

Virus-resistant plants have been created by the production of transgenic plants expressing recombinant virus proteins or nucleic acids which interfere with virus replication without producing the pathogenic consequences of infection, e.g:

- Virus coat proteins which inhibit uncoating of incoming virus particles
- Intact or partial virus replicases which interfere with genome replication
- Antisense RNAs
- Defective viral genomes
- Satellite sequences (see Chapter 8)
- Catalytic RNA sequences (ribozymes)
- Modified movement proteins.

This is a very promising technology which offers the possibility of substantial increases in agricultural production without the use of expensive, toxic and ecologically damaging chemicals (fertilizers, herbicides, or pesticides), but it is still in its infancy.

Immune Responses to Virus Infections in Animals

The most significant response to virus infection in vertebrates is the activation of both the cellular and humoral arms of the immune system. A thorough description of all the events involved in the immune response to the presence of foreign antigens in the body is beyond the scope of this chapter. Readers should refer to the books cited under Further Reading to ensure that they are familiar with all the immune mechanisms (and jargon!) described below. However, a brief summary of some of the more pertinent aspects is well worth considering, beginning with the humoral immune response, which results in the production of antibodies.

The major impact of the humoral immune response is the eventual clearance of virus from the body, i.e. serum neutralization stops the spread of virus to uninfected cells and allows other defence mechanisms to mop up the infection. Figure 6.2 shows a grossly simplified version of the mammalian humoral response to infection. Virus infection induces at least three classes of antibody: immunoglobulin G (IgG), IgM, and IgA. IgM is a large, multivalent molecule which is most effective at cross-linking large targets (e.g. bacterial cell walls or flagella), but is probably less important in combating virus infections. In contrast, the production of IgA is very important for initial protection from virus infection. Secretory IgA is produced at mucosal surfaces and results in 'mucosal immunity', an important factor in preventing infection from occurring. Induction of mucosal immunity depends to a large extent on the way in which antigens are presented to and recognized by the immune system. Similar antigens incorporated into different vaccine delivery systems (see Prevention and Therapy of Virus Infection) can lead to very different results in this respect and mucosal immunity is such an important factor that similar vaccines may vary considerably in their efficacy. IgG is probably the most important class of antibody for direct neutralization of virus particles in serum and other body fluids (into which it diffuses).

Direct virus neutralization by antibodies results from a number of mechanisms, including conformational changes in the virus capsid caused by antibody binding, or blocking of the function of the virus target molecule (e.g. receptor binding) by steric hindrance. A secondary consequence of antibody binding is phagocytosis of antibody-coated ('opsonized') target molecules by mononuclear cells or polymorphonuclear leukocytes. This process is mediated by the presence of the Fc receptor on the surface of these cells, but as has already been noted in Chapter 4, in some cases opsonization of virus by the binding of non-neutralizing antibodies can result in enhanced virus uptake. This has been shown to occur

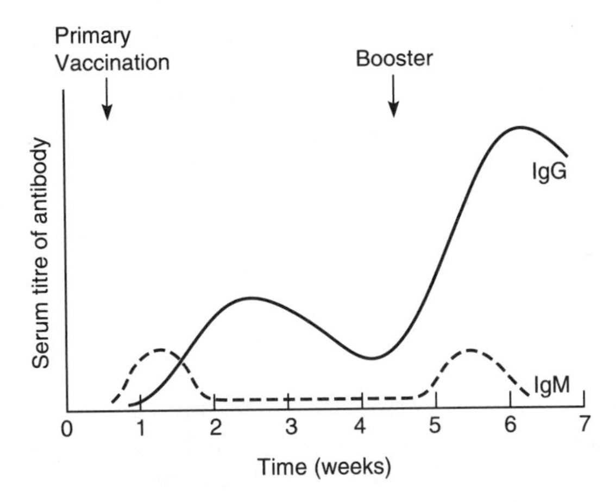

Figure 6.2 Simplified version of the kinetics of the mammalian humoral response to a 'typical' foreign virus (or other) antigen.

with rabies virus and in the case of HIV may promote uptake of the virus by macrophages. Antibody binding also leads to the activation of the complement cascade, which assists in the neutralization of virus particles. The morphological alterations induced by the disruption of viruses by complement can sometimes be visualized directly by electron microscopy. Complement is particularly important early in virus infection when limited amounts of low-affinity antibody are made – complement potentiates the action of these.

Despite all the above mechanisms, in overall terms, cell-mediated immunity is probably more important than humoral immunity in the control of virus infections. This is demonstrated by the following observations:

- Congenital defects in cell-mediated immunity tend to result in predisposition to viral (and parasitic) infections, rather than to bacterial infections
- The functional defect in acquired immune deficiency syndrome (AIDS) is a reduction in the ratio of T-helper ($CD4^+$):T-suppressor ($CD8^+$) cells from the normal value of about 1.2 to 0.2. AIDS patients commonly suffer many opportunistic virus infections (e.g. various herpesviruses such as HSV, CMV and EBV), which may have been present before the onset of AIDS but were previously suppressed by the intact immune system.

Cell-mediated immunity is effected through three main mechanisms (Figure 6.3):

- Non-specific cell killing (mediated by 'natural killer' (NK) cells)
- Specific cell killing (mediated by cytotoxic T-lymphocytes, CTL)
- Antibody-dependent cellular cytotoxicity (ADCC).

NK cells mediate cell lysis independently of conventional immunological specificity, i.e. clonal antigen recognition. They are not MHC-restricted (i.e. only able to recognize a specific antigen in the context of MHC antigens plus the T-cell receptor/CD3 complex). The advantages of this are that they have broad specificity and are active without the requirement for sensitizing antibodies. They are therefore the first line of defence against virus infection. NK cells are most active in the early stages of infection (i.e. the first few days) and their activity is stimulated by interferon-α (see below). NK cells are not directly induced by virus infection – they exist even in immunologically naive individuals and are 'revealed' in the presence of interferon-α. Their function is complementary to and is later taken over by CTL (below). The target for NK cells on the surface of infected cells is not known, but their action is inhibited by MHC class I antigens (which are present on all nucleated cells), allowing recognition of 'self' and preventing total destruction of the body. It is known that some virus infections disturb normal cellular MHC class I expression and this is one possible mechanism of NK recognition of infected cells. The cell killing is achieved by the release of perforin (or 'cytolysin'), a protein found in NK cell cytoplasmic granules. Perforin is a peptide related to the complement protein C9 which, after release from the NK cells, polymerizes to form polyperforin. This forms transmembrane channels in the target cells, resulting in permeability of the target cell membrane and death.

Cytotoxic T-lymphocytes (CTL) are usually of $CD8^+$ (suppressor) phenotype.

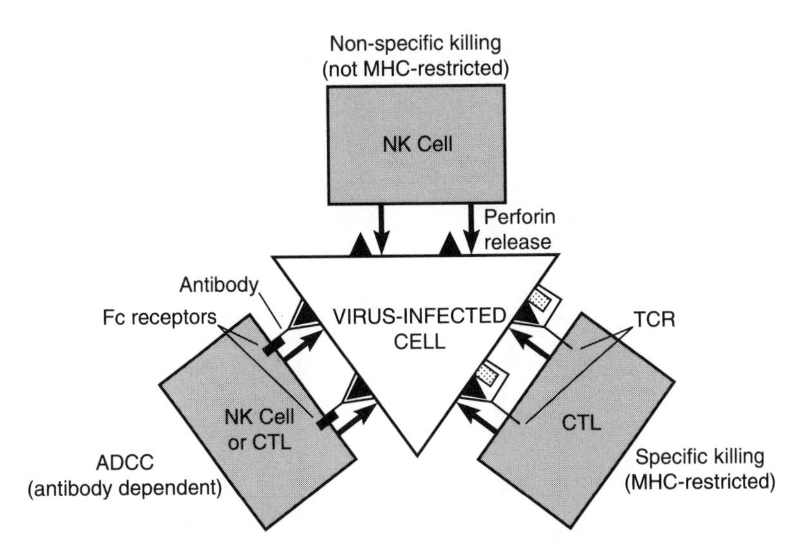

Figure 6.3 Diagram illustrating the three main mechanisms by which cell-mediated immunity is effected.

CTL are the major cell-mediated immune response to virus infections and are MHC-restricted, i.e. clones of cells recognize a specific antigen only when presented by MHC class I antigen on the target cell to the T-cell receptor–CD3 complex on the surface of the CTL. (MHC class I antigens are expressed on all nucleated cells in the body; MHC class II antigens are expressed only on the surface of the antigen-presenting cells of the immune system – T-cells, B-cells and macrophages.) This process requires 'help' (i.e. lymphokine production) from T-helper cells. The CTL themselves recognize foreign antigens through the TCR–CD3 complex, which 'docks' with antigen presented by MHC class I on the surface of the target cell (Figure 6.4). However, the mechanism of cell killing by CTL is similar to that of NK cells, the production and release of perforin. The induction of a CTL response also results in the release of many different lymphokines from T-helper cells, some of which result in clonal proliferation of antigen-specific CTL and others which have direct antiviral effects, for example, interferons (see below). The kinetics of the CTL response (which peak at about 7 days after infection) are somewhat slower than the NK response (e.g. 1–3 days) and therefore these are complementary systems.

The induction of a CTL response is dependent on recognition of specific T-cell epitopes by the immune system. These are distinct from the B-cell epitopes recognized by the humoral arm of the immune system. T-cell epitopes are frequently more highly conserved (less variable) than B-cell epitopes, which are frequently able to mutate quickly to escape immune pressure. These are important considerations in the design of antiviral vaccines. The specificity of killing by CTL is not absolute. Although they are better 'behaved' than NK cells, diffusion of perforin and local lymphokine production frequently results in

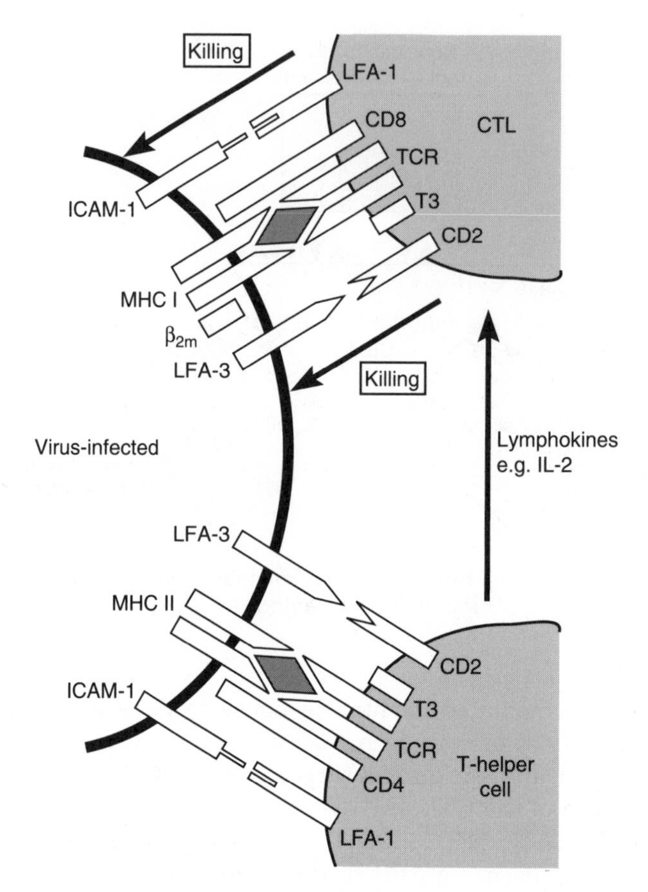

Figure 6.4 Cell surface proteins involved in immune recognition. Close contact between cells results in cell-to-cell signalling which regulates the immune response.

inflammation and bystander cell damage. This is a contributory cause of the pathology of many virus diseases (see Chapter 7), but the less attractive alternative is to allow virus replication to proceed unchecked.

ADCC is less well understood than either of the above two mechanisms. It can be mediated either by NK cells or by CTL. The mechanism of cell killing is the same as above, although complement may also be involved in ADCC. This mechanism is dependent on the recognition of antigen on the surface of the target cell by antibody on the surface of the effector cell. The antibody involved is usually IgG, which is bound to Fc receptors on the surface of the T-cell. ADCC therefore requires a pre-existing antibody response and hence does not occur early during primary virus infections. The overall contribution of ADCC to the control of virus infections is not clear, although it is now believed that it plays a significant part in their control.

Another important factor in protection from virus infection is the production of interferons (IFNs). By the 1950s, interference (i.e. the blocking of a virus infection by a competing virus) was a well-known phenomenon in virology. In some cases, the mechanism responsible is just that; for example, avian retroviruses are grouped into nine interference groups, A–I, based on their ability to infect various strains of chickens, pheasants, partridges, quail, etc., or cell lines derived from these species. In this case, the inability of particular viruses to infect the cells of some strains is due to the expression of the envelope glycoprotein of an endogenous provirus present in the cells, which sequesters the cellular receptor needed by the exogenous virus for infection. In other cases, the mechanism of viral interference is less clear.

In 1957, Alick Issacs and Jean Lindenmann were studying this phenomenon and performed the following experiment. Pieces of chick chorioallantoic membrane were exposed to u.v.-inactivated (non-infectious) influenza virus in tissue culture. The 'conditioned' medium from these experiments (which did not contain infectious virus) was found to inhibit the infection of fresh pieces of chick chorioallantoic membrane by (infectious) influenza virus in separate cultures. Their conclusion was that a soluble factor, which they called 'interferon', was produced by cells as a result of virus infection and that this factor could prevent the infection of other cells. As a result of this provocative observation interferon became the great hope for virology and was thought to be directly equivalent to the use of antibiotics to treat bacterial infections.

The true situation has turned out to be far more complex than was first thought. Interferons do have antiviral properties, but by and large their effects are exerted indirectly via their major function as cellular regulatory proteins. Interferons are immensely potent; less than 50 molecules per cell shows evidence of antiviral activity. Hence, following Isaacs and Lindenmann's initial discovery, many fairly fruitless years were spent trying to purify minute amounts of naturally produced interferon. This situation changed with the development of molecular biology and the cloning and expression of interferon genes, which has led to rapid advances in our understanding over the last 15 years. There are three types of interferon: α, β, and γ.

(1) *Interferon-α*: There are at least 15 molecular species of interferon-α, all of which are closely related; some species differ by only one amino acid. They are synthesized predominantly by lymphocytes. The mature proteins contain 143 amino acids, with a minimum homology of 77% between the different types. All the genes encoding interferon-α are located on human chromosome 9 and gene duplication is thought to be responsible for this proliferation of genes.

(2) *Interferon-β*: There is a single gene for interferon-β, also on human chromosome 9. The mature protein contains 145 amino acids and, unlike interferon-α, is glycosylated, with approximately 30% homology to other interferons. It is synthesized predominantly by fibroblasts.

(3) *Interferon-γ*: There is a single gene for interferon-γ, on human chromosome 12. The mature protein contains 146 amino acids, is glycosylated, has very low sequence homology to other interferons. It is synthesized predominantly by lymphocytes.

Induction of interferon synthesis results from upregulation of transcription from the interferon gene promoters. There are three main mechanisms involved:

(1) *Virus infection*: This mechanism is thought to act by the inhibition of cellular protein synthesis which occurs during many virus infections, resulting in a reduction in the concentration of intracellular repressor proteins and hence in increased interferon gene transcription. In general, RNA viruses are potent inducers of interferon while DNA viruses are relatively poor inducers. However, there are exceptions to this rule (e.g. poxviruses are very potent inducers). The molecular events in the induction of interferon synthesis by virus infection are not clear. In some cases (e.g. influenza virus), u.v.-inactivated virus is a potent inducer. Therefore, virus replication is not necessarily required. Induction by viruses might involve perturbation of the normal cellular environment and/or production of small amounts of double-stranded RNA (see below).

(2) *Double-stranded (d/s) RNA*: All naturally occurring double-stranded RNAs (e.g. reovirus genomes) are potent inducers of interferon, as are synthetic molecules (e.g. poly I:C). Therefore, this process is independent of nucleotide sequence. Single-stranded RNA and double-stranded DNA are not inducers, hence the mechanism of induction is thought to depend on the secondary structure of the RNA rather than any particular nucleotide sequence.

(3) *Metabolic inhibitors*: Compounds which inhibit cellular transcription (e.g. actinomycin D) or translation (e.g. cycloheximide) result in induction of interferon. Tumour promoters such as TPA (tetradecanoyl phorbol acetate) or DMSO (dimethyl sulfoxide) are also inducers. Their mechanism of action remains unknown but they almost certainly act at the level of transcription.

The effects of interferons are exerted via specific receptors on the surface of target cells. The main action is on cellular regulatory activities and is rather complex. Interferon affects both cellular proliferation and immunomodulation. These effects result from the induction of transcription of a wide variety of cellular genes, including other lymphokines. The net result is complex regulation of a cell's ability to proliferate, differentiate and communicate. This cell regulatory activity itself has indirect effects on virus replication (see below). All interferons have similar antiviral capacities, but IFN-γ is by far the most potent cellular regulator.

The effect of interferons on virus infections *in vivo* is extremely important. Animals experimentally infected with viruses and injected with anti-interferon antibodies experience much more severe infections than control animals infected with the same virus. This is because interferons protect cells from damage and death. However, they do not appear to play a major role in the clearance of virus infections – the other parts of the immune response are necessary for this. Interferon is a 'firebreak' which inhibits virus replication in its earliest stages by several mechanisms. Two of these are understood in some detail (below), but there are a number of others (in some cases specific to certain viruses) which are less well understood.

Interferons induce transcription of a cellular gene encoding 2′,5′-oligo A synthetase (Figure 6.5). There are at least four molecular species of 2′,5′-oligo A, induced by different forms of interferon. This compound activates an RNA-

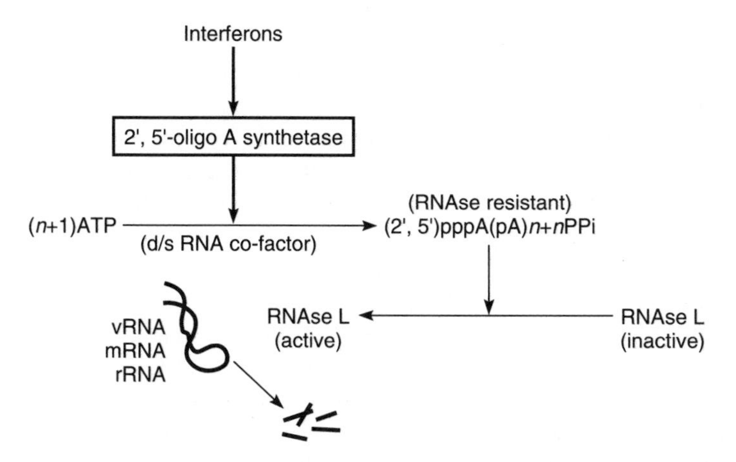

Figure 6.5 Mechanism of induction of 2′,5′-oligo A synthetase by interferons.

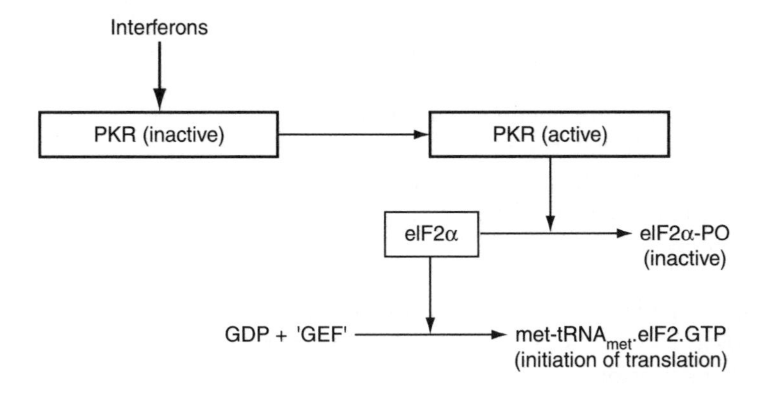

Figure 6.6 Mechanism of induction of PKR by interferons.

digesting enzyme, RNAse L, which digests viral genomic RNAs, viral and cellular mRNAs and cellular ribosomal RNAs. The end result of this mechanism is a reduction in protein synthesis (due to the degradation of mRNAs and rRNAs) and therefore, the cell is protected from virus damage. The second method relies on the activation of a 68 kDa protein called PKR (RNA-activated protein kinase) (Figure 6.6). PKR phosphorylates a cellular factor, eIF2α, which is required by ribosomes for the initiation of translation. Therefore, the net result of this mechanism is also the inhibition of protein synthesis and this reinforces the 2′,5′-oligo A mechanism. A third, well-established mechanism depends on the M_x gene, a single-copy gene located on human chromosome 21, whose transcription is induced by IFN-α and IFN-β but not IFN-γ. The product of this gene inhibits the primary transcription of influenza virus but not of other viruses. Its method of action is unknown. In addition to the three mechanisms above, there are many additional recorded effects of interferons. They inhibit the penetration and uncoating of SV40 and some other viruses, possibly by altering the composition/structure of the cell membrane; they inhibit the primary transcription of many virus genomes (e.g. SV40, HSV) and also cell transformation by retroviruses. None of the molecular mechanisms by which these effects are mediated have been fully explained.

In conclusion, it can be stated that interferons are a powerful weapon against virus infection, but that they act as a blunderbuss, rather than a 'magic bullet'! The severe side-effects (fever, nausea, malaise) which result from the powerful cell-regulatory action of interferons means that they will never be widely used for the treatment of trivial virus infections – they are not the cure for the common cold. However, as the cell-regulatory potential of interferons is becoming better understood, they are finding increasing use as a treatment for certain cancers (e.g. the use of IFN-α in the treatment of hairy cell leukaemia). Current therapeutic uses of interferons are summarized in Table 6.1. The long-term prospects for their use as antiviral compounds are less certain, except for possibly in life-threatening infections where there is no alternative therapy, e.g. chronic viral hepatitis.

Interferons are an effective means of curbing the worst effects of virus infections. Part of their wide-ranging efficacy results from their generalized, non-specific effects (e.g. the inhibition of protein synthesis in virus-infected cells). This lack of specificity means that it is very difficult for viruses to evolve strategies to counteract their effects. Nevertheless, there are instances where this has happened. The anti-interferon effect of adenovirus VA RNAs has already been described in Chapter 5. Other mechanisms of virus resistance to interferons include:

- Epstein–Barr virus EBER RNAs are similar in structure and function to the adenovirus VA RNAs. The EBER-2 protein also blocks interferon-induced signal transduction
- Vaccinia virus is known to show resistance to the antiviral effects of interferons. One of the early genes of this virus, K3L, encodes a protein which is homologous to eIF-2α which inhibits the action of PKR. In addition, the E3L protein also binds dsRNA and inhibits PKR activation

Table 6.1 Therapeutic uses of interferons

Condition	Virus
Chronic active hepatitis	HBV, HCV
Condylomata acuminata (Genital warts)	Papilloma viruses
Tumours	
Hairy cell leukaemia	–
Kaposi's sarcoma (in AIDS patients)	HHV-8 (?)
Congenital diseases	
Chronic granulomatous disease – IFN-γ reduces bacterial infections	–

- Poliovirus infection activates a cellular inhibitor of PKR in virus-infected cells
- Reovirus capsid protein σ3 is believed to sequester dsRNA and therefore prevent activation of PKR.

Virus–Host Interactions

At this point, it would be as well to state once again that viral pathogenesis is an abnormal situation of no value to the virus – the vast majority of virus infections are asymptomatic. However, for pathogenic viruses, there are a number of critical stages in replication which determine the nature of the disease they produce. For all viruses, pathogenic or non-pathogenic, the first factor which influences the course of infection is the mechanism and site of entry into the body (Figure 6.7):

- *The skin*: Mammalian skin is a highly effective barrier against viruses. The outer layer (epidermis) consists of dead cells and therefore does not support virus replication. Very few viruses infect directly by this route unless there is prior injury such as minor trauma or puncture of the barrier, such as insect or animal bites or subcutaneous injections. Some viruses which do use this

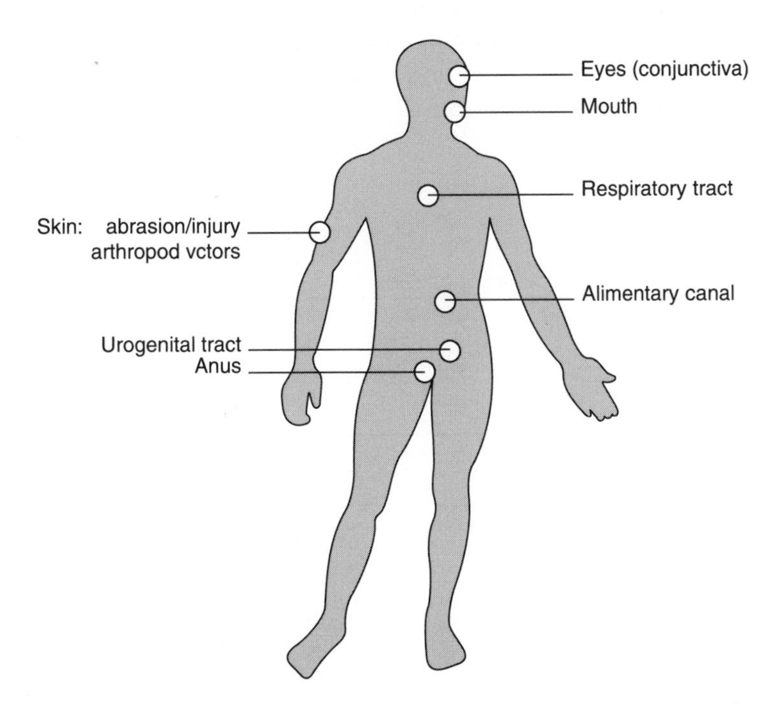

Figure 6.7 Schematic diagram to illustrate possible sites of virus entry into the body.

route are herpes simplex virus and papillomaviruses, although these viruses probably still require some form of disruption of the skin such as small abrasions or eczema

- *Mucosal membranes*: The mucosal membranes of the eye and genitourinary (GU) tract are much more favourable routes of access for viruses to the tissues of the body. This is reflected by the number of viruses which can be sexually transmitted; virus infections of the eye are also quite common (Table 6.2)
- *Alimentary canal*: Viruses may infect the alimentary canal via the mouth, oropharynx, gut, or rectum, although viruses which infect the gut via the oral route must survive passage through the stomach, an extremely hostile environment with a very low pH and high concentrations of digestive enzymes. Nevertheless, the gut is a highly valued prize for viruses – the intestinal epithelium is constantly replicating and there is a good deal of lymphoid tissue associated with the gut which provides many opportunities for virus replication. Moreover, the constant intake of food and fluids provides ample opportunity for viruses to infect these tissues (Table 6.3). To counteract this problem, the gut has many specific (e.g. secretory antibodies) and non-specific (e.g. stomach acids and bile salts) defence mechanisms
- *Respiratory tract*: The respiratory tract is probably the most frequent site of virus infection. As with the gut, it is constantly in contact with external virus particles which are taken in during respiration. As a result, the respiratory tract also has defences aimed at virus infection – filtering of particulate matter in the sinuses, and cells and antibodies of the immune system present

Table 6.2 Viruses which infect via mucosal surfaces

Virus	Site of infection
Adenoviruses	Conjunctiva
Picornaviruses: enterovirus 70	Conjunctiva
Papillomaviruses	Genitourinary tract
Herpesviruses	Genitourinary tract
Retroviruses: HIV, HTLV	Genitourinary tract

Table 6.3 Viruses which infect via the alimentary canal

Virus	Site of infection
Herpesviruses	Mouth and oropharynx
Adenoviruses	Intestinal tract
Caliciviruses	Intestinal tract
Coronaviruses	Intestinal tract
Picornaviruses: enteroviruses	Intestinal tract
Reoviruses	Intestinal tract

Table 6.4 Viruses which infect via the respiratory tract

Virus	Localized infection	Systemic infection
Adenoviruses	Upper respiratory tract	–
Coronaviruses	Upper respiratory tract	–
Orthomyxoviruses	Upper respiratory tract	–
Picornaviruses: rhinoviruses	Upper respiratory tract	–
Paramyxoviruses: parainfluenza, respiratory syncytial virus (RSV)	Lower respiratory tract	–
Herpesviruses	–	Varicella–Zoster (VZV)
Paramyxoviruses	–	Measles, mumps
Poxviruses	–	Smallpox
Togaviruses	–	Rubella

in the lower regions. Viruses which infect the respiratory tract usually come directly from the respiratory tract of others, since aerosol spread is very efficient: 'coughs and sneezes spread diseases' (Table 6.4).

The natural environment is a considerable barrier to virus infections. Most viruses are relatively sensitive to heat, drying, ultraviolet light (sunlight), etc., although a few types are quite resistant to these factors. This is particularly important for viruses which are spread via contaminated water or foodstuffs – not only must they be able to survive in the environment until they are ingested by another host, but as most are spread by the faecal–oral route, they must also be able to pass through the stomach to infect the gut before being shed in the faeces. One way of overcoming environmental stress is to take advantage of a secondary vector for transmission between the primary hosts (Figure 6.8). As with plant viruses (above), the virus may or may not replicate while in the vector. Viruses without a secondary vector must rely on continued host-to-host transmission, and have evolved various strategies to do this (Table 6.5):

- *Horizontal transmission*: The direct host-to-host transmission of viruses. This strategy relies on a high rate of infection to maintain the virus population
- *Vertical transmission*: The transmission of the virus from one generation of hosts to the next. This may occur by infection of the foetus before, during, or shortly after birth (e.g. during breastfeeding). More rarely, it may involve direct transfer of the virus via the germ line itself, e.g. retroviruses. In contrast to horizontal transmission, this strategy relies on long-term persistence of the virus in the host rather than rapid propagation and dissemination of the virus.

Having gained entry to a potential host, the virus must initiate an infection by entering a susceptible cell (primary replication). This initial interaction frequently determines whether the infection will remain localized at the site of entry or spread to become a systemic infection (Table 6.6). In some cases, virus spread is controlled by infection of polarized epithelial cells and the preferential

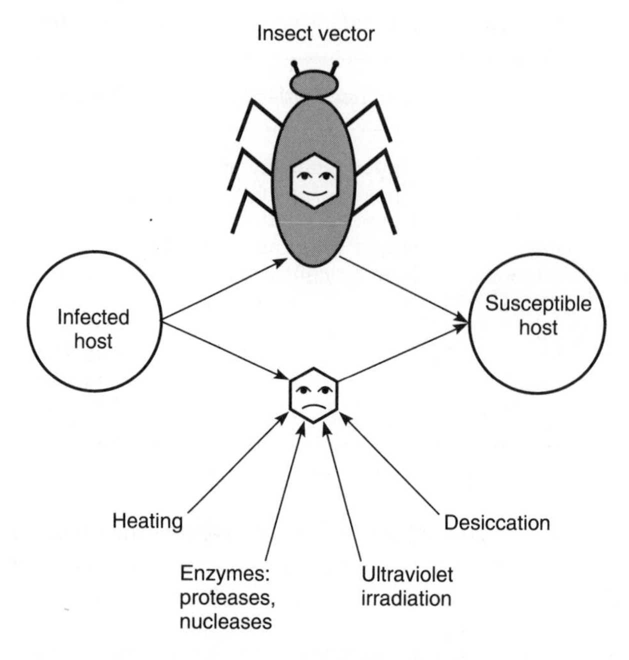

Figure 6.8 Transmission of viruses through the environment. Some viruses have adopted the use of vectors such as insects or other arthropods to avoid environmental stress.

Table 6.5 Virus transmission patterns

Pattern	Example
Horizontal transmission	
Human–human (aerosol)	Influenza
Human–human (faecal-oral)	Rotaviruses
Animal–human (direct)	Rabies
Animal–human (vector)	Bunyaviruses
Vertical transmission	
Placental–foetal	Rubella
Mother–child (birth)	HSV, HIV
Mother–child (breastfeeding)	HIV, HTLV
Germ line	In mice, retroviruses; In humans ?

release of virus from either the apical (e.g. influenza virus – a localized infection in the upper respiratory tract) or basolateral (e.g. rhabdoviruses – a systemic infection) surface of the cells (Figure 6.9). Following primary replication at the site of infection, the next stage may be spread throughout the host. In addition to direct cell–cell contact, there are two main mechanisms for spread throughout the host:

Table 6.6 Examples of localized and systemic virus infections

Virus	Primary replication	Secondary replication
Localized infections		
Papillomaviruses	Dermis	–
Rhinoviruses	Upper respiratory tract	–
Rotaviruses	Intestinal epithelium	–
Systemic infections		
Enteroviruses	Intestinal epithelium	Lymphoid tissues, CNS
Herpesviruses	Oropharynx or genitourinary tract	Lymphoid cells, CNS

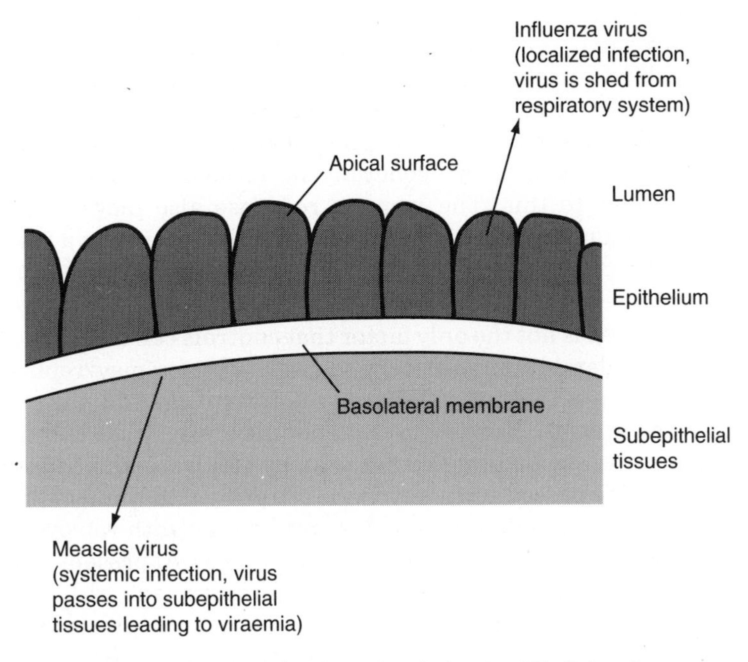

Figure 6.9 Virus infection of polarized epithelial cells.

(1) *Via the bloodstream*: Viruses may get into the bloodstream by direct inoculation, for example, by arthropod vectors, blood transfusion, or intravenous drug abuse (sharing of non-sterilized needles). The virus may travel free in the plasma (e.g. togaviruses, enteroviruses), or in association with red cells (orbiviruses), platelets (HSV), lymphocytes (EBV, CMV), or monocytes (lentiviruses). Primary viraemia usually precedes and is necessary for the spread of virus to other parts of the body via the bloodstream and is followed by a more generalized, higher titre secondary viraemia as the virus reaches the other target tissues or replicates directly in blood cells.

(2) *Via the nervous system*: As above, spread of virus to the nervous system is usually preceded by primary viraemia. In some cases, spread occurs directly by contact with neurones at the primary site of infection, in other cases it

occurs via the bloodstream. Once in peripheral nerves, the virus can spread to the CNS by axonal transport along neurones. The classic example of this is herpes simplex virus (see Latent Infection, below). Viruses can cross synaptic junctions since these frequently contain virus receptors, allowing the virus to 'jump' from one cell to another.

The spread of the virus to various parts of the body is controlled to a large extent by its cell or tissue tropism. Tissue tropism is controlled partly by the route of infection, but largely by the interaction of a virus attachment protein with a specific receptor molecule on the surface of a cell (as discussed in Chapter 4), and has considerable effect on pathogenesis.

At this stage, following significant virus replication and the production of viral antigens, the host immune response comes into play. This has already been discussed (above) and obviously has a major impact on the outcome of an infection. To a large extent, the efficiency of the immune response determines the amount of secondary replication which occurs and hence, the spread to other parts of the body. If a virus can be prevented from reaching tissues where secondary replication can occur, generally no disease results, although there are some exceptions to this. The immune response also plays a large part in determining the amount of cell and tissue damage that occurs as a result of virus replication. As described above, the production of interferons is a major factor in preventing virus-induced tissue damage.

The immune system is not the only factor that controls cell death, the amount of which varies considerably for different viruses. Viruses may replicate widely throughout the body without any disease symptoms if they do not cause significant cell damage or death. Retroviruses do not generally cause cell death, being released from the cell by budding rather than by cell lysis, and cause persistent infections, even being passed vertically to the offspring if they infect the germ line. All vertebrate genomes, including humans, are littered with retrovirus genomes which have been with us for millions of years (Chapter 3). At present, these ancient virus genomes are not known to cause any disease in humans, although there are examples of tumours caused by them in rodents. Conversely, picornaviruses cause lysis and death of the cells in which they replicate, leading to fever and increased mucus secretion in the case of rhinoviruses, and paralysis or death (usually due to respiratory failure due to damage to the central nervous system resulting, in part, from viral replication in these cells) in the case of poliovirus.

The eventual outcome of any virus infection depends on a balance between two processes:

$$\begin{array}{ccc} \text{CLEARANCE} & \leftharpoondown & \text{PERSISTENCE} \\ \text{(by the host organism)} & \rightharpoondown & \text{(by the virus)} \end{array}$$

Clearance is mediated by the immune system (as discussed previously). However, the virus is a moving target which rapidly responds to pressure from the immune system by altering its antigenic composition (whenever possible). The classic example of this phenomenon is influenza virus, which displays two genetic mechanisms that allow the virus to alter its antigenic constitution:

(1) *Antigenic drift*: This involves the gradual accumulation of minor mutations (e.g. nucleotide substitutions) in the virus genome which result in subtly altered coding potential and therefore altered antigenicity, leading to decreased recognition by the immune system. This process occurs in all viruses all the time, but at greatly different rates; for example, it is much more frequent in RNA viruses than in DNA viruses. In response, the immune system constantly adapts by recognition of and response to novel antigenic structures – but it is always one step behind. In most cases, however, the immune system is eventually able to overwhelm the virus, resulting in clearance.

(2) *Antigenic shift*: In this process a sudden and dramatic change in the antigenicity of a virus occurs owing to reassortment of the segmented virus genome with another genome of a different antigenic type (see Chapter 3). This results initially in the failure of the immune system to recognize a new antigenic type, giving the virus the upper hand (Figure 6.10).

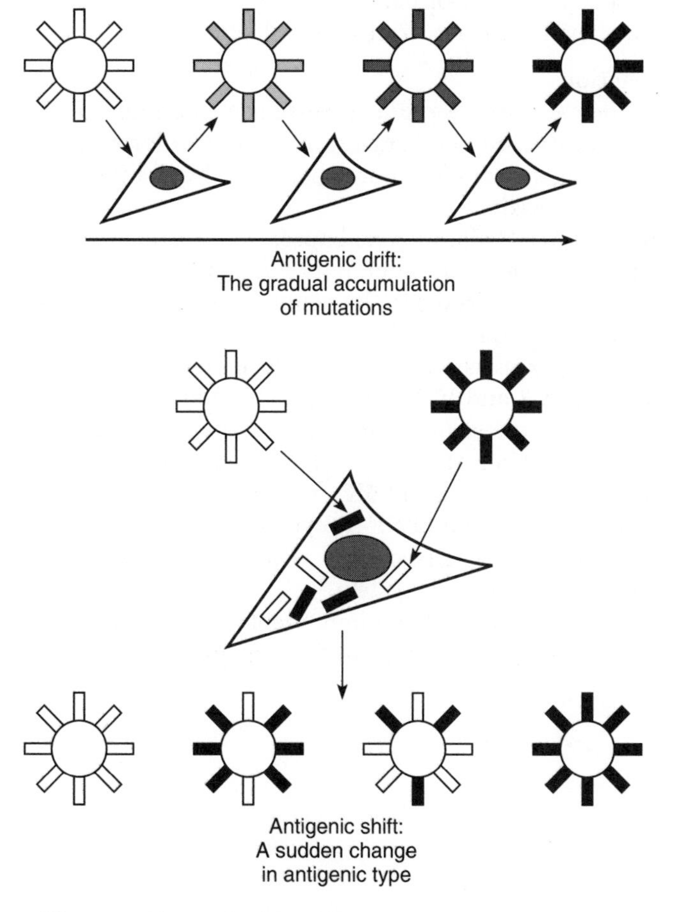

Figure 6.10 Antigenic shift and drift in influenza virus.

The occurrence of past antigenic shifts in influenza virus populations is recorded by **pandemics** (worldwide epidemics; Figure 6.11). These events are marked by the sudden introduction of a new antigenic type of haemagglutinin and/or neuraminidase into the circulating virus, overcoming previous immunity in the human population. Previous haemagglutinin/neuraminidase types become resurgent when a sufficiently high proportion of the people who have 'immunological memory' of that type have died, thus overcoming the effect of 'herd immunity'.

The other side of the relationship that determines the eventual outcome of a virus infection is the ability of the virus to persist in the host. Long-term persistence of viruses results from two main mechanisms. The first is the regulation of lytic potential. The strategy followed here is to achieve the continued survival of a critical number of virus infected cells, i.e. sufficient to continue the infection without killing the host organism. For viruses which do not usually kill the cells in which they replicate, this is not usually a problem; hence these viruses tend naturally to cause persistent infections, e.g. retroviruses. For viruses that undergo lytic infection, e.g. herpesviruses, it is necessary to develop mechanisms which restrict virus gene expression, and consequently, cell damage (see below). The second aspect of persistence is the evasion of immune surveillance. Viruses have evolved many tricks to fool the immune system:

- Antigenic variation
- Immune tolerance, causing a reduced response to an antigen, which may be due to genetic factors, pre-natal infection, or molecular mimicry by the virus of a host 'self' antigen, which is not recognized by the immune system
- Restricted gene expression
- Downregulation of MHC class I gene expression, resulting in lack of recognition of infected cells, e.g. the adenovirus E3 gp19K protein prevents CTL recognition of infected fibroblasts by sequestering MHC class I proteins in the endoplasmic reticulum
- Downregulation of expression of accessory molecules involved in immune recognition, e.g. LFA-3 and ICAM-1 (Figure 6.4) by EBV
- Overcoming the immunostimulatory effects of cytokines. Ways in which viruses escape the action of interferons have already been described. The adenovirus E3 proteins 10.4K, 14.5K and 14.7K protect infected cells from tumour necrosis factor alpha (TNF-α) cytolysis
- Infection of immunocompromised sites within the body, e.g. HSV 'hides' in sensory ganglia in the nervous system
- Direct infection of the cells of the immune system itself, e.g. herpesviruses, retroviruses (HIV), often resulting in immunosuppression.

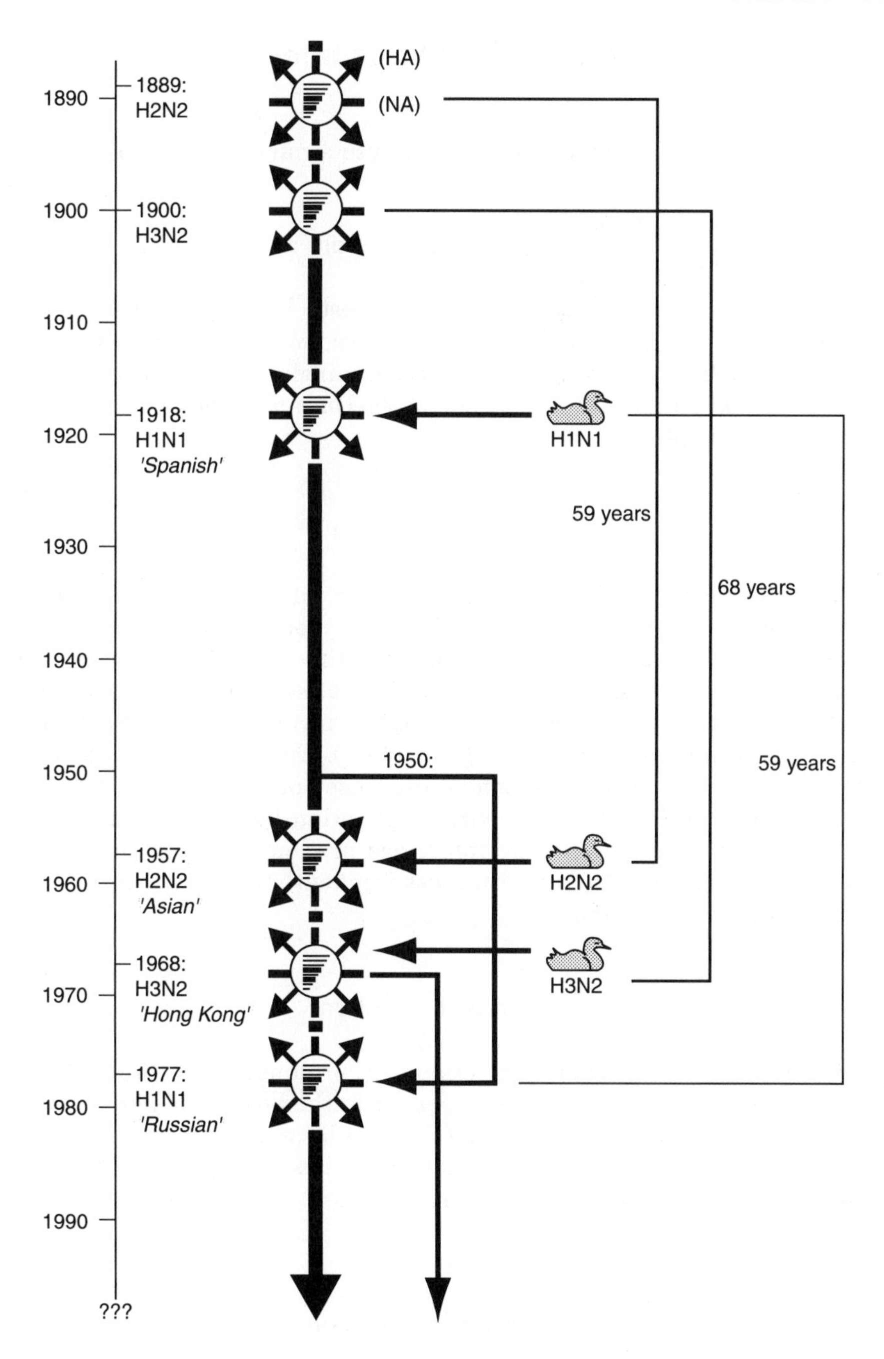

Figure 6.11 Influenza pandemics.

The Course of Virus Infections

Patterns of virus infection can be divided into a number of different types.

Abortive infection

This occurs when a virus infects a cell (or host), but cannot complete the full replication cycle. Therefore, this is a non-productive infection. However, the outcome of such infections is not necessarily insignificant; for example, SV40 infection of non-permissive rodent cells sometimes results in transformation of the cells (see Chapter 7).

Acute infection

This pattern is familiar for many common virus infections (e.g. 'colds'). In these relatively brief infections, the virus is usually eliminated completely by the immune system, but occasionally can have lasting effects – for example, herpes zoster infection causes chicken pox (an acute, self-limited disease), but also shingles (a chronic disease with neurological involvement caused by reactivation of the latent virus; see Latent Infection). Typically, in acute infections, much viral replication occurs before the onset of any symptoms (e.g. fever), which are the result not only of viral replication but of the activation of the immune system. Therefore, acute infections present a serious problem for the epidemiologist and are the pattern most frequently associated with epidemics (e.g. influenza, measles, etc.).

Chronic infection

These are the converse of acute infections, i.e. prolonged and stubborn. To cause this type of infection, the virus must persist in the host for a significant period. To the clinician, there is no clear distinction between chronic, persistent and latent infections and the terms are often used interchangeably. They are listed separately here because to virologists, there are significant differences in the events which occur during these infections.

Persistent infection

These infections result from a delicate balance between the virus and the host organism, in which ongoing virus replication occurs, but the virus adjusts its replication and pathogenicity to avoid killing the host. They differ from chronic infections in that whereas in chronic infections the virus is usually eventually cleared by the host (unless the infection proves fatal), in persistent infections, the virus may continue to be present and to replicate in the host for its entire lifetime.

The best studied example of such a system is lymphocytic choriomeningitis virus (LCMV, an arenavirus) infection in mice (Figure 6.12). Mice can be experimentally infected with this virus either at a peripheral site (e.g. a footpad or the tail) or by direct inoculation into the brain. Adult mice infected in the latter way are killed by the virus, but if infected by a peripheral route, there are two possible outcomes to the infection: some mice die, whereas others survive, having cleared the virus from the body completely. It is not clear what factors determine the survival or death of LCMV-infected mice, but other evidence shows that this is related to the immune response to the virus. In immunosuppressed adult mice which are infected via the CNS route, a persistent infection is established in which the virus is not cleared (due to the non-functional immune system), but remarkably, these mice are not killed by the virus. If, however, syngeneic LCMV-specific T-lymphocytes (i.e. of the same MHC type) are injected into these persistently infected mice, the animals develop the full pathogenic symptoms of LCMV infection and are killed. Newborn mice, whose immune systems are immature, infected via the CNS route also develop a persistent infection, but in this case, if subsequently injected with syngeneic LCMV-specific T-lymphocytes, they clear the virus and survive the infection. The mechanisms which control these events are not completely understood, but evidently there is a delicate balance between the virus and the host animal and that the immune response to the virus is partly responsible for the pathology of the disease and the death of the animals.

Not infrequently, persistent infections may result from the production of **defective-interfering (D.I.) particles** (see Chapter 3). Such particles contain

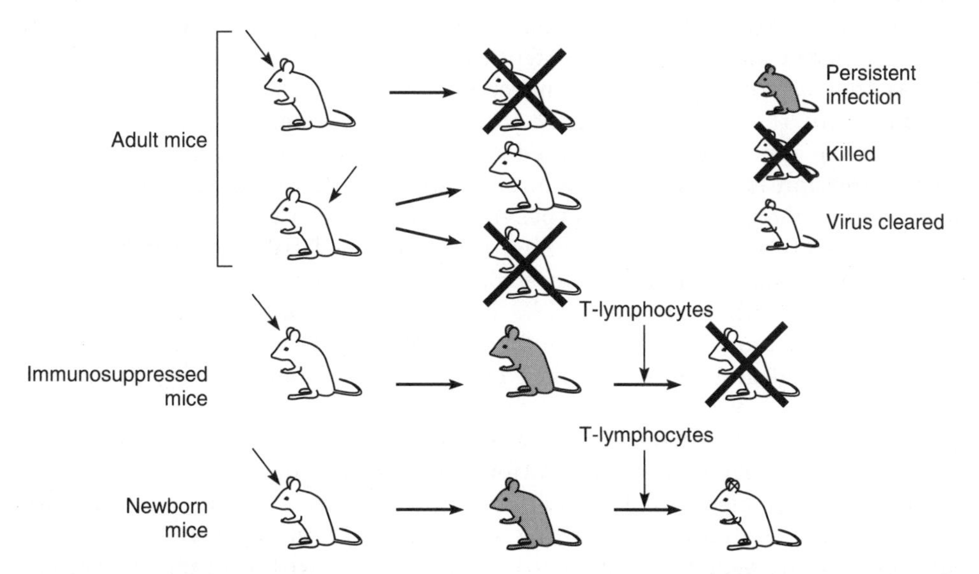

Figure 6.12 Persistent infection of mice by lymphocytic choriomeningitis virus (LCMV).

a partial deletion of the virus genome and are replication defective, but are maintained and may even tend to accumulate during infections because they can replicate in the presence of replication-competent helper virus. The production of D.I. particles is a common consequence of virus infection of animals, particularly by RNA viruses, but also occurs with DNA viruses and plant viruses and can be mimicked *in vitro* by continuous high titre passage of virus. Although not able to replicate themselves independently, D.I. particles are not necessarily genetically inert and may alter the course of an infection by recombination with the genome of a replication-competent virus. The presence of D.I. particles can profoundly influence the course and the outcome of a virus infection. In some cases, they appear to moderate pathogenesis, whereas in others they potentiate it, making the symptoms of the disease much more severe. Moreover, as D.I. particles effectively cause restricted gene expression (because they are genetically deleted), they may also result in a persistent infection by a virus which normally causes an acute infection and is rapidly cleared from the body.

Latent infection

This is the ultimate infection! In latent infection the virus is able to down-regulate its gene expression and establish an inactive state, i.e. with strictly limited gene expression and without ongoing virus replication. Latent virus infections typically persist for the entire life of the host. An example of such an infection in man is herpes simplex virus (HSV). Infection of sensory nerves serving the mucosa results in localized primary replication. Subsequently, the virus travels via axonal transport mechanisms further into the nervous system. There, it 'hides' in dorsal root ganglia, such as the trigeminal ganglion, establishing a truly latent infection. The nervous system is an immunologically privileged site and is not patrolled by the immune system in the same way as the rest of the body, but nevertheless, the major factor in latency is the ability of the virus to restrict its gene expression. This eliminates the possibility of recognition of infected cells by the immune system. Restricted gene expression is achieved by tight regulation of α-gene expression, which is an essential control point in herpesvirus replication (Chapter 5). There has been much interest in mRNAs which are made by HSV during latent infection; these are known as latency-associated transcripts or LATs and raise the possibility of parallels between HSV latency and lysogeny in phage λ. However, the importance of the LATs in controlling latency has more recently been questioned and the mechanism by which they might act is not clear at present.

When reactivated by some provocative stimulus, HSV travels down the sensory nerves to cause peripheral manifestations such as cold sores or genital ulcers. It is not altogether clear what constitutes a provocative stimulus but there are many possible alternatives, including psychological and physical factors. Periodic reactivation establishes the pattern of infection, with sporadic, sometimes very painful reappearance of disease symptoms for the rest of the host's life.

Even worse than this, immunosuppression later in life can cause the latent infection to flare up (which indicates that the immune system normally has a role in helping to suppress these latent infections), resulting in a very severe, systemic and often life-threatening infection.

In a manner somewhat similar to herpesviruses, infection by retroviruses may result in a latent infection. Integration of the provirus into the host genome certainly results in the persistence of the virus for the lifetime of the host organism and may lead to an episodic pattern of disease. In some ways, acquired immunodeficiency syndrome (AIDS) which results from HIV infection, shows aspects of this pattern of infection. The pathogenesis of AIDS is discussed in detail in Chapter 7.

Prevention and Therapy of Virus Infection

There are two aspects of the response to the threat of virus diseases: first, prevention of infection, and second, treatment of the disease. The former strategy relies on two approaches: public and personal hygiene, which perhaps plays the major role in preventing virus infection (e.g. provision of clean drinking water and disposal of sewage; good medical practice such as the sterilization of surgical instruments) and vaccination, which makes use of the immune system to combat virus infections. Most of the damage to cells during virus infections occurs very early, often before the clinical symptoms of disease appear. This makes the treatment of virus infection very difficult, and therefore, in addition to being cheaper, prevention of virus infection is undoubtedly better than cure.

To design effective vaccines, it is important to understand both the immune response to virus infection (above) and the stages of virus replication that are appropriate targets for immune intervention. To be effective, vaccines must stimulate as many of the body's defence mechanisms as possible. In practice, this usually means trying to mimic the disease, without of course causing pathogenesis – for example, the use of nasally administered influenza vaccines and orally administered poliovirus vaccines. To be effective, it is not necessary to get 100% uptake of vaccine. 'Herd immunity' results from the break in transmission of a virus which occurs when a sufficiently high proportion of a population has been vaccinated. This strategy is most effective where there is no alternative host for the virus, e.g. measles, and in practice is the situation that usually occurs since it is impossible to achieve 100% coverage with any vaccine. However, this is a risky business. If protection of the population falls below a critical level, epidemics can easily occur. There are three basic types of vaccine: subunit vaccines, inactivated vaccines, and live virus vaccines.

Subunit vaccines

These are the newest type of vaccine and consist of only some components of the virus, sufficient to induce a protective immune response but not enough to allow any danger of infection. In general terms, they are completely safe, except for very rare cases in which adverse immune reactions may occur. Unfortunately, at present, they are also the least effective and most expensive type of vaccines. The major technical problems associated with subunit vaccines are their relatively poor antigenicity and the need for new delivery systems, such as improved carriers and adjuvants. There are several categories of such vaccines:

- *Synthetic vaccines*, such as short, chemically synthesized peptides. The major disadvantage with these molecules is that they are not usually very effective immunogens and are very costly to produce. However, because they can be made to order for any desired sequence, they have great potential for the future. None are currently in use
- *Recombinant vaccines*, produced by genetic engineering. Such vaccines have been already produced and are better than the above type because they tend to give rise to a more effective immune response. Some practical success has already been achieved with this type of vaccine. For example, vaccination against hepatitis B virus (HBV) used to rely on the use of 'Australian antigen' (HBsAg) obtained from the serum of chronic HBV carriers. This was a very risky practice indeed (because HBV carriers are often also infected with HIV). A completely safe recombinant HBV vaccine produced in yeast is now widely used
- *Virus vectors*, i.e. recombinant virus genomes genetically manipulated to express protective antigens from (unrelated) pathogenic viruses. The idea here is to utilize the genome of a well-understood, **attenuated** virus to express and present antigens to the immune system. Many different viruses offer possibilities for this type of approach, but the most highly developed system so far is based on the vaccinia virus (VV) genome. This virus has been used to vaccinate millions of people worldwide in the campaign to eradicate smallpox (see below), and is generally a safe and effective vehicle for antigen delivery. Such vaccines are difficult to produce and no human example is clearly successful yet, although many different trials are currently underway. VV-rabies recombinants have been used to eradicate rabies in European fox populations. VV-based vaccines have advantages and disadvantages for use in humans:
  - the high percentage of the human population which has already been vaccinated during the smallpox eradication campaign – lifelong protection may result in poor response to recombinant vaccines (?)
  - although generally safe, VV is dangerous in immunocompromised hosts, thus cannot be used in HIV-infected individuals.

  A possible solution to these problems may be to use Avipoxvirus vectors, e.g. fowlpox or canarypox, as 'suicide vectors' which can only establish **abortive infections** of mammalian cells:

- can express high levels of foreign proteins
- no danger of pathogenesis (abortive infection)
- no natural immunity in humans (avian virus).

Inactivated vaccines

These are produced by exposing the virus to a denaturing agent under precisely controlled conditions. The objective is to cause loss of virus infectivity without loss of antigenicity. Obviously, this involves a delicate balance. However, inactivated vaccines have certain advantages, such as generally being effective immunogens (if properly inactivated), being relatively stable, and carrying little or no risk of vaccine-associated virus infection (if properly inactivated, but accidents can and do occur). The disadvantage of these vaccines is that it is not possible to produced inactivated vaccines for all viruses, since denaturation of virus proteins may lead to loss of antigenicity, e.g. measles virus. Although relatively effective, 'killed' vaccines are sometimes not as effective at preventing infection as 'live' virus vaccines (see below), often because they fail to stimulate protective mucosal immunity to the same extent. A more recent concern is that these vaccines contain viral nucleic acids, which may themselves be a source of infection, either of their own accord (e.g. (+)sense RNA virus genomes) or after recombination with other viruses.

Live (attenuated) virus vaccines

This strategy relies on the use of viruses with reduced pathogenicity to stimulate an immune response without causing disease. The vaccine strain may be a naturally occurring virus (e.g. the use of cowpox virus by Edward Jenner to vaccinate against smallpox) or artificially **attenuated** *in vitro* (e.g. the oral poliomyelitis vaccines produced by Albert Sabin). The advantage of attenuated vaccines is that they are good immunogens and induce long-lived, appropriate immunity. Set against this are their many disadvantages. They are often biochemically and genetically unstable and may either lose infectivity (when they become worthless) or revert to virulence unexpectedly. Despite intensive study, it is not possible to produce an attenuated vaccine to order, and there appears to be no general mechanism by which different viruses can be reliably and safely attenuated. Contamination of the vaccine stock with other, possibly pathogenic viruses is also possible – this was the way in which SV40 was first discovered in oral poliovirus vaccine in 1960. Inappropriate use of live virus vaccines, for example, in immunocompromised hosts or during pregnancy, may lead to vaccine-associated disease, whereas the same vaccine given to a healthy individual may be perfectly safe.

Despite these difficulties, vaccination against virus infection has been one of the great triumphs of medicine during the twentieth century. Most of the success stories result from the use of live attenuated vaccines; for example, the use of vaccinia virus against smallpox. On the 8th May 1980, the World Health Organization (WHO) officially declared smallpox to be completely eradicated, the first virus disease to be eliminated from the world. The WHO Expanded Programme on Immunization currently focuses on the prevention of selected childhood diseases and aims to eradicate poliomyelitis by the year 2000, reduce measles deaths and incidence, eliminate neonatal tetanus (a bacterial infection) and introduce hepatitis B vaccine worldwide. Cases of poliomyelitis have declined by 75% since the campaign was launched in 1988. Only 8254 cases were reported in 1994, although the true figure may be as high as 100000 cases. Poliomyelitis has essentially been eliminated from the Western Hemisphere – the last case was reported from Peru in 1991, although small numbers of cases are imported from regions where the virus remains endemic. Poliomyelitis was last reported in China in 1994 and in the Philippines in 1993. However, the Indian subcontinent remains heavily affected, reporting more than two-thirds of all cases. Poliomyelitis remains endemic in West and Central Africa and some countries in the Middle East and the Horn of Africa. The vaccination campaign against measles virus is estimated to have reduced the total reported annual number of cases from 6 million to about half a million (still an underestimate) and considerably reduced the death total from the previous level of one million annually. It is certain that these advances will be continued by the development and application of vaccines produced by recombinant DNA technology.

Virus vaccines do not have to be based on virion structural proteins. For example, human papillomavirus (HPV) infection is associated with cervical carcinoma (Chapter 7). Two HPV regulatory proteins, E6 and E7, are consistently expressed in tumour cells and vaccines based on live recombinant vaccinia viruses expressing the E6 and E7 proteins of HPV 16 and 18 are under development for the prevention and treatment of cervical cancer. This raises another important point. Although prevention of infection by prophylactic vaccination is much the preferred option, post-exposure therapeutic vaccines can be of great value in modifying the course of some virus infections. Examples of this are rabies virus, where the course of infection may be very long and there is time for post-exposure vaccination to generate an effective immune response and prevent the virus from carrying out secondary replication in the CNS, which is responsible for the pathogenesis of rabies. Other potential examples are in virus-associated tumours, such as HPV-induced cervical carcinoma, as described above.

Virus Vectors and Gene Therapy

Viruses are being developed as gene delivery systems for the treatment of inherited and also acquired diseases. The first human trial to treat children with immunodeficiency resulting from a lack of the enzyme adenosine deaminase (ADA) began in September 1990 and showed encouraging although not completely successful results. Like most of the initial attempts, this trial used recombinant retrovirus genomes as vectors. A variety of different viruses are being

Table 6.7 Virus vectors in gene therapy

Virus	Advantages	Possible disadvantages
Adenoviruses	Relatively easily manipulated *in vitro* (c.f. Retroviruses); genes coupled to the major late promoter (MLP) are efficiently expressed in large amounts.	Possible pathogenesis associated with partly attenuated vectors (especially in the lungs); immune response makes multiple doses ineffective if gene needs to be administered repeatedly (virus does not integrate).
Parvoviruses (AAV)	Integrate into cellular DNA at high frequency to establish a stable latent state; not associated with any known disease; vectors can be constructed which will not express any viral gene products.	Only *c.* 5kb of DNA can be packaged into the parvovirus capsid and some viruses sequences must be retained for packaging; integration into host cells DNA may potentially have damaging consequences.
Herpesviruses	Relatively easy to manipulate *in vitro*; grows to high titres; long-term persistence in neuronal cells without integration.	(Long-term) pathogenic consequences?
Retroviruses	Integrate into cell genome giving long-lasting (life-long?) expression of recombinant gene.	Difficult to grow to high titre and purify for direct administration (patient cells need to be cultured *in vitro*); cannot infect non-dividing cells – most somatic cells (except Lentiviruses?); insertional mutagenesis/ activation of cellular oncogenes.

tested as potential vectors (Table 6.7) and a large number of different trials are underway, e.g. the first human trial of gene therapy for cystic fibrosis began in 1993. Non-viral methods of gene delivery including liposome/DNA complexes, peptide/DNA complexes and direct injection of recombinant DNA are also under active investigation. It is important to note that such experiments are aimed at augmenting defective cellular genes in the somatic cells of patients to alleviate the symptoms of the disease and not at manipulating the human germ line, which is a different issue. There can be no question that carefully applied, this new application of virology will change the treatment of inherited diseases during the next century.

Chemotherapy of Virus Infections

The alternative to vaccination is to attempt to treat virus infections using drugs which block virus replication (Table 6.8). Historically, discovery of antiviral drugs has been largely fortuitous. Spurred on by successes in the treatment of bacterial infections with antibiotics, drug companies launched huge blind-screening programmes to identify chemical compounds with antiviral activity, with relatively little success. The key to the success of any antiviral drug lies in its specificity. Almost any stage of virus replication can be a target for a drug, but the drug must be more toxic to the virus than the host. This is measured by the chemotherapeutic index, given by:

$$\frac{\text{Dose of drug which inhibits virus replication}}{\text{Dose of drug which is toxic to host}}$$

The smaller the value of the chemotherapeutic index, the better. In practice, a difference of several orders of magnitude between the two toxicity values is

Table 6.8 Antiviral drugs

Drug	Viruses	Chemical type	Target
Acyclovir	Herpes simplex	Nucleoside analogue	Virus polymerase (needs virus thymidine kinase for activation)
Amantadine/ Rimantadine	Influenza A viruses	Tricyclic amines	Matrix protein/ haemagglutinin
Dideoxthiacytidine (3TC)	HIV	Nucleoside analogue	Reverse transcriptase
Dideoxyinosine (ddI), dideoxycytosine (ddC)	HIV	Nucleoside analogue	Reverse transcriptase
Foscarnet	Herpesviruses	Pyrophosphate analogue	Virus polymerase
Ganciclovir	Cytomegalovirus	Nucleoside analogue	Virus polymerase (needs virus kinase for activation)
Ribavirin	HSV, Measles, Mumps, Lassa fever (broad spectrum)	Triazole carboxamide	Virus replicase/ transcriptase
Vidarabine	Herpesviruses	Nucleoside analogue	Virus polymerase
Zidovudine (AZT)	HIV	Nucleoside analogue	Reverse transcriptase

usually required to produce a safe and clinically useful drug. Modern technology, including molecular biology and computer-aided design of chemical compounds, allows the deliberate design of drugs, but to do this, it is necessary to 'know your enemy', i.e. to understand the key steps in virus replication which might be inhibited. Any of the stages of viral replication can be a target for antiviral intervention. The only requirements are:

(1) That the process targeted be essential for replication.
(2) That the drug is active against the virus but has 'acceptable toxicity' to the host organism.

What degree of toxicity is 'acceptable' clearly varies considerably, for example, between a cure for the common cold, which might be sold over the counter and taken by millions of people, and a drug used to treat fatal virus infections such as AIDS.

The attachment phase of replication can be inhibited in two ways: by agents which mimic the virus-attachment protein (VAP) and bind to the cellular receptor, or by agents which mimic the receptor and bind to the VAP. Synthetic peptides are the most logical class of compound to use for this purpose. While this is a promising line of research, there are considerable problems with clinical use of these substances, mostly the high cost of synthetic peptides and the poor pharmacokinetic properties of many of these synthetic molecules.

It is difficult to target specifically the penetration/uncoating stages of virus replication as relatively little is known about them. Uncoating in particular is largely mediated by cellular enzymes and is therefore a poor target for intervention, although like penetration, it is often influenced by one or more virus proteins. Amantadine and rimantadine are two drugs which are active against influenza A viruses. The action of these closely related agents is to block cellular membrane ion channels. The target for both drugs is the matrix protein (M_2), but resistance to the drug may also map to the haemagglutinin (HA) gene. This biphasic action results from the inability of drug-treated cells to lower the pH of the endosomal compartment (a function normally controlled by the M_2 gene product), which is essential to induce conformational changes in the HA protein to permit membrane fusion (see Chapter 4).

Many viruses have evolved their own specific enzymes to replicate virus nucleic acids preferentially at the expense of cellular molecules. There is often sufficient specificity in virus polymerases to provide a target for an antiviral agent, and this method has produced the majority of the specific antiviral drugs currently in use. The majority of these drugs function as polymerase substrates (i.e. nucleoside/nucleotide) analogues, and their toxicity varies considerably from some which are well tolerated (e.g. acyclovir) to others which are quite toxic (e.g. AZT). There is a problem with the pharmacokinetics of these nucleoside analogues in that their typical serum half-life is 1–4 h. Nucleoside analogues are, in fact, pro-drugs, since they need to be phosphorylated before becoming effective. This is the key to their selectivity:

- Acyclovir is phosphorylated by HSV thymidine kinase 200 times more efficiently than by cellular enzymes

- Ganciclovir is 10 times more effective against CMV than acyclovir, but must be phosphorylated by a kinase encoded by CMV gene *UL97* before it becomes pharmaceutically active.

A series of other nucleoside analogues derived from these drugs and active against herpesviruses have been developed, e.g. valciclovir and famciclovir. These compounds have improved pharmacokinetic properties such as better oral bioavailability and longer half lives. In addition to these, there are a number of non-nucleoside analogues which inhibit viral polymerases. Foscarnet is an analogue of pyrophosphate which interferes with the binding of incoming nucleotide triphosphates by viral DNA polymerases. Ribavirin is an interesting compound with a very wide spectrum of activity against many different viruses, especially against many (−)sense RNA viruses. The way in which these effects are caused are not entirely clear, but may include mechanisms such as:

- Ribavirin 5′ monophosphate inhibits inosine monophosphate dehydrogenase and decreases intracellular pools of guanosine triphosphate
- Ribavirin 5′ triphosphate inhibits guanylyl transferase and the 5′ capping of mRNAs
- Ribavirin 5′ triphosphate inhibits initiation and elongation by RNA polymerases.

Ribavirin is thus quite unlike the other nucleoside analogues described above and its use is likely to become much more widespread in future.

Virus gene expression is less amenable to chemical intervention than genome replication, because viruses are much more dependent on the cellular machinery for transcription, mRNA splicing, cytoplasmic export, and translation than for replication. To date, no clinically useful drugs which discriminate between viral and cellular gene expression have been developed. As with penetration and uncoating, for the majority of viruses, the processes of assembly, maturation and release are poorly understood, and therefore, have not yet become targets for antiviral intervention.

The most striking aspect of antiviral chemotherapy is how few clinically useful drugs are available. As if this were not bad enough, there is also the problem of drug resistance to consider. In practice, the speed and frequency with which resistance arises when drugs are used to treat virus infections varies considerably and depends largely on the biology of the virus involved rather than on the chemistry of the compound. To illustrate this, two extreme cases are described.

Acyclovir used to treat herpes simplex virus (HSV) infections is easily the most widely used antiviral drug. This is particularly true in the case of genital herpes, which causes painful recurrent ulcers on the genitals. It is estimated that 40–60 million people suffer from this condition in the USA. Fortunately, resistance to acyclovir arises infrequently. This is partly due to the high fidelity with which the DNA genome of HSV is copied (Chapter 3). Mechanisms which give rise to acyclovir resistance include:

- HSV *pol* gene mutants which do not incorporate acyclovir
- HSV thymidine kinase (TK) mutants in which TK activity is absent (TK^-), reduced, or shows altered substrate specificity.

Strangely, it is possible to find mutations which give rise to each of these phenotypes with a frequency of 10^{-3}–10^{-4} in clinical HSV isolates. The discrepancy between this and the very low frequency with which resistance is recorded clinically is probably explained by the observation that most *pol*/TK mutants appear to be **attenuated** (e.g. TK^- mutants of HSV do not reactivate from the latent state).

Conversely, azidothymidine (AZT) treatment of HIV infection is much less effective. In untreated HIV-infected individuals, AZT produces a rise in the numbers of $CD4^+$ cells within 2–6 weeks. However, this beneficial effect is transient; after 20 weeks, $CD4^+$ T-cell counts generally revert to baseline. This is due partly to the development of AZT resistance in treated HIV populations and to the toxicity of AZT on haematopoesis, since the chemotherapeutic index of AZT is much worse than that of acyclovir. AZT resistance is initiated by the acquisition of a mutation in the HIV reverse transcriptase (RT) gene at codon 215. In conjunction with two to three additional mutations in the RT gene, a fully AZT-resistant phenotype develops. After 20 weeks of treatment, 40–50% of AZT-treated patients develop at least one of these mutations. This high frequency is due to the error-prone nature of reverse transcription (Chapter 3). Because of the large number of replicating HIV genomes in infected patients (Chapter 7), many mistakes occur continuously. It has been shown that the mutations which confer resistance already exist in untreated virus populations. Thus, treatment with AZT does not cause but merely selects these resistant viruses from the total pool. With other anti-RT drugs such as ddI, a resistant phenotype can result from a single base pair change. However, ddI has an even lower therapeutic index than AZT and relatively low levels of resistance can potentially render this drug useless. However, some combinations of resistant mutations may make it difficult for HIV to replicate and resistance to one RT-inhibitor may counteract resistance to another. Current therapy for HIV infection therefore employs combinations of different drugs. Molecular mechanisms of resistance are important to consider in designing combination regimes:

- Combinations such as AZT+ddI or AZT+3TC that have antagonistic patterns of resistance are effective
- Combinations such as ddC+3TC that show cross-reactive resistance should be avoided.

Other potential benefits of combination antiviral therapy include lower toxicity profiles and the use of drugs that may have different tissue distributions or cell tropisms. Combination therapy may also prevent or delay the development of drug resistance. Combinations of drugs which can be employed include not only small synthetic molecules but also ‘biological response modifiers’ such as interleukins and interferons.

Summary

Virus infection is a complex, multi-stage interaction between the virus and the host organism. The course and eventual outcome of any infection is the result of a balance between host and virus processes. Host factors involved include exposure to different routes of virus transmission and the control of virus replication by the immune response. Virus processes include the initial infection of the host, spread throughout the host and the regulation of gene expression to evade the immune response. Medical intervention against virus infections includes the use of vaccines to stimulate the immune response and drugs to inhibit virus replication. Molecular biology is stimulating the production of a new generation of antiviral drugs and vaccines.

Further Reading

Beachy, R.N. (Ed.). (1993). Transgenic resistance to plant viruses. *Sem. Virol.* **4**: 327–416.

Brochier, B. *et al.* (1991). Large scale eradication of rabies using recombinant vaccinia–rabies vaccine. *Nature* **354**: 520–2.

Brown, F., Dougan, G., Hoey, E.M., Martin, S.J., Rima, B.K. and Trudgett, A. (Eds). (1993). *Vaccine Design*. Wiley, Chichester.

Fraser, R. (1990). The genetics of resistance to plant viruses. *Annu. Rev. Plant Phytopathol.* **28**: 179–200.

Garcia-Blanco, M.A. and Cullen, B.R. (1991). Molecular basis of latency in pathogenic human viruses. *Science* **254**: 815–20.

Ho, D.Y. (1992). Herpes simplex virus latency – molecular aspects. *Prog. Med. Virol.* **39**: 76–115.

Miller, A.D. (1992). Human gene therapy comes of age. *Nature* **357**: 455–60.

Mims, C.A., Dimmock, N., Nash, A. and Stephen, A. (1995). *Mims' Pathogenesis of Infectious Disease* (4th edition). Academic Press, London.

Oldstone, M.A. (1991). Molecular anatomy of viral persistence. *J. Virol.* **65**: 6381–6.

Roitt I. *Essential Immunology* (8th edition). (1994). Blackwell Scientific, Oxford.

Sen, G.C. and Lengyel, P. (1992). The interferon system: a birds eye view of its biochemistry. *J. Biol. Chem.* **267**: 5017–5020.

Smith, A.E. (1995). Viral vectors in gene therapy. *Annu. Rev. Microbiol.* **49**: 807–838.

Self-Assessment Questions

(6.1) Plant viruses (true or false?):
- (a) Are responsible for an estimated US$60 000 000 worth of damage each year.
- (b) Can be transmitted by infected seeds.
- (c) May be transmitted to new hosts as wind-blown dust or as rain-splashed mud.
- (d) Never replicate in insect vectors.
- (e) Always cause a systemic infection involving the whole plant.

(6.2) Are the following statements true or false?
- (a) Movement proteins allow plant viruses to be transmitted from insect vectors to plants.
- (b) Damage to virus-infected plants occurs through necrosis of cells, hypoplasia and hyperplasia.
- (c) The hypersensitive response in plants is functionally analogous to the production of interferons in animals.
- (d) The hypersensitive response in plants tends to limit the occurrence of systemic infection.
- (e) Virus-resistant plants have been created by genetic manipulation resulting in the endogenous expression of virus genes.

(6.3) Are the following statements true or false?
- (a) The major impact of the humoral immune response on virus infections is the eventual clearance of virus from the body.
- (b) Unlike bacteria, virus particles are not directly neutralized by antibodies or phagocytosed by white blood cells.
- (c) Overall, cell-mediated immunity is more important than humoral immunity in controlling virus infections.
- (d) Natural killer (NK) cells are the first cell-mediated defence mechanism to be activated against virus infection.
- (e) Cytotoxic T-lymphocytes (CTL) are not MHC-restricted.

(6.4) Are the following statements true or false?
- (a) Interferon was discovered in chickens in 1975.
- (b) Biochemical purification of interferons from natural sources was difficult because interferons were so abundant.
- (c) There are three distinct types of interferon: α, β, and γ.
- (d) Interferons are synthesized predominantly by fibroblasts.
- (e) Double-stranded DNA is a potent inducer of interferon synthesis.

(6.5) Are the following statements true or false?
- (a) All interferons have similar antiviral capacities, but act differently as cellular regulators.
- (b) 2′,5′-oligo A activates RNAse L, which digests all RNAs.

(c) PKR phosphorylates eIF2α, which is required for the initiation of translation.
(d) Interferons are used to treat chronic viral hepatitis.
(e) Several viruses have evolved mechanisms which enable them to resist the effects of interferons.

(6.6) Are the following statements true or false?
(a) Mucosal membranes are favourable routes of access for viruses to the tissues of the body.
(b) ‘Coughs and sneezes spread diseases.’
(c) The natural environment is a considerable barrier to virus infections.
(d) Primary systemic infection by viruses is frequently followed by localized secondary infection.
(e) Spread of virus to the nervous system usually precedes primary viraemia.

(6.7) Viruses have evolved many tricks to fool the immune system (true or false?):
(a) Antigenic variation.
(b) Upregulation of MHC class I gene expression.
(c) Upregulation of accessory molecules involved in immune recognition.
(d) Infection of immunocompromised sites within the body.
(e) Direct infection of the cells of the skin.

(6.8) Are the following statements true or false?
(a) Abortive infection of cells can have significant pathological consequences.
(b) In acute infections, much viral replication occurs before the onset of any symptoms.
(c) In persistent infections, the virus adjusts its replication and pathogenicity to avoid killing the host.
(d) Persistent infections may result from the production of defective-interfering (D.I.) particles.
(e) Latent infections are typified by strictly limited gene expression without ongoing virus replication.

(6.9) Are the following statements true or false?
(a) Subunit vaccines are the most effective and least expensive virus vaccines.
(b) No recombinant virus vaccines produced by genetic engineering are currently in use.
(c) Inactivated virus vaccines are sometimes not as effective as ‘live’ virus vaccines because they fail to stimulate mucosal immunity.
(d) The majority of successful virus vaccines are based on attenuated viruses.
(e) Retroviruses have been used successfully as vectors for gene therapy.

(6.10) Are the following statements true or false?

(a) Prevention of virus infection by drug treatment is preferable to vaccination.

(b) Any of the stages of viral replication can be a potential target for antiviral chemotherapy.

(c) The majority of antiviral drugs in use are nucleoside/nucleotide analogues.

(d) Ribavirin is active only against influenza A viruses.

(e) The frequency with which viruses become resistant to antiviral drugs depends on the drug and the virus concerned.

Answers to Self-Assessment Questions are given in Appendix 3.

Chapter 7

Pathogenesis

Pathogenicity, which is the capacity of one organism to cause disease in another, is a complex and variable phenomenon. For one thing, it is rather difficult to define. At the simplest level there is the question, what is disease? An all-embracing definition would be that it is a departure from the normal physiological parameters of an organism. This could range from a transient and very minor condition such as a slightly elevated temperature or rather subjective feelings of lethargy to chronic pathologic conditions which eventually result in death. Any of these conditions may result from a tremendous number of internal or external sources. However, there is rarely one single factor which 'causes' a disease. Most disease states are multi-factorial at one level or another.

Considering viral diseases only, there are two components involved: the direct effects of virus replication and the effects of bodily responses to the infection. The course of any virus infection is determined by a delicate and dynamic balance between the host and the virus, as described in Chapter 6. The extent and severity of viral pathogenesis is determined similarly. In some virus infections, most of the pathologic symptoms observed are attributable not to viral replication, but to the side-effects of the immune response. Inflammation, fever, headaches and skin rashes are not usually caused by viruses themselves, but by the cells of the immune system due to the release of potent chemicals such as interferons and interleukins. In the most extreme cases, it is possible that none of the pathologic effects of certain diseases are caused directly by the virus, excepting that its presence stimulates the activation of the immune system.

Viral pathogenesis is an abnormal and fairly rare situation. The vast majority of virus infections are silent and do not result in outward signs of disease. It is sometimes said that viruses would disappear if they killed their hosts. This is not necessarily true. It is perfectly possible to envisage viruses with a hit-and-run strategy, moving quickly from one dying host to the next and relying on continuing circulation for their survival. Nevertheless, there is a clear tendency for viruses not to injure their hosts if possible. A good example of this is the rabies virus. The symptoms of human rabies virus infections are truly dreadful, but thankfully very rare. However, in its normal hosts (e.g. foxes), rabies virus infection produces a much milder disease which does not usually kill the animal. Man is an unnatural, dead-end host for this virus and the severity of human rabies is extreme as the condition is rare. Ideally, a virus would not even provoke an immune response from its host, or at least be able to hide to avoid the effects. Herpesviruses and some retroviruses have evolved complex lifestyles which enable them to get close to this objective. Of course, fatal infections such as rabies and AIDS always grab the headlines. Much less effort has been devoted to isolating and studying the myriad viruses which have not (yet) caused well-defined diseases in man, his domestic animals or economically valuable crop plants.

In the past few decades, molecular genetic analysis has contributed enormously to our understanding of viral pathogenesis. Nucleotide sequencing and site-directed mutagenesis have been used to explore molecular determinants of virulence in many different viruses. Specific sequences and structures which are found only in disease-causing strains of virus and not in closely related attenuated or avirulent strains have been identified. Sequence analysis has also led to the identification of T-cell and B-cell epitopes on virus proteins responsible for their recognition by the immune system. Unfortunately, these advances do not automatically lead to an understanding of the mechanisms responsible for pathogenicity.

Unlike the rest of this book, this chapter is specifically about viruses which cause disease in animals. It does not discuss bacteriophages, for which the terms 'pathogenesis' and 'disease' are hardly appropriate, nor viruses which cause disease in plants, which has already been considered in Chapter 6. Three major aspects of viral pathogenesis are considered: direct cell damage resulting from virus replication, damage resulting from immune activation or suppression, and cell transformation caused by viruses.

Mechanisms of Cellular Injury

Virus infection results in a number of changes which are detectable by visual or biochemical examination of infected cells. These changes result from the production of viral proteins and nucleic acids, but also from alterations to the biosynthetic capabilities of cells. Intracellular parasitism by viruses sequesters cellular apparatus such as ribosomes and raw materials which would normally be devoted to synthesizing molecules required by the cell. Eukaryotic cells must carry out constant macromolecular synthesis, whether they are growing and dividing or in a state of quiescence. A growing cell clearly needs to manufacture more proteins, more nucleic acids and more of all of its myriad components to increase its size before dividing. However, there is a much more fundamental requirement for such activity. The function of all cells is regulated by controlled expression of their genetic information and the subsequent degradation of the molecules produced. Such control relies on a delicate and dynamic balance between synthesis and decay which determines the intracellular levels of all the important molecules in the cell. This is particularly true of the control of the cell cycle, which determines the behaviour of dividing cells (see Cellular Transformation by DNA Viruses). Any interruption of the synthesis of these vital messages, or the addition of any virus-encoded protein which interferes with this process, has severe consequences for the cell. Even non-dividing cells must constantly synthesize a variety of macromolecules which require continual replacement. This activity is known as 'housekeeping' and all cells contain many thousands of these housekeeping genes. In general terms, a number of common phenotypic changes can be recognized in virus-infected cells. These changes are often referred to as the **cytopathic effects (c.p.e.)** of a virus, and include:

- *Altered shape*: Adherent cells which are normally attached to other cells (*in vivo*) or an artificial substrate (*in vitro*) may assume a rounded shape different from their normal flattened appearance. The extended 'processes' (extensions of the cell surface resembling tendrils) involved in attachment or mobility are withdrawn into the cell
- *Detachment from the substrate*: For adherent cells, this is the stage of cell damage which follows that above. Both of these effects are caused by partial degradation or disruption of the cytoskeleton which is normally responsible for maintaining the shape of the cell
- *Lysis*: This is the most extreme case, where the entire cell breaks down. Membrane integrity is lost and the cell may swell due to the absorption of extracellular fluid and finally, break open. This is an extreme case of cell damage and it is important to realize that not all viruses induce this effect, although they may cause other cytopathic effects. Lysis is beneficial to a virus in that it provides an obvious method of releasing new virus particles from an infected cell. However, there are alternative ways of achieving this, such as release by budding (see Chapter 4)
- *Membrane fusion*: The membranes of adjacent cells fuse, resulting in a mass of cytoplasm containing more than one nucleus, known as a syncytium, or, depending on the number of cells which merge, a giant cell. Fused cells are short lived and subsequently lyse – apart from direct effects of the virus, they cannot tolerate more than one non-synchronized nucleus per cell
- *Membrane permeability*: A number of viruses cause an increase in membrane permeability, allowing an influx of extracellular ions such as sodium. Translation of some viral mRNAs is resistant to high concentrations of sodium ions, permitting the expression of virus genes at the expense of cellular messages
- *Inclusion bodies*: These are areas of the cell where virus components have accumulated. They are frequently sites of virus assembly, and some cellular inclusions consist of crystalline arrays of virus particles. It is not clear how these structures damage the cell, but they are frequently associated with viruses which cause cell lysis, such as herpesviruses and rabies virus
- *Apoptosis*: Virus infection may trigger **apoptosis** ('programmed cell death'), a highly specific mechanism involved in the normal growth and development of organisms (see Viruses and Immunodeficiency).

In some cases, a great deal of detail is known about the molecular mechanisms of cell injury. A number of viruses which cause cell lysis exhibit a phenomenon known as **shutoff** early in infection. Shutoff is the sudden and dramatic cessation of most host cell macromolecular synthesis. In poliovirus-infected cells, this is the result of production of the virus 2A protein. This molecule is a protease which cleaves the p220 component of eIF–4F, a complex of proteins required for cap-dependent translation of messenger RNAs by ribosomes. Since poliovirus RNA does not have a 5′ methylated cap but is modified by the addition of the VPg protein, virus RNA continues to be translated. In poliovirus-infected cells, the dissociation of mRNAs and polyribosomes from the cytoskeleton can be observed and this is the reason for the inability of the cell to translate its own messages. A few hours after translation ceases, lysis of the cell occurs.

In other cases, shutoff results from a different molecular mechanism. For many viruses, the sequence of events which occurs is not known. In the case of adenoviruses, the penton protein (part of the virus capsid) has a toxic effect on cells. Although its precise action on cells is not known, addition of purified penton protein to cultured cells results in their rapid death. Toxin production by pathogenic bacteria is a common phenomenon, but this is the only well-established case of a virus-encoded molecule with a toxin-like action. However, some of the normal contents of cells released on lysis may have toxic effects on other cells, and antigens which are not recognized as 'self' by the body (e.g. nuclear proteins) may result in immune activation and inflammation. The adenovirus E3-11.6K protein is synthesized in small amounts from the E3 promoter at early stages of infection and in large amounts from the major late promoter at late stages of infection (see Chapter 5). It has recently been shown that E3-11.6K is required for the lysis of adenovirus-infected cells and the release of virus particles from the nucleus.

Membrane fusion is the result of virus-encoded proteins required for infection of cells (see Chapter 4), typically, the glycoproteins of enveloped viruses. One of the best known examples of such a protein comes from Sendai virus (a paramyxovirus), which has been used to induce cell fusion during the production of monoclonal antibodies (Chapter 1). Herpes simplex virus (HSV) encodes at least nine glycoproteins (Table 7.1). A number of these proteins are involved in fusion

Table 7.1 The glycoproteins of herpes simplex virus

Protein	Known functions
gB	Active in viral entry; major neutralization antigen.
gC	Active in attachment.
gD	Active in post-attachment viral entry.
gE	Fc receptor; required for basolateral spread of virus in polarized cells.
gG	Active in viral entry, egress and cell-to-cell spread.
gH	Transported to plasma membrane as a complex with gL; active in viral entry, egress and cell-to-cell spread.
gJ	Predicted from nucleotide sequence, protein not yet found.
gK	Required for efficient exocytosis.
gL	Forms complexes with gH promoting transport of gH to plasma membrane; required for viral entry mediated by gH.
gM	?

of the virus envelope with the cell membrane and also in cell penetration. Production of fused syncytia is a common feature of HSV infection.

Another virus which causes cell fusion is human immunodeficiency virus (HIV). Infection of $CD4^+$ cells with some but not all isolates of HIV causes cell–cell fusion and the production of syncytia or giant cells (Figure 7.1). The protein responsible for this is the transmembrane envelope glycoprotein of the virus and the domain near the amino terminus responsible for this fusogenic activity has been identified by molecular genetic analysis. Because HIV infects $CD4^+$ cells and it is the reduction in the number of these crucial cells of the immune system which is the most obvious defect in AIDS, it was initially believed that direct killing of these cells by the virus was the basis for the pathogenesis of AIDS. Although direct cell fusion and killing by the virus undoubtedly occurs *in vivo*, it is now believed that the pathogenesis of AIDS is

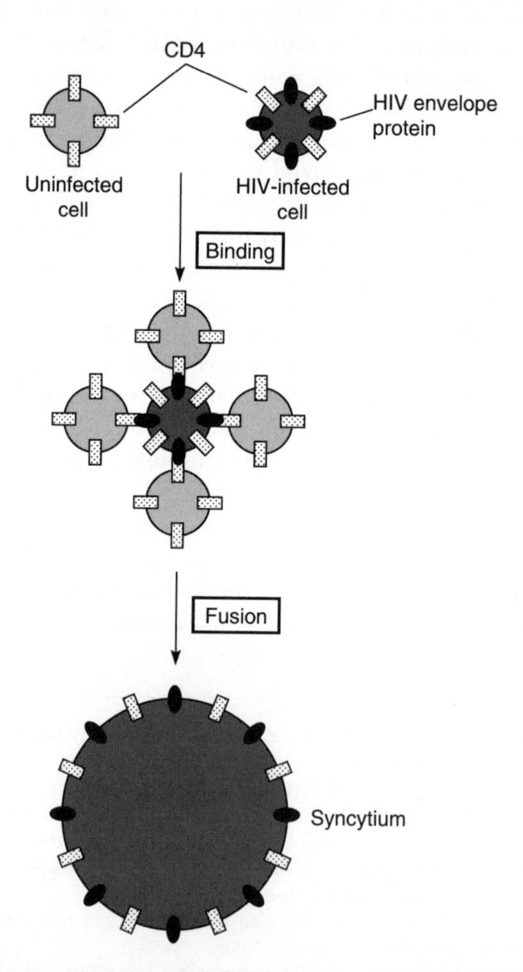

Figure 7.1 Mechanism of HIV-induced cell fusion. The virus envelope glycoprotein, whose role on virus particles is in receptor-binding and membrane fusion, is expressed on the surface of infected cells. Uninfected $CD4^+$ cells coming into contact with these infected cells are fused together to form a multi-nucleate **syncytium**.

considerably more complex (see Viruses and Immunodeficiency). Many animal retroviruses also cause cell killing and in most cases, it appears that the envelope protein of the virus is required, although there may be more than one mechanism involved.

Viruses and Immunodeficiency

At least two groups of viruses, herpesviruses and retroviruses, directly infect the cells of the immune system. This has important consequences for the outcome of the infection and for the immune system of the host. Herpes simplex virus (HSV) establishes a systemic infection, spreading via the bloodstream in association with platelets, but does not show particular tropism for cells of the immune system. However, herpes saimirii and Marek's disease virus are herpesviruses which cause lymphoproliferative diseases (but not clonal tumours) in monkeys and chickens, respectively. The most recently discovered human herpesviruses, human herpesvirus-6 (HHV-6), HHV-7 and HHV-8 all infect lymphocytes (Chapter 8).

Epstein–Barr virus (EBV) infection of B-cells leads to their immortalization and proliferation, resulting in 'glandular fever' or mononucleosis, a debilitating but benign condition. EBV was first identified in a lymphoblastoid cell line derived from Burkitt's lymphoma and, in rare instances, EBV infection may lead to the formation of a malignant tumour (see Cellular Transformation by DNA Viruses). While some herpesviruses such as HSV are notably cytopathic, most of the lymphotropic herpesviruses do not cause a significant degree of cellular injury. However, infection of the delicate cells of the immune system may perturb their normal function. Because the immune system is internally regulated by complex networks of interlinking signals, relatively small changes in cellular function can result in its collapse. Alteration of the normal pattern of production of cytokines could have profound effects on immune function. The *trans*-regulatory proteins involved in the control of herpesvirus gene expression may also affect the transcription of cellular genes. Therefore, the effects of herpesviruses on immune cells are more complex than just cell killing.

Retroviruses cause a variety of pathogenic conditions including paralysis, arthritis, anaemia and malignant cellular transformation. A significant number of retroviruses infect the cells of the immune system. Although these infections may lead to a diverse array of diseases and haematopoetic abnormalities such as anaemia and lymphoproliferation, the most commonly recognized consequence of retrovirus infection is the formation of lymphoid tumours (see Cellular Transformation by DNA Viruses). However, some degree of immunodeficiency, ranging from very mild to quite severe, is a common consequence of the interference with the immune system resulting from the presence of a lymphoid or myeloid tumour.

In the last decade, the most prominent aspect of virus-induced immunodeficiency has been acquired immunodeficiency syndrome (AIDS), which is a consequence of infection with human immunodeficiency virus (HIV), a member of the lentivirus group of the retroviruses. There are a number of similar lentiviruses which cause immunodeficiency diseases in primates and cats as well as man. Unlike infection by other types of retrovirus, HIV infection does not directly result in the formation of tumours. Some tumours such as B-cell lymphomas are sometimes seen in AIDS patients but are probably a consequence of the lack of immune surveillance which is responsible for the destruction of tumours in healthy individuals. The clinical course of AIDS is long and very variable. A great number of different abnormalities of the immune system are seen in AIDS (Table 7.2). As a result of the biology of lentivirus infections, the pathogenesis of AIDS is highly complex. Some of the relevant factors to consider are:

- *Latency*: Lentiviruses do not show true latency (unlike herpesviruses or phage λ), but they do have the capacity to control the expression of their genomes by means of the virus-encoded *trans*-acting regulatory proteins tat and rev (Chapter 5)
- *Antigenic variation*: New antigenic variants of lentiviruses continually arise in infected hosts because of the inaccuracy of reverse transcriptase, which results in a type of antigenic drift. In caprine arthritis encephalitis virus (CAEV) infection of goats, each new antigenic variant has been shown to cause a flare up of inflammatory arthritis. Although HIV does not cause arthritis directly, a similar mechanism may contribute to the decline of immune system in AIDS (see below)
- *The 'Trojan horse' mechanism*: HIV and other lentiviruses escape immune recognition by infecting monocytes, which not only serve as a reservoir for the virus but also help them to spread to other tissues (e.g. the brain) and to other hosts.

It is not clear how much of the pathology of AIDS is caused by the virus and how much is caused by the immune system. There are numerous models that have been suggested to explain how HIV causes immunodeficiency. These mechanisms are not mutually exclusive and indeed it is probable that the underlying loss of $CD4^+$ cells in AIDS is multifactorial.

Table 7.2 Immune abnormalities in HIV infection

Altered cytokine expression
Decreased CTL and NK cell function
Decreased humoral and proliferative response to antigens and mitogens
Decreased MHC-II expression
Decreased monocyte chemotaxis
Depletion of $CD4^+$ cells
Impaired delayed type hypersensitivity (DTH) reactions
Lymphopaenia
Polyclonal B-cell activation

Direct cell killing

This was the earliest mechanism suggested, based on the behaviour of certain laboratory isolates of HIV. Cell fusion resulting in syncytium formation is one of the major mechanisms of cell killing by HIV (Figure 7.1). However, different isolates of HIV vary considerably in the extent to which they promote the fusion of infected cells. Subsequent experiments suggested there may not be sufficient virus present in AIDS patients to account for all the damage seen, although killing of $CD4^+$ cells may contribute to the overall pathogenesis of AIDS. Recently, this hypothesis has been resurrected as a result of more accurate quantitation of virus load and replication kinetics in infected individuals (see below).

Indirect killing of HIV-infected cells

Indirect effects of infection, e.g. disturbances in cell biochemistry and lymphokine production may also affect the regulation of the immune system. However, the expression of virus antigens on the surface of infected cells leads to indirect killing by the immune system (NK/CTL/ADCC) – effectively a type of autoimmunity. The extent of this activity is dependent on the virus load and replication kinetics in infected individuals (see below).

Antigenic diversity

This theory proposes that the continual generation of new antigenic variants eventually swamps and overcomes the immune system, leading to its collapse. There is no doubt that new antigenic variants of HIV constantly arise during the long course of AIDS because of the low fidelity of reverse transcription (Chapter 3). It is envisaged that there might be a 'ratchet' effect, with each new variant contributing to the slight but irreversible decline in immune function as described in 'T-cell Anergy' and 'Apoptosis' below (Figure 7.2). Because of the way virus infections are handled by the immune system, it is probable that variation of T-cell epitopes on target proteins recognized by CTL are more important than B-cell epitopes which generate the antibody response to a foreign antigen (Chapter 6). A mathematical model has been constructed which simulates antigenic variation during the course of infection. When primed with known data about the state of immune system during HIV infection, it provides a startlingly accurate depiction of the course of AIDS.

T-cell anergy

Anergy is an immunologically unresponsive state in which lymphocytes are present but not functionally active. This is usually due to incomplete activation signals and may be an important regulatory mechanism in the immune system, e.g. tolerance of 'self' antigens. In AIDS, anergy could be induced due to HIV

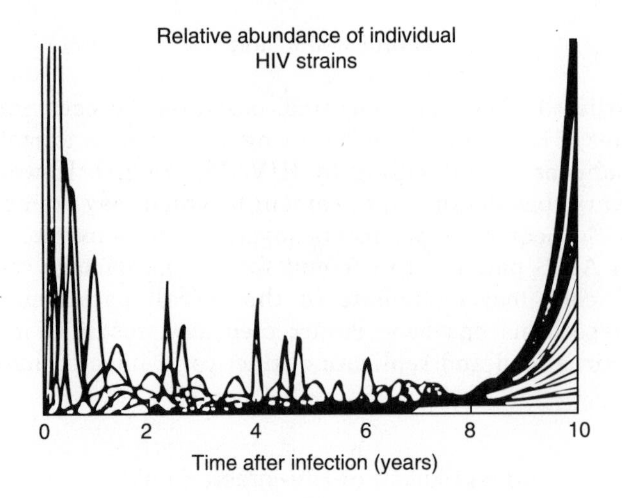

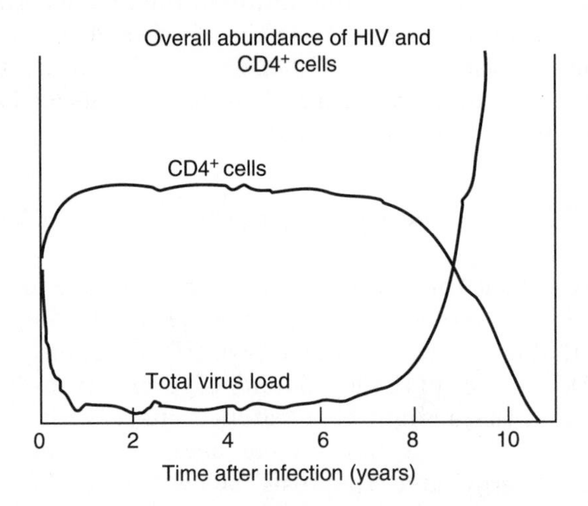

Figure 7.2 Antigenic diversity threshold theory of AIDS pathogenesis. Each line on the upper graph shows the (theoretical) relative abundance of an individual HIV strain (antigenic type) in a single case of AIDS. The lower graph shows the total virus load (i.e. total number of HIV strains) and number of $CD4^+$ cells predicted by a computer model.

infection, e.g. interference with cytokine expression. There is experimental *in vitro* evidence that gp120-CD4 interactions result in anergy due to interference with signal transduction. However, there is no strong evidence that this occurs *in vivo*.

Apoptosis

Apoptosis or 'programmed cell death' is believed to be a normal part of T-cell maturation, e.g. the elimination of self-responsive clones. A number of unrelated viruses are known to induce apoptosis in infected cells. Like T-cell anergy, apoptosis could potentially be induced in large numbers of uninfected cells by factors released from a much smaller number of HIV-infected cells. In addition to clonal deletion as a normal part of the evolution of the T-cell repertoire, apoptosis may be induced following T-cell activation as a negative regulatory mechanism to control the strength and duration of the immune response. This is relevant, since HIV infection of T-cells induces an activated phenotype, e.g. surface expression of CD45 and HLA-DR markers, which suggests that these cells may be inevitably doomed due to activation of the apoptosis pathway. Because HIV establishes a persistent infection, it is by no means clear that apoptosis has an entirely negative effect – induction of cell death may well limit virus production and slow down the course of infection. Several mechanisms have been suggested by which apoptosis of uninfected cells could be induced:

- Crosslinking of CD4 by gp120
- Continuous immune activation (see 'Antigenic diversity' above)
- Defective activation signals: induction of an activated phenotype without appropriate co-signals from antigen presenting cells (e.g. via CD28 or cytokine production) may induce apoptosis immediately, or when cells subsequently respond to their corresponding antigen
- Activation of T-cells by **superantigens** (below).

Superantigens

Superantigens are molecules which short-circuit the immune system, resulting in massive activation of T-cells rather than the usual, carefully controlled response to foreign antigens. It is believed that they do this by binding to both the variable region of the β-chain of the T-cell receptor (Vβ) and to MHC class II molecules, cross-linking them in a non-specific way (Figure 7.3). This results in polyclonal T-cell activation rather than the usual situation where only the few clones of T-cells responsive to a particular antigen presented by the MHC class II molecule are activated. The over-response of the immune system produced results in autoimmunity as whole families of T-cells which bind superantigens are activated, and immunosuppression as the activated cells are killed by other activated T-cells or undergo apoptosis.

It has been reported that in some AIDS patients, certain clones of T-cells bearing particular Vβ T-cell receptor rearrangements are depleted or absent. This is precisely what would be expected if some clones of cells were being eliminated by the presence of a superantigen. However, unlike other retroviruses (e.g. mouse mammary tumour virus (MMTV) and the murine leukaemia virus (MLV) responsible for murine acquired immunodeficiency syndrome (MAIDS)) no superantigen has been conclusively identified in HIV, despite intensive investigation. Thus the practical relevance of superantigens in AIDS remains in some doubt. However, it is possible that exposure to superantigens produced by opportunistic infection(s) might play an important role in AIDS.

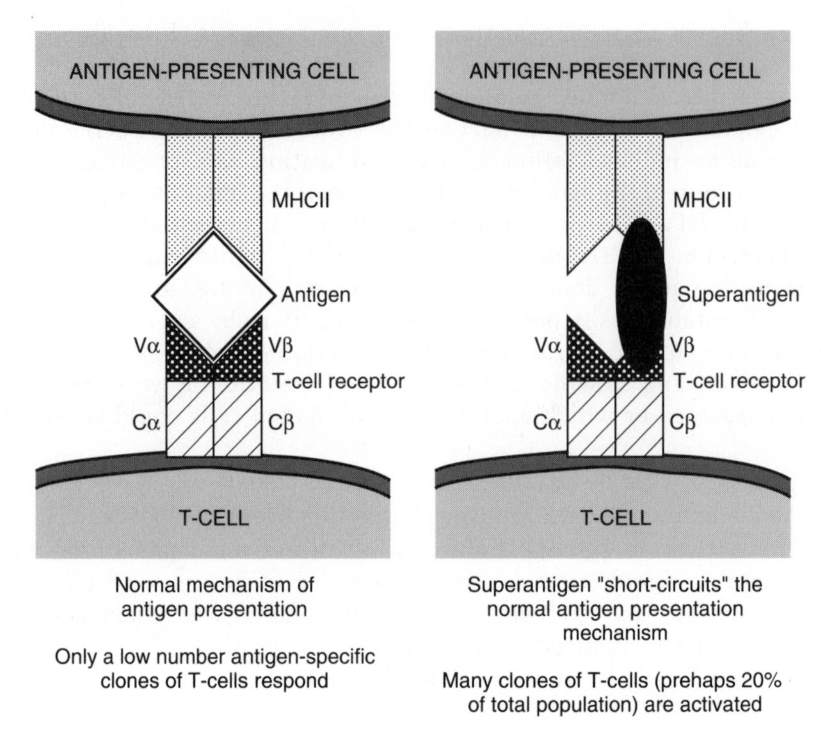

Figure 7.3 Mechanism of action of superantigens.

T_H1/T_H2 imbalance

Regulation of the immune system depends on a complex network of cells, but central to the process is the role of $CD4^+$ T-helper (T_H) cells. Current immunological theory suggests that there are two types of these, T_H1 cells which promote the cell mediated response and T_H2 cells which promote the humoral response (Figure 7.4). This theory suggests that early in HIV infection, T_H1–responsive T-cells predominate and are effective in controlling (but not eliminating) the virus. At some point, a (relative) loss of the T_H1 response occurs and T_H2 HIV-responsive cells predominate. It has recently been reported that at least some virus variants can inhibit the CTL response to HIV. The hypothesis is therefore that the T_H2–dominated humoral response is not effective at maintaining HIV replication at a low level and the virus load builds up, resulting in AIDS. Although this is largely a theoretical proposal which has not yet been proved, this thinking is shaping our understanding of the immune response to many different pathogens, not just HIV.

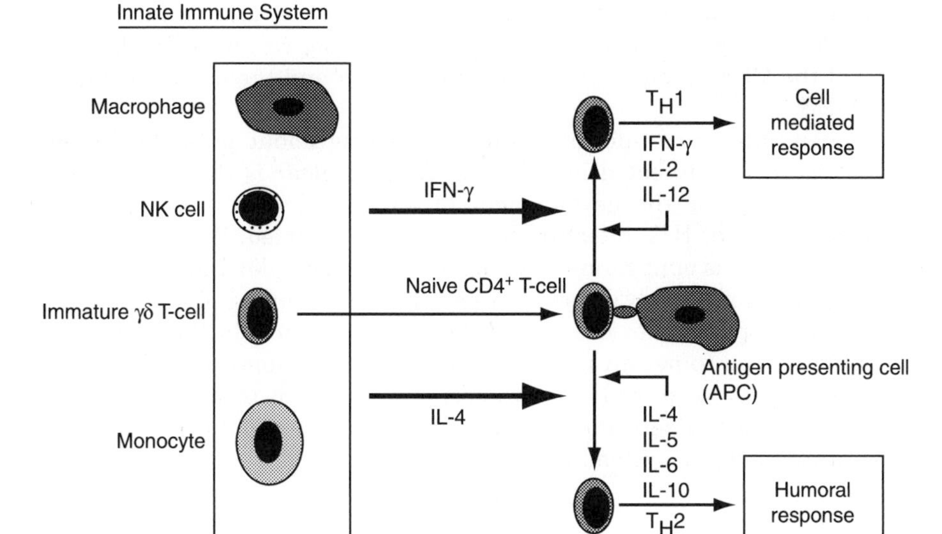

Figure 7.4 Regulation of the cell-mediated and humoral immune responses. IFN, interferon; IL, interleukin;, T_H, T-helper cell.

Virus load and replication kinetics

Recent reports involving accurate quantitation of the amount of virus in infected individuals have revealed that much higher virus loads are present than was originally measured by less sensitive techniques. Using quantitative polymerase chain reaction (PCR) methods to accurately measure the amount of virus present and determine how these levels change when patients are treated with drugs which inhibit HIV replication, it has been shown that:

- Continuous and highly productive replication of HIV occurs in all infected individuals, although the rates of virus production vary by up to 70–fold in different individuals
- The average half-life of an HIV particle/infected cell *in vivo* is 2.1 days
- Up to 2×10^9 HIV particles are produced each day
- An average of 2.6×10^9 new $CD4^+$ cells are produced each day.

Thus contrary to what was previously believed, there is a very dynamic situation in HIV-infected subjects involving continuous infection, destruction and replacement of $CD4^+$ cells. Billions of new $CD4^+$ cells are produced, infected and killed each day. These data suggest a return to cellular killing (although predominantly immune-mediated rather than virus-mediated) as a direct cause of the $CD4^+$ cell decline in AIDS. For reasons which are not yet clear, AIDS is a (marathon) race between virus production, destruction and cellular regeneration which, after many years, most individuals lose, resulting in the absolute decline of the CD4 segment of the immune system and the development of full-blown AIDS.

These new ideas are informing future thinking about possible therapeutic intervention in HIV-infected individuals. What is clear is that the presence of HIV is necessary for the development of AIDS and that it is vital that the worldwide spread of HIV infection is halted and reversed. Work on developing anti-HIV vaccines is continuing but because of the complex biology of the virus, is proving to be formidably difficult. A better understanding of the pathogenesis of AIDS and, in particular, the role of the immune system in the early stages of the disease is vital to permit the development of more appropriate therapies for AIDS. This might well involve a two-pronged approach incorporating both an attack on the virus itself and therapeutic immunomodulation to prevent the decline of the immune system.

Virus-Related Diseases

There are a number of human disease syndromes for which virus infections are believed to be a necessary prerequisite. In some instances, the link between a particular virus and a pathological condition is well established, but it is clear that the pathogenesis of the disease is complex and also involves the immune system of the host. In other cases, the pathogenic involvement of a particular virus is less certain, and in a few instances, rather speculative.

Although the incidence of measles virus infection has been reduced sharply by vaccination (Chapter 6), measles still causes hundreds of thousands of deaths each year. The normal course of measles virus infection is an acute febrile illness during which the virus spreads throughout the body, infecting many tissues. The vast majority of people spontaneously recover from the disease without any lasting harm. In rare cases (about 1 in 2000), measles may progress to a severe encephalitis. This is still an acute condition which either regresses or kills the patient within a few weeks. There is however, another, much rarer late consequence of measles virus infection which occurs many months or years after initial infection of the host. This is the condition known as subacute sclerosing panencephalitis (SSPE). Evidence of prior measles virus infection (antibodies or direct detection of the virus) is found in all patients with SSPE, whether they can recall having a symptomatic case of measles or not. In about 1 in 300 000 cases of measles, the virus is not cleared from the body by the immune system, but establishes a persistent infection in the CNS. In this condition, virus replication continues at a low level, but defects in the envelope genes prevent the production of extracellular infectious virus particles. The lack of envelope protein production causes the failure of the immune system to recognize and eliminate infected cells. However, the virus is able to spread directly from cell to cell, bypassing the usual route of infection. It is not known to what extent damage to the cells of the brain is caused directly by virus replication, or whether there is any contribution by the immune system to the pathogenesis of SSPE. Vaccination against measles virus and the prevention of primary infection should ultimately eliminate this condition.

Another well-established case where the immune system is implicated in pathogenesis concerns dengue virus infections. Dengue virus is a flavivirus which is transmitted from one human host to another via mosquitoes. The primary infection may be asymptomatic, or may result in dengue fever. Dengue fever is normally a self-limited illness from which patients recover after 7–10 days without further complications. Following primary infections, patients carry antibodies to the virus. Unfortunately, there are four serotypes of dengue virus (DEN-1, -2, -3 and -4) and the presence of antibody directed against one type does not give cross-protection against the other three; worse still is the fact that antibodies can enhance the infection of peripheral blood mononuclear cells by Fc-receptor mediated uptake of antibody-coated dengue virus particles (see Chapter 4). In a few cases, the consequences of dengue virus infection are much more severe than the usual fever. Dengue haemorrhagic fever (DHF) is a life-threatening disease. In the most extreme cases, so much internal haemorrhaging occurs that hypovolemic shock (dengue shock syndrome, DSS) occurs. DSS is frequently fatal. The cause of shock in dengue and other haemorrhagic fevers is partly due to the virus, but largely due to immune-mediated damage of virus-infected cells (Figure 7.5). DHF and DSS following primary dengue virus infections occur in approximately 1 in 14 000 and 1 in 500 patients respectively. However, after secondary dengue virus infections, the incidence of DHF is 1 in 90 and DSS 1 in 50, since cross-reactive but non-neutralizing antibodies to the virus are now present. These figures show the problems of cross-infection with different serotypes of dengue virus, and the difficulties which must be faced in developing a safe vaccine against the virus. Dengue virus is discussed further later in this chapter (New and Emergent Viruses).

Another instance where virus vaccines have resulted in increased pathology rather than the prevention of disease is the occurrence of post-vaccination Reye's syndrome. Reye's syndrome is a neurological condition involving acute cerebral oedema and occurs almost exclusively in children. It is well known as a rare post-infection complication of a number of different viruses, but most commonly influenza virus and VZV (chicken pox). Symptoms include frequent vomiting, painful headaches, behavioural changes, extreme tiredness and disorientation. The chances of contracting Reye's syndrome are increased if aspirin is administered during the initial illness. The basis for the pathogenesis of this condition is completely unknown, but some of the most unfortunate cases have followed the administration of experimental influenza virus vaccines.

Guillain–Barré syndrome is a similarly mysterious condition in which demyelination of nerves results in partial paralysis and muscle weakness. The onset of Guillain–Barré syndrome usually follows an acute 'virus-like' infection, but no single agent has ever been firmly associated with this condition. Kawasaki disease is similar to Reye's syndrome in that it occurs in children, but distinct

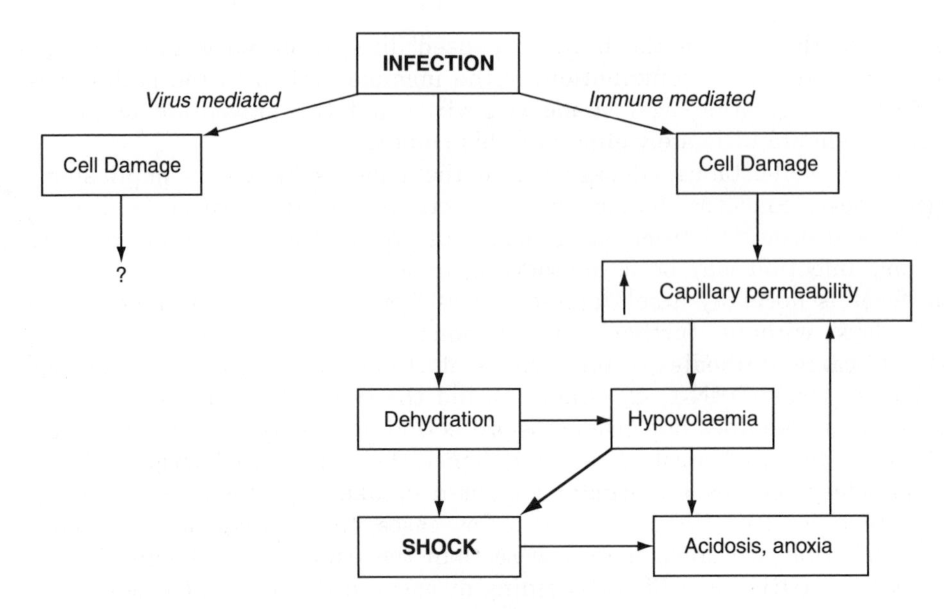

Figure 7.5 Causes of shock in haemorrhagic fevers.

in that it results in arthritis and/or serious damage to the heart. Like Guillain–Barré syndrome, Kawasaki disease appears to follow acute infections. The disease itself is not infectious but does appear to occurs in epidemics, which suggests an infectious agent as the cause. A large number of bacterial and viral pathogens have been suggested to be associated with the induction of Kawasaki disease, but once again the underlying cause of the pathology is unknown. It would appear that acute infection rather than a particular pathogen is responsible for the onset of these diseases.

In recent years, there has been a search for an agent responsible for a newly diagnosed disease called chronic fatigue syndrome (CFS) or myalgic encephalomyelitis (ME). Unlike the other conditions described above, CFS is a rather ill-defined disease and is not recognized by all physicians. Recent research has discounted the earlier idea that EBV might cause CFS but a variety of other possible viral causes including other herpes viruses, enteroviruses and retroviruses have also been suggested. Although it is attractive to believe that a virus might be responsible for this condition this has not been proved. All these various conditions and syndromes illustrate the complexity of viral pathogenesis and show that the direct effects of virus replication and self-inflicted damage resulting from poor control of the immune system are sometimes hard to differentiate.

Cellular Transformation by Viruses

Transformation is a change in the morphological, biochemical, or growth parameters of a cell. Transformation may or may not result in cells able to produce tumours in experimental animals, which is properly known as neoplastic transformation. Therefore, transformed cells do not automatically result in the development of 'cancer'. Carcinogenesis (or more properly, oncogenesis) is a complex, multi-step process in which cellular transformation may be only the first, although essential, step along the way. Transformed cells have an altered phenotype, which is displayed as one (or more) of the following characteristics:

- *Loss of anchorage dependence*: Normal (i.e. non-transformed) adherent cells such as fibroblasts or epithelial cells require a surface to which they can adhere. In the body, this requirement is supplied by adjacent cells or structures. *In vitro*, it is met by the glass or plastic vessels in which the cells are cultivated. Some transformed cells lose the ability to adhere to solid surfaces and float free (or in clumps) in the culture medium without loss of viability
- *Loss of contact inhibition*: Normal adherent cells in culture divide and grow until they have coated all the available surface for attachment. At this point, when adjacent cells are touching each other, cell division stops – the cells do not continue to grow and pile up on top of one another. Many transformed cells have lost this characteristic. Single transformed cells in a culture dish become visible as small thickened areas of growth called 'transformed foci' – clones of cells all derived from a single original cell
- *Colony formation in semi-solid media*: Most normal cells (both adherent and non-adherent cells such as lymphocytes) will not grow in media that are partially solid owing to the addition of substances such as agarose or hydroxymethyl cellulose. However, many transformed cells will grow under these conditions, forming colonies since movement of the cells is restricted by the medium
- *Decreased requirements for growth factors*: All cells require multiple factors for growth. In a broad sense, these include compounds such as ions, vitamins, and hormones which cannot be manufactured by the cell. More specifically, it includes regulatory peptides such as epidermal growth factor (EGF) and platelet-derived growth factor (PDGF) which regulate the growth of cells. These are potent molecules which have powerful effects on cell growth. Some transformed cells may have decreased or may even have lost their requirement for particular factors. The production by a cell of a growth factor which is required for its own growth is known as **autocrine** stimulation and is one route by which cells may be transformed.

Cell transformation is a single-hit process, i.e. a single virus transforms a single cell (c.f. oncogenesis, i.e. the formation of tumours, which is a multi-step process). All or part of the virus genome persists in the transformed cell and is usually (but not always) integrated into the host cell chromatin. Transformation is usually accompanied by continued expression of a limited repertoire of virus genes, or rarely by productive infection. Virus genomes found in transformed cells are frequently replication-defective and contain substantial deletions.

Transformation is mediated by proteins encoded by **oncogenes**. These regulatory genes can be grouped in several ways, for example, by their origins, biochemical function or subcellular locations. Oncogenes are involved in the control of cell growth and are found in all cells. About 30 such genes have now been identified, and these normal cellular sequences are known as cellular oncogenes or *c-oncs*. Cell-transforming viruses may have RNA or DNA genomes, but all have at least a DNA stage in their replication cycle, i.e. the only RNA viruses directly capable of cell transformation are the retroviruses (Table 7.3). Certain retroviruses carry homologues of *c-oncs* derived originally from the cellular genes and known as *v-oncs*. In contrast, the oncogenes of cell-transforming DNA viruses are unique to the virus genome – there are no homologous sequences present in normal cells. Oncogenes can be grouped by their common biochemical functions:

- Extracellular growth factors (homologues of normal growth factors)
- Membrane receptors responsible for the capture of extracellular signals
- Tyrosine kinases associated with the inner surface of the cell membrane
- G proteins involved in signal transduction
- Cytoplasmic protein kinases
- Transcription factors (see Chapter 5)
- Nuclear hormone receptors.

The function of each class of oncogene product depends on their cellular location (Figure 7.6). Several classes of oncogenes are associated with the process of signal transduction, i.e. the transfer of information derived from the binding of extracellular ligands to cellular receptors to the nucleus (Figure 7.7). Many of the kinases in these groups have a common type of structure with conserved functional domains representing the hydrophobic transmembrane and hydrophilic intracellular kinase regions (Figure 7.8). These proteins are associated with the cell membranes or are present in the cytoplasm. Other classes of oncogenes located in the nucleus are normally involved with the control of the cell cycle (Figure 7.9). The products of these genes overcome the restriction between the G_1 and S phases of the cell cycle, which is the key control point in preventing uncontrolled cell division. Some viral oncogenes are not sufficient on their own to produce a fully transformed phenotype in cells. However, in some instances, they may cooperate with another oncogene of complementary function to produce a fully transformed phenotype, for example, the adenovirus E1A gene plus either the E1B gene or the cellular *ras* gene transforms NIH3T3 cells (a mouse fibroblast cell line). This further underlines the fact that oncogenesis is a complex, multi-step process.

Cell transformation by retroviruses

Not all retroviruses are capable of transforming cells, for example, the lentiviruses such as HIV do not transform cells, although they are cytopathic. The retroviruses

Table 7.3 Known oncogenes transduced by retroviruses

Oncogene	Protein function (cellular homologue)	Type of tumour
abl	Non-receptor tyrosine kinase, signal transduction	Pre-B-cell leukaemia/ sarcoma
crk	Adaptor protein, signal transduction	
*erb*A	Hormone receptor (thyroid hormone receptor)	
*erb*B	Tyrosine kinase, growth-factor receptor (EGF receptor)	Erythroleukaemia/ fibrosarcoma
ets	Nuclear protein, transcription factor	
eyk	Tyrosine kinase, growth-factor receptor	
fes	Non-receptor tyrosine kinase, signal transduction	Sarcoma
fgr	Non-receptor tyrosine kinase, signal transduction	Sarcoma
fms	Tyrosine kinase, growth-factor receptor (CSF-1 receptor)	Sarcoma
fos	Nuclear protein, transcription factor (AP-1 complex)	Osteosarcoma
fps	Non-receptor tyrosine kinase, signal transduction	Sarcoma
H-*ras*	G proteins (GTPase)	Sarcoma/ erythroleukaemia
jun	Nuclear protein, transcription factor (AP-1 complex)	Sarcoma
K-*ras*	G proteins (GTPase)	Sarcoma/ erythroleukaemia
kit	Tyrosine kinase, growth-factor receptor	Sarcoma
maf	Nuclear protein, transcription factor	
mil	Serine–threonine kinase, signal transduction	Myeloblastosis
mos	Serine–threonine kinase	Sarcoma
mpl	Tyrosine kinase, growth-factor receptor	Leukaemia
myb	Nuclear protein, transcription factor	Myeloblastosis
myc	Nuclear protein, transcription factor	Sarcoma/ myelocytoma/ carcinoma
qin	Nuclear protein, transcription factor (HNF-3 family)	
raf	Serine–threonine kinase, signal transduction	Sarcoma
rel	Nuclear protein, transcription factor	Reticuloendotheliosis
ros	Tyrosine kinase, growth-factor receptor	Sarcoma
sea	Tyrosine kinase, growth-factor receptor	Sarcoma/ leukaemia
sis	Growth factor (PDGF)	Sarcoma
ski	Nuclear protein, transcription factor	Carcinoma
src	Non-receptor tyrosine kinase, signal transduction	Sarcoma
yes	Non-receptor tyrosine kinase, signal transduction	Sarcoma

which can transform cells fall into three groups: transducing, *cis*-activating and *trans*-activating. The characteristics of these groups are given in Table 7.4.

If oncogenes are present in all cells, why does transformation occur as a result of virus infection? The reason is that oncogenes may become activated in one of

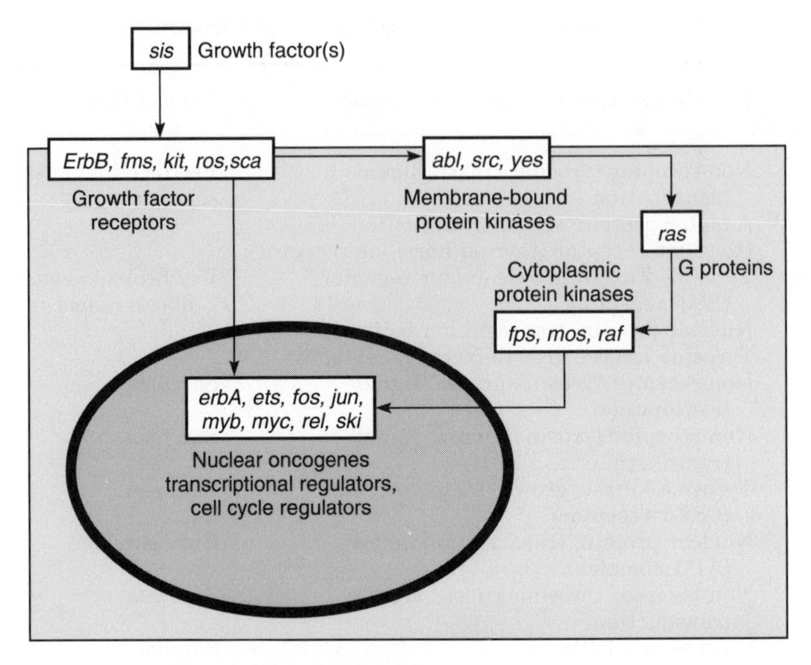

Figure 7.6 Subcellular location of oncoproteins.

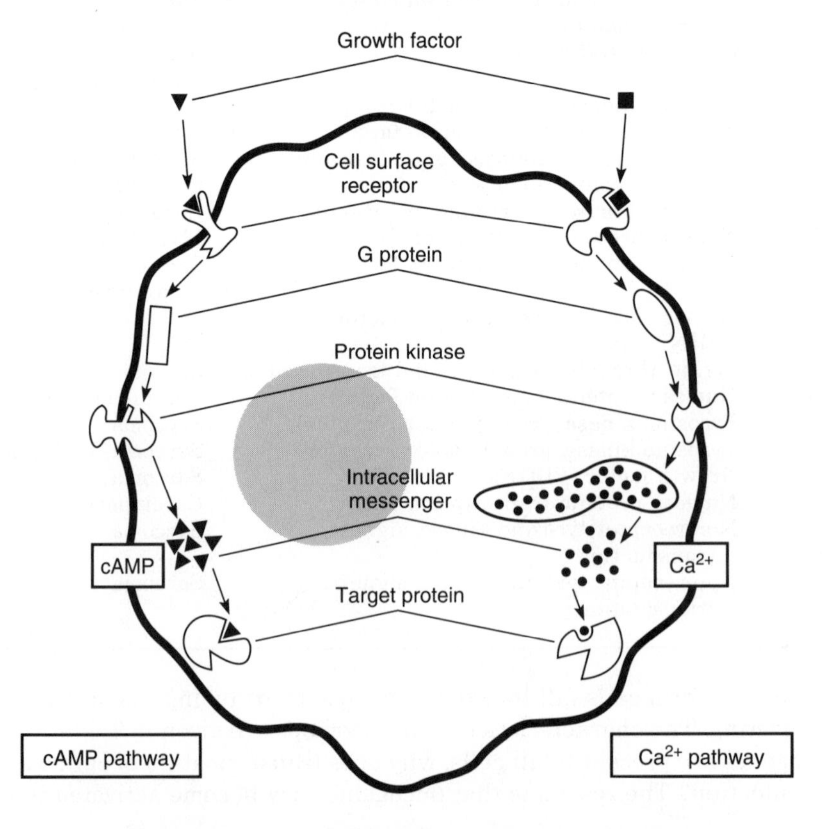

Figure 7.7 Cellular mechanism of signal transduction.

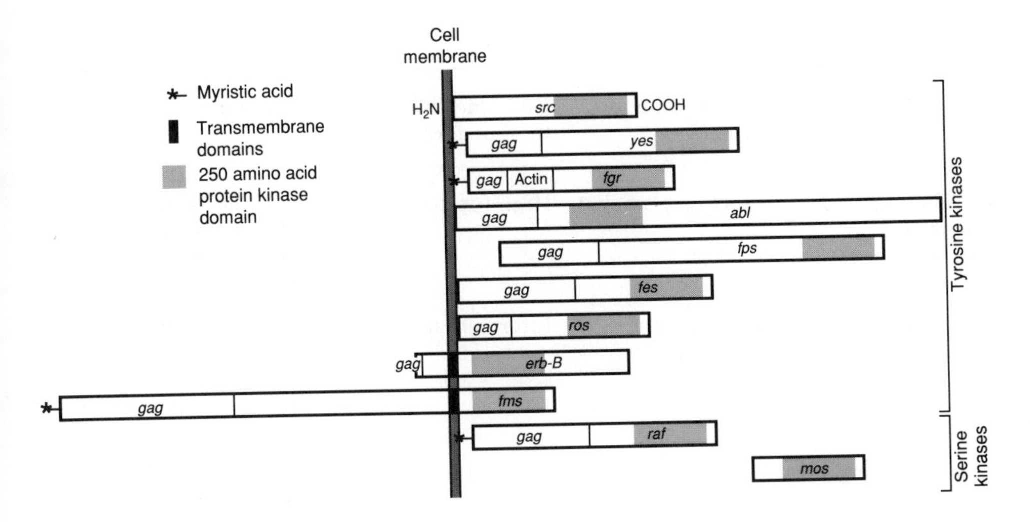

Figure 7.8 The family of retrovirus protein kinases involved in cell transformation. Many of these molecules are fusion proteins containing amino-terminal sequences derived from the viruses *gag* gene. Most of this type contain the fatty acid myristate which is added to the *N*-terminus of the protein after translation and which links the protein to the inner surface of the host cell cytoplasmic membrane. In a number of cases, it has been shown that this post-translational modification is essential to the transforming action of the protein.

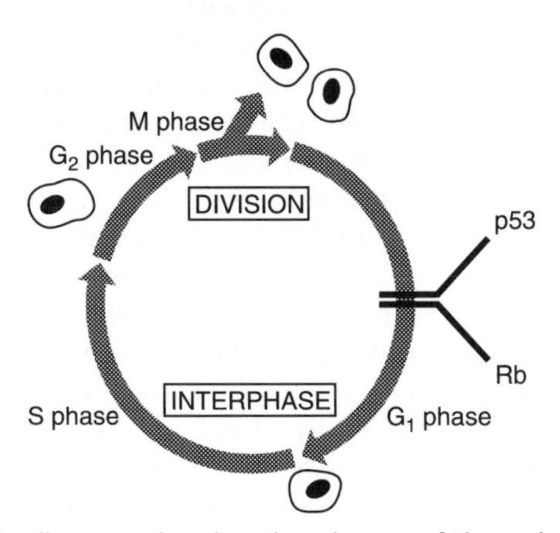

Figure 7.9 Schematic diagram showing the phases of the eukaryotic cell cycle.

two ways, either by subtle changes to the normal structure of the gene, or by interruption of the normal control of expression. The transforming genes of the acutely transforming retroviruses (*v-oncs*) are derived from and are highly homologous to *c-oncs* and are believed to have been transduced by viruses (Table 7.3). However, most *v-oncs* possess slight alterations from their *c-onc* progenitors. Many contain minor sequence alterations which alter the structure and the function of the oncoprotein produced. Others contain short deletions of part of the gene. Most oncoproteins from replication-defective, acutely transforming

Table 7.4 Cell-transforming retroviruses

Virus type	Time to tumour formation	Efficiency of tumour formation	Type of oncogene
Transducing (acutely transforming)	Short (e.g. weeks)	High (up to 100%)	*c-onc* transduced by virus, i.e. *v-onc* present in virus genome (usually replication defective)
cis-Activating (chronic transforming)	Intermediate (e.g. months)	Intermediate	*c-onc* in cell genome activated by provirus insertion – no oncogene present in virus genome (replication competent)
trans-Activating	Long (e.g. years)	Low (<1%)	Activation of cellular genes by *trans*-acting virus proteins (replication competent)

retroviruses are fusion proteins, containing additional sequences derived from virus genes, most commonly viral *gag* sequences at the amino terminus of the protein. These additional sequences may alter the function or the cellular localization of the protein and these abnormal attributes result in transformation.

Alternatively, viruses may result in abnormal expression of an unaltered oncoprotein. This might be either the overexpression of an oncogene under the control of a virus promoter rather than its normal promoter in the cell, or it may be the inappropriate temporal expression of an oncoprotein which disrupts the cell cycle. Chronic transforming retrovirus genomes do not contain oncogenes. These viruses activate *c-oncs* by a mechanism known as insertional activation. A provirus which integrates into the host cell genome close to a *c-onc* sequence may indirectly activate the expression of the gene in a way analogous to that in which *v-oncs* have been activated by transduction (Figure 7.10). This can occur if the provirus is integrated upstream of the *c-onc* gene, which might be expressed via a read-through transcript of the virus genome plus downstream sequences. However, insertional activation can also occur when a provirus integrates downstream of a *c-onc* sequence or upstream but in an inverted orientation. In these cases, activation results from enhancer elements in the virus promoter (see Chapter 5). These can act even if the provirus integrates at a distance of several kilobases from the *c-onc* gene. The best-known examples of this phenomenon occur in chickens, where insertion of avian leukosis virus (ALV) activates the *myc* gene, and in mice, where mouse mammary tumour virus (MMTV) insertion activates the *int* gene.

Transformation by the third class of retroviruses operates by quite a different mechanism. Human T-cell leukaemia virus (HTLV) and related animal viruses encode a transcriptional activator protein in the virus *tax* gene. The tax protein acts *in trans* to stimulate transcription from the virus LTR. It is believed that the

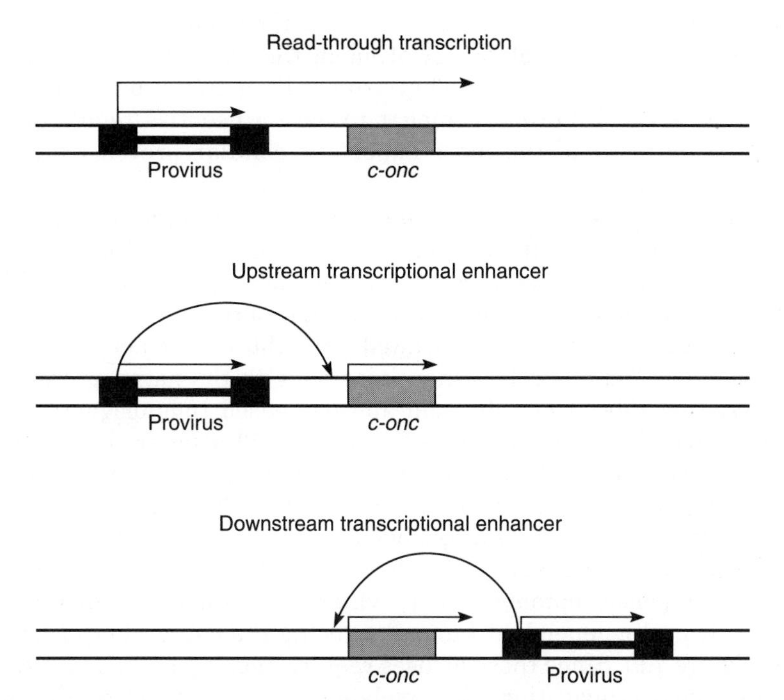

Figure 7.10 Mechanisms by which cellular oncogenes can be transcriptionally activated by retrovirus insertional mutagenesis.

protein also activates transcription of some cellular genes, notably those encoding the T-cell growth factor interleukin-2 (IL-2) and its receptor (IL-2R), by interacting with cellular transcription factors (Chapter 5). This could result in the formation of a cycle of positive feedback involving autocrine stimulation of T-cells by these two genes. However, HTLV oncogenesis, i.e. the formation of a leukaemic tumour, has a latent period of some 20–30 years. Therefore, cell transformation (which can be mimicked *in vitro*) and tumour formation (which cannot) are not one and the same – they are additional events required for the development of leukaemia. It is thought that chromosomal abnormalities which may occur in the population of HTLV-transformed cells are also required to produce a malignant tumour, although because of the difficulties of studying this lengthy process, this is not completely understood.

Cell transformation by DNA viruses

In contrast to the oncogenes of retroviruses, the transforming genes of DNA tumour viruses have no cellular counterparts. Several families of DNA viruses are capable of transforming cells (Table 7.5). In general terms, the functions of their oncoproteins are much less diverse than those encoded by retroviruses. They are mostly nuclear proteins involved in the control of DNA replication which directly affect the cell cycle. They achieve their effects by interacting with cellular proteins which normally appear to have a negative regulatory role in cell proliferation. Two of the most important cellular proteins involved are known as p53 and Rb.

p53 was originally discovered by virtue of the fact that it forms complexes with SV40 T-antigen. It is now known that it also interacts with other DNA virus oncoproteins, including those of adenoviruses and papillomaviruses. The gene encoding p53 is mutated or altered in the majority of tumours, implying that loss of the normal gene product is associated with the emergence of malignantly transformed cells. Tumour cells, when injected with the native protein *in vitro* show a decreased rate of cell division and decreased tumourigenicity *in vivo*. Transgenic mice which do not possess an intact p53 gene are developmentally normal but are susceptible to the formation of spontaneous tumours. Therefore, it is clear that p53 plays a central role in controlling the cell cycle. It is believed to be a tumour suppressor or 'anti-oncogene' and has been called 'the guardian of the genome'. p53 is a transcription factor which activates the expression of certain cellular genes, notably WAF1, which encodes a protein which is an inhibitor of G_1 cyclin-dependent kinases, causing the cell cycle to arrest at the G_1 phase (Figure 7.9). Since these viruses require ongoing cellular DNA replication for their own propagation, this explains why their transforming proteins target p53.

Rb was discovered when it was noticed that the gene which encodes this protein is always damaged or deleted in a tumour of the optic nerve known as retinoblastoma. Therefore, the normal function of this gene is also thought to be that of a tumour suppressor. The Rb protein forms complexes with a transcription factor called E2F. This factor is required for the transcription of adenovirus genes, but E2F is also involved in the transcription of cellular genes which drive quiescent cells into S phase. The formation of inactive E2F–Rb complexes thus has the same overall effect as the action of p53 – arrest of the cell cycle at G_1.

Table 7.5 Transforming proteins of DNA tumour viruses

Virus	Transforming protein(s)	Cellular target
Adenoviruses	E1A	Rb
	E1b	p53
Polyomaviruses (SV40)	T-antigen	p53, Rb
Papillomaviruses		
BPV-1	E5	PDGF receptor
	E6	p53
HPV-16, 18	E7	Rb

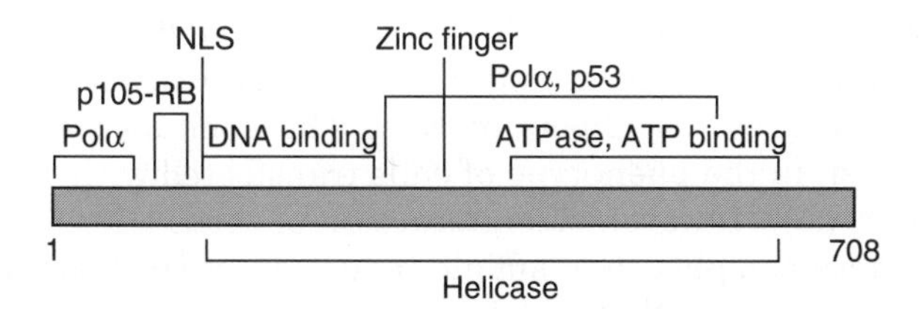

Figure 7.9 Regions of SV40 T-antigen involved in protein–protein interactions. Other functional domains of the protein involved in virus DNA replication are also shown, e.g. the helicase, ATPase, and nuclear location signal (NLS) domains.

Release of E2F by replacement of E2F–Rb complexes with E1A–Rb, T-antigen-RB or E7–RB complexes therefore stimulates cellular and virus DNA replication.

SV40 T-antigen is one of the known virus proteins which binds p53. Chapter 5 describes the role of large T-antigen in the regulation of SV40 transcription. Infection of cells by SV40 or other polyomaviruses can result in two possible outcomes:

- Productive (lytic) infection
- Non-productive (abortive) infection.

The outcome of infection appears to be determined primarily by the cell type infected; for example, mouse polyomavirus establishes a lytic infection of mouse cells but an abortive infection of rat or hamster cells, while SV40 shows lytic infection of monkey cells but abortive infection of mouse cells. However, in addition to transcription, T-antigen is also involved in genome replication. SV40 DNA replication is initiated by binding of large T-antigen to the origin region of the genome (Figure 5.12). The function of T-antigen is controlled by phosphorylation, which decreases the ability of the protein to bind to the SV40 origin.

The SV40 genome is very small and does not encode all the information necessary for DNA replication. Therefore, it is essential for the host cell to enter S phase, when cell DNA and the virus genome are replicated together. Protein–protein interactions between T-antigen and DNA polymerase α directly stimulate replication of the virus genome. The precise regions of the T-antigen involved in binding to DNA, DNA polymerase α, p53 and Rb are all known (Figure 7.11). Inactivation of tumour suppressor proteins bound to T-antigen causes G_1-arrested cells to enter S phase and divide and this is the mechanism which results in transformation. However, the frequency with which abortively infected cells are transformed is low (about 1×10^{-5}). Therefore, the function of T-antigen is to alter the cellular environment to permit virus DNA replication. Transformation is a rare and accidental consequence of the sequestration of tumour suppressor proteins.

The immediate early proteins of adenoviruses are analogous in many ways to SV40 T-antigen. E1A is a *trans*-acting transcriptional regulator of the adenovirus early genes (see Chapter 5). Like T-antigen, the E1A protein binds to Rb, inactivating the regulatory effect of this protein, permitting viral DNA replication and accidentally stimulating cellular DNA replication (see above). E1B binds p53 and reinforces the effects of E1A. The combined effect of the two proteins can be seen in the phenotype of cells transfected with DNA containing these genes (Table 7.6). However, the interaction of these transforming proteins with the cell is more complex than simple induction of DNA synthesis. Expression of E1A alone causes cells to undergo **apoptosis**. Expression of E1A and E1B together overcomes this response and permits transformed cells to survive and grow.

Human papillomavirus (HPV) genital infections are very common, occurring in more than 50% of young, sexually active adults, and are usually asymptomatic. Certain serotypes of HPV appear to be associated with a low risk of subsequent development of anogenital cancers such as cervical carcinoma, after an incubation period of several decades. Of the 70 HPV types currently recognized, only four are associated with a high risk of tumour formation (HPV-16, 18, 31, 45). Once again, transformation is mediated by the early gene products of the virus. However, the transforming proteins appear to vary from one type of papillomavirus to another as shown in Table 7.5. In bovine papillomavirus, it is the E5 protein which is responsible for transformation. In HPV-16 and HPV-18, the E6 and E7 proteins are involved. The E6 protein is known to bind p53 and to target it for accelerated degradation, decreasing the intracellular level of the protein. The E7 protein has structural and functional similarities (e.g. binding to Rb) with adenovirus E1A. In general terms, it appears that two or more early proteins often cooperate to give a transformed phenotype. Although some papillomaviruses can transform cells on their own (e.g. BPV-1), others appear to require the cooperation of an activated cellular oncogene (e.g. HPV-16/*ras*). More confusingly, in most cases, all or part of the papillomavirus genome, including the putative transforming genes, is maintained in the tumour cells, whereas in some cases (e.g. BPV-4), the virus DNA may be lost after transformation, which may indicate a possible hit-and-run mechanism of transformation. Different papillomaviruses appear to use slightly different mechanisms to achieve genome replication and consequently, cell transformation may proceed via a slightly different route. It is imperative that a better understanding of these processes is obtained. There is no positive evidence that adenoviruses or polyomaviruses are involved in the formation of human tumours. In contrast, the evidence that papillomaviruses may be involved in the formation of malignant penile and cervical carcinomas is now strong.

Table 7.6 Role of the adenovirus E1A and E1B proteins in cell transformation

DNA	Cell phenotype
E1A	Immortalized but morphologically unaltered; not tumorigenic in animals.
E1B	Not transformed
E1A + E1B	Immortalized and morphologically altered; tumorigenic in animals.

Viruses and Cancer

There are numerous examples of viruses which cause tumours in experimental animals. This stimulated a long search for viruses which might be the cause of cancer in man. For many years, this search was unsuccessful, so much so that a few scientists categorically stated that viruses did not cause human tumours. Like all rash statements, this was wrong. We currently know of at least six viruses which are associated with the formation of tumours in infected humans (EBV, HBV, HCV, HHV-8, HPVs, HTLV). However, the relationship between virus infection and tumorigenesis is indirect and complex. The role of the HTLV tax protein in leukaemia has already been described (see Cell Transformation by Retroviruses). The evidence that papillomaviruses may be involved in human tumours is also growing. There are almost certainly many more viruses which cause human tumours, but the remainder of this chapter will describe two examples which have been intensively studied, Epstein–Barr virus (EBV) and hepatitis B virus (HBV).

EBV was first identified in 1964 in a lymphoblastoid cell line derived from an African patient with Burkitt's lymphoma. In 1962, Dennis Burkitt described a highly malignant lymphoma whose distribution in Africa paralleled that of malaria. Burkitt recognized that this tumour was rare in India, but occurred in Indian children living in Africa, and therefore looked for an environmental cause. Initially, he thought that the tumour might be caused by a virus spread by mosquitoes (which is wrong). The association between EBV and Burkitt's lymphoma is not entirely clear cut:

- EBV is widely distributed worldwide but Burkitt's lymphoma is rare.
- EBV is found in many cell types in Burkitt's lymphoma patients, not just in the tumour cells.
- Rare cases of EBV-negative Burkitt's lymphoma are sometimes seen in countries where malaria is not present, suggesting there may be more than one route to this tumour.

EBV has a dual cell tropism for human B-lymphocytes (generally a non-productive infection) and epithelial cells, in which a productive infection occurs. The usual outcome of EBV infection is polyclonal B-cell activation and a benign proliferation of these cells which is frequently asymptomatic but sometimes produces a relatively mild disease known as infectious mononucleosis or glandular fever. In 1968, it was shown that EBV could efficiently transform (i.e. immortalize) human B-lymphocytes *in vitro*. This observation clearly strengthens the case that EBV is involved in the formation of tumours. There is now epidemiological and/or molecular evidence that EBV infection is associated with at least four human tumours:

- Burkitt's lymphoma
- Nasopharyngeal carcinoma (NPC), a highly malignant tumour seen most frequently in China. There is a strong association between EBV and NPC: unlike Burkitt's lymphoma, the virus has been found in all the tumours which

have been studied. Environmental factors, such as the consumption of nitrosamines in salted fish, are also believed to be involved in the formation of NPC (c.f. the role of malaria in the formation of Burkitt's lymphoma)
- B-cell lymphomas in immunosuppressed individuals, e.g. AIDS patients
- Some clonal forms of Hodgkin's disease.

There are three possible explanations to explain the link between EBV and Burkitt's lymphoma:

(1) EBV immortalizes a large pool of B-lymphocytes. Concurrently, malaria causes T-cell immunosuppression. There is thus a large pool of target cells in which a third event (e.g. a chromosomal translocation) results in the formation of a malignantly transformed cell. Most Burkitt's lymphoma tumours contain translocations involving chromosome 8, resulting in activation of the *c-myc* gene, which supports this hypothesis.
(2) Malaria results in polyclonal B-cell activation. EBV subsequently immortalizes a cell containing a pre-existing *c-myc* translocation. This mechanism would be largely indistinguishable from the above.
(3) EBV is just a passenger virus! Burkitt's lymphoma also occurs in Europe and North America although it is very rare in these regions; however, 85% of these patients are not infected with EBV, which implies that there are other causes for Burkitt's lymphoma.

Although it has not been formally proved, it seems likely that either (1) and/or (2) are the true explanations for the origin of Burkitt's lymphoma.

Another case where a virus appears to be associated with the formation of a human tumour is that of HBV and hepatocellular carcinoma (HCC). Hepatitis is an inflammation of the liver and as such is not a single disease. Because of the central role of the liver in metabolism, many virus infections may involve the liver. However, there are at least seven viruses which seem specifically to infect and damage hepatocytes. No two of these belong to the same family (see Chapter 8). HBV is the prototype member of the family *Hepadnaviridae* and causes the disease formerly known as 'serum hepatitis'. This disease was distinguished clinically from 'infectious hepatitis' (caused by other types of hepatitis virus) in the 1930s. HBV infection formerly was the result of inoculation with human serum (e.g. blood transfusions, organ transplants), but is still common among intravenous drug abusers where it is spread by the sharing of needles and syringes. However, the virus is also transmitted sexually, by oral ingestion, and from mother to child, which accounts for familial clusters of HBV infection. All blood, organ, and tissue donations in developed countries are now tested for HBV and risk of transmission is extremely low. The virus does not replicate in tissue culture and this has seriously hindered investigations into its pathogenesis. HBV infection has three possible outcomes:

- An acute infection followed by complete recovery and immunity from reinfection (> 90% of cases)
- Fulminant hepatitis, developing quickly and lasting a short time, causing liver failure and a mortality rate of approximately 90% (< 1% of cases)

- Chronic infection, leading to the establishment of a carrier state with virus persistence (about 10% of cases).

There are approximately 250 million chronic HBV carriers worldwide. The total population of the world is approximately 5.5 billion, therefore about 5% of the world population is infected with HBV. All of these chronic carriers of the virus are at 100–200 times the risk of non-carriers of developing HCC. HCC is a rare tumour in the West, where it represents < 2% of fatal cancers. Most cases which do occur in the West are alcohol-related and this is an important clue to the pathogenesis of the tumour. However, in South-East Asia and in China, HCC is the most common fatal cancer, causing about half a million deaths every year. The virus might cause the formation of the tumour by three different pathways: direct activation of a cellular oncogene(s), *trans*-activation of a cellular oncogene(s), or indirectly via tissue regeneration (Figure 7.12). As with EBV and Burkitt's lymphoma, the relationship between HBV and HCC is not clear cut:

- Cirrhosis (a hardening of the liver which may be the result of infections or various toxins, e.g. alcohol) appears to be a prerequisite for the development of HCC. It would appear that chronic liver damage induces tissue regeneration and that faulty DNA repair mechanisms result eventually in malignant cell transformation. Unrelated viruses which cause chronic active hepatitis, such as the flavivirus hepatitis C virus (HCV), are also associated with HCC after a long latent period
- A number of co-factors, such as aflatoxins and nitrosamines, can induce HCC-

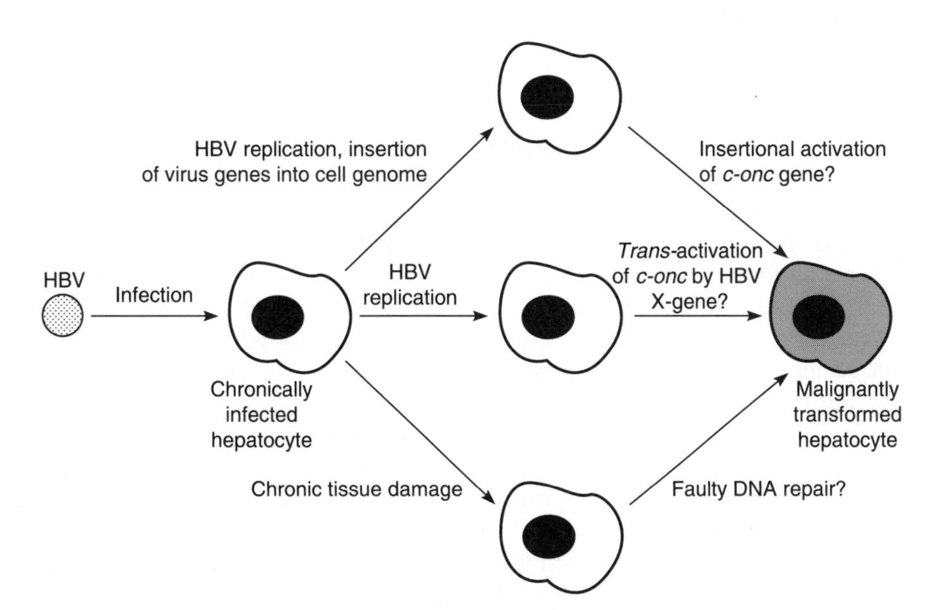

Figure 7.12 Possible mechanisms of hepatocellular carcinoma formation due to hepatitis B virus infection.

like tumours in experimental animals without virus infection. Therefore, such substances may also be involved in human HCC (c.f. nitrosamines and NPC, above)
- There is no consistent evidence for the integration of the HBV genome or even the persistence of particular HBV genes (e.g. the *X* gene, which encodes a *trans*-activator protein functionally analogous to the HTLV tax protein) in tumour cells.

It is possible that all the mechanisms shown in Figure 7.12 might operate *in vivo*. The key risk factor is the development of a chronic as opposed to an acute HBV infection. This in itself is determined by a number of other factors:

- *Age*: The frequency of chronic infections declines with increasing age at the time of infection
- *Sex*:

Chronic infection:	male/female ratio	1.5:1
Cirrhosis:	male/female ratio	3:1
HCC:	male/female ratio	6:1

- *Route of infection*: Oral or sexual infections give rise to fewer cases of chronic infection than serum infection.

Until there is a much better understanding of the pathogenesis and normal course of HBV infection it is unlikely that the reasons for these differences will be understood. There may be a happy ending to this story. A safe and effective vaccine which prevents HBV infection is now available and widely used in the areas of the world where HBV infection is endemic as part of the WHO Expanded Programme on Immunization. This will in future prevent half a million deaths annually from HCC.

New and Emergent Viruses

What constitutes a 'new' infectious agent? Are these just viruses which have never been encountered before, or are there previously known viruses which appear to have changed their behaviour? This section will describe and attempt to explain current understanding of a number of agents which meet the above criteria. Much of the information presented here is not the result of many years of study, but is derived from very recent investigations into these 'new' infectious agents. In the last few decades, there have been massive and unexpected epidemics caused by certain viruses. For the most part, these are not caused by completely new (i.e. previously unknown) viruses, but by viruses that were well known in the geographical areas in which they may be currently causing epidemic outbreaks of disease. Such viruses are known as **emergent viruses** (Table 7.7). There are numerous examples of such viruses which appear to have mysteriously altered their behaviour with time, with significant effects on their pathogenesis.

One of the better known examples of this phenomenon is poliovirus. It is known that poliovirus and poliomyelitis have existed in human populations for at least 4000 years. For most of this time, the pattern of disease was **endemic** rather than **epidemic**, i.e. a low, continuous level of infection in particular geographical areas. During the first half of this century, the pattern of occurrence of poliomyelitis in Europe, North America and Australia changed to an epidemic one, with vast annual outbreaks of infantile paralysis. Although we do not have samples of polioviruses from earlier centuries, the clinical symptoms of the disease give no reason to believe that the virus changed substantially. Why then did the pattern of disease change so dramatically? It is believed that the reason is as follows. In rural communities with primitive sanitation facilities, poliovirus circulated freely. Serological surveys in similar contemporary situations reveal that more than 90% of children of 3 years of age have antibodies to at least one of the three serotypes of poliovirus. (Even the most virulent strains of poliovirus cause 100–200 subclinical infections for each case of paralytic poliomyelitis seen.) In such communities, infants experience subclinical immunizing infections while still protected by maternal antibodies – a form of natural vaccination. The relatively few cases of paralysis and death which do occur are likely to be overlooked, especially in view of high infant mortality rates. During the nineteenth century, industrialization and urbanization changed the pattern of poliovirus transmission. Dense urban populations and increased travelling afforded opportunities for rapid transmission of the virus. In addition, improved sanitation broke the natural pattern of virus transmission. Children were likely to encounter the virus for the first time at a later age and without the protection of maternal antibodies. These children were at far greater risk when they did eventually become infected and it is believed that these social changes account for the altered pattern of disease. Fortunately, the widespread use of poliovirus vaccines has since brought the situation under control in industrialized countries, and poliovirus is well on the way to being eradicated (Chapter 6).

There are many examples of the epidemic spread of viruses caused by movement of human populations. Measles and smallpox were not known to the ancient Greeks. Both of these viruses are maintained by direct person-to-person transmission and have no known alternative hosts. Therefore, it has been suggested that it was not until human populations in China and the Roman Empire reached a critical density that these viruses were able to propagate in an epidemic pattern and cause recognizable outbreaks of disease. Before this time, the few cases that did occur could easily have been overlooked. Smallpox reached Europe from the Far East in AD 710 and in the eighteenth century, achieved plague proportions – five reigning European monarchs died from smallpox. However, the worst effects occurred when these viruses were transmitted to the New World. Smallpox was (accidentally) transferred to the Americas by Hernando Cortés in 1520. In the next two years, 3.5 million Aztecs died from the disease and the Aztec empire was decimated by disease rather than conquest. Although not as highly pathogenic as smallpox, epidemics of measles subsequently finished off the Aztec and Inca civilizations. More recently, the first contacts with isolated groups of Eskimos and tribes in New Guinea and South America have had similarly devastating results, although on a smaller scale.

Table 7.7 New and emergent viruses

Virus	Date	Family	Comments
New viruses			
Human herpesvirus 6 (HHV-6)	1986	Herpesvirus	See text
Hepatitis C virus (HCV)	1989	Flavivirus	See text
Human herpesvirus 7 (HHV-7)	1990	Herpesvirus	See text
GB viruses (hepatitis)	1994	Flavivirus	See text
Human herpesvirus 8 (HHV-8)	1995	Herpesvirus	See text
Emergent viruses			
Cocoa swollen shoot		Badnavirus	Has caused destruction of 200 million cocoa trees in West Africa. Transmitted by mealybugs
Dengue		Flavivirus	See text
Ebola		Filovirus	See text
Equine Morbillivirus		Paramyxovirus	Emerged in Brisbane, Australia, September 1994. Causes acute respiratory disease in horses with high mortality. Believed to cause a fatal encephalitis in humans.
Hantaan group		Bunyaviruses	See text
Phocine distemper		Paramyxovirus	Emerged in 1987 and caused high mortalities in seals in the Baltic and North Sea. Similar viruses subsequently recognized as responsible for cetacean (porpoise and dolphin) deaths in Irish Sea and Mediterranean
Rabbit calicivirus disease (RCD)/ Viral haemorrhagic disease (VHD)		Calicivirus	Emerged in China in 1985, spread naturally through UK and Europe. Introduced to Wardang Island off the coast of South Australia to test potential for rabbit population control, accidentally spread to Australian mainland causing huge kill in rabbit populations. Vaccine available to protect domestic and farmed rabbits
Rift Valley Fever		Bunyaviruses	See text
Tomato spotted wilt		Bunyavirus	See text
Whitefly-transmitted geminiviruses (group III geminiviruses)		Geminivirus	See text

These historical incidents illustrate the way in which a known virus can suddenly cause illness and death on a catastrophic scale following a change in human behaviour.

Measles and smallpox viruses are transmitted exclusively from one human host to another. For viruses with more complex cycles of transmission, e.g. those with secondary hosts and insect vectors, control of infection becomes much more difficult (Figure 7.13). This is particularly true of the families of viruses known collectively as 'arboviruses' (arenaviruses, bunyaviruses, flaviviruses and togaviruses). As man's territory has expanded, this has increasingly brought people into contact with the type of environment where these viruses are found – warm, humid, vegetated areas where insect vectors occur in high densities, such as swamps and jungles. In addition to changes in agricultural practices, many emergent virus diseases are zoonose i.e. transmitted from animals to humans. This emphasises the importance of the 'species barrier' in preventing transmission of infectious diseases.

A classic example is the mortality caused by yellow fever virus during the building of the Panama Canal at the end of the nineteenth century. More recently, the increasing pace of ecological alteration in tropical areas has resulted in the resurgence of yellow fever in Central America, particularly an urban form of the disease transmitted directly from one human to another by mosquitoes. Dengue fever is also primarily an urban disease of the tropics, transmitted by *Aedes aegypti*, a domestic, day-biting mosquito that prefers to feed on humans. Some outbreaks of dengue fever have involved more than a million cases with attack rates of up to 90% of the population. There are believed to be over 40 million cases of Dengue virus infection worldwide each year. This disease was first described in 1780 and by 1906 it was known that the virus was transmitted by mosquitoes. The virus was isolated in 1944. Therefore,

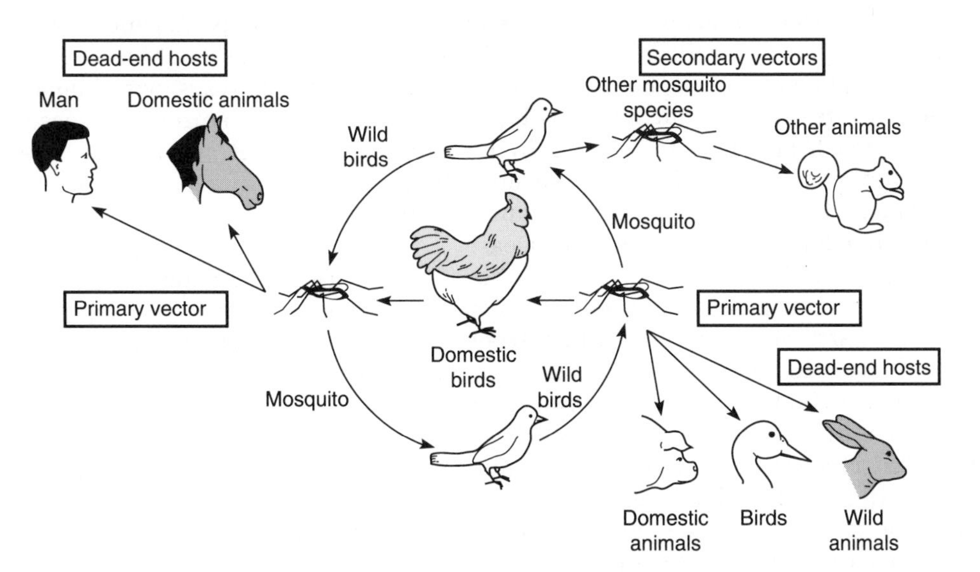

Figure 7.13 Complex transmission pattern of an 'arbovirus'.

this is not a new virus, but the frequency of Dengue virus infection has increased dramatically in the last twenty years. Murphy (see Further Reading) suggested this is due to factors such as:

- Population movements and the intrusion of humans and domestic animals into new arthropod habitats, particularly tropical forests
- Deforestation, with development of new forest–farmland margins and exposure of farmers and domestic animals to new arthropods
- Irrigation, especially primitive irrigation systems which are oblivious to arthropod control
- Uncontrolled urbanization, with vector populations breeding in accumulations of water (tin cans, old tyres, etc.) and sewage
- Increased long-distance air travel, with potential for transport of arthropod vectors
- Increased long-distance livestock transportation, with potential for carriage of viruses and arthropods (especially ticks)
- New routing of long-distance bird migration brought about by new man-made water impoundments
- Cessation of mosquito control control programmes and political upheaval resulting in degraded primary medical services.

Of more than 520 arboviruses known, at least 100 are pathogenic for man and perhaps 20 would meet the criteria for emergent viruses. Attempts to control these diseases rely on twin approaches involving both the control of insect vectors responsible for transmission of the virus to man and the development of vaccines to protect human populations. However, both of these approaches present considerable difficulties, the former in terms of avoiding environmental damage and the latter in terms of understanding viral pathogenesis and developing appropriate vaccines (see discussion of Dengue virus pathogenesis earlier). Rift Valley Fever virus was first isolated from sheep in 1930 but has caused repeated epidemics in sub-Saharan Africa during the last 20 years, with human infection rates in epidemic areas as high as 35%. This is an epizootic disease, transmitted from sheep to humans by a number of different mosquitoes. The construction of dams which increase mosquito populations, increasing numbers of sheep and the movement of sheep and human populations are believed to be responsible for the upsurge in this disease.

The Hantaan Virus group (*Bunyavirus* family) are a particular cause for concern. Hantaviruses cause millions of cases of haemorrhagic fever each year in many parts of the world. Unlike arboviruses, hantaviruses are transmitted directly from rodent hosts to humans (e.g. via faeces) rather than by an invertebrate host. Hantaviruses cause two acute diseases: haemorrhagic fever with renal syndrome (HFRS) and hantavirus pulmonary syndrome (HPS). HFRS was first recognized in 1951 after an outbreak among US troops stationed in Korea. In May 1993, HPS was first recognized in the USA and a new virus, Sin Nombre, identified as the cause. It is now known that at least three different hantaviruses cause HFRS and four different viruses cause HPS. By January 1995, HPS had been recognized in 102 patients in 21 states of the USA, seven in

Canada and three in Brazil with an overall mortality rate of approximately 40%. This illustrates the disease-causing potential of emerging viruses.

Ebola virus has received massive attention in recent years and so will only be dealt with briefly here. The virus was first described in 1976 and since then has reappeared several times (Table 7.8). The extreme pathogenicity of this virus has severely inhibited investigations, most of which have been carried out using molecular biological techniques. However, this predominantly molecular approach has left important questions unanswered, for example:

- *Strain variation*: Some strains of Ebola virus are highly pathogenic, whereas other strains are not (Table 7.8). Relatively non-pathogenic strains have been isolated from monkeys imported into the USA from the Philippines. The molecular basis for these differences is unknown
- *Natural hosts and distribution*: Most human infections appear to be associated with contact with infected primates. However, extensive ecological surveys in Central Africa have failed to show any evidence that primates (or any of the thousands of animals, plants and invertebrate species examined) are the natural reservoir for infection. Isolates from Central Africa appear to be highly pathogenic whereas those from the Philippines are apathogenic for humans.

Plant viruses can also be responsible for emergent diseases. Group III geminiviruses are transmitted by insect vectors (whiteflies) and their genomes consist of two circular, single-stranded DNA molecules (Chapter 3). These viruses cause a great deal of crop damage in plants such as tomatoes, beans, squash, cassava and cotton and their spread may be directly linked to the inadvertent worldwide dissemination of a particular biotype of the whitefly *Bemisia tabaci*. This vector is an indiscriminate feeder, encouraging rapid and efficient spread of viruses from indigenous plant species to neighbouring crops.

Occasionally, there appears an example of an emergent virus which has acquired extra genes and as a result of this new genetic capacity, becomes capable of infecting new species. A possible example of this phenomenon is seen in tomato spotted wilt virus (TSWV). TSWV is a Bunyavirus with a very wide plant host range, infecting over 600 different species from 70 families. In

Table 7.8 Ebola virus outbreaks

Date	Place	Number of cases	Fatalities (%)
1976	Zaire	276	93
1976	Sudan	280	53
1977	Zaire	1	100
1979	Sudan	34	65
1989	Virginia, USA	4	0
1995	Zaire	315	77
1996	Gabon	19	53
1996	Texas, USA	0	0

recent decades, this virus has been a major agricultural pest in Asia, the Americas, Europe and Africa. Its rapid spread has been the result of dissemination of its insect vector (the thrip *Frankinellia occidentalis*) and diseased plant material. TSWV is the type species of the Tospovirus genus and has a similar morphology and genomic organization to the other bunyaviruses (Chapter 3). However, TSWV undergoes **propagative transmission** and it has been suggested that it may have acquired an extra gene in the M segment via recombination, either from a plant or from another plant virus. This new gene encodes a movement protein (Chapter 6), conferring the capacity to infect plants and cause extensive damage.

In addition to viruses whose ability to infect their host species appears to have changed, 'new' viruses are being discovered continually. In recent years, three new human herpesviruses have been discovered:

(1) *Human herpesvirus 6 (HHV-6)*: First isolated in 1986 in lymphocytes of patients with lymphoreticular disorders; tropism for $CD4^+$ lymphocytes. HHV-6 is now recognized as being an almost universal human infection. Discovery of the virus solved a longstanding mystery: the primary infection in childhood causes 'roseola infantum' or 'fourth disease', a common childhood rash whose cause was previously unknown. Antibody titres are highest in children and decline with age. The consequences of childhood infection appear to be mild. Primary infections of adults are rare but have more severe consequences, mononucleosis or hepatitis, and infections may be a severe problem in transplant patients.
(2) *Human herpesvirus 7 (HHV-7)*: First isolated from human $CD4^+$ cells in 1990. Its genome organization is similar to but distinct from that of HHV-6 and there is limited antigenic cross-reactivity between the two viruses. At present, there is no clear evidence for the direct involvement of HHV-7 in any human disease, but might it be a co-factor in HHV-6-related syndromes.
(3) *Human herpesvirus 8 (HHV-8)*: In 1995, sequences of a unique herpesvirus were identified in DNA samples from AIDS patients with Kaposi's sarcoma (KS) and in some non-KS tissue samples from AIDS patients. There is a strong correlation (>95%) with KS in both HIV^+ and HIV^- patients. HHV-8 can be isolated from lymphocytes and from tumour tissue and appears to have a less ubiquitous world distribution than other HHVs, i.e. may only be associated with a specific disease state (c.f. HSV, EBV). However, the virus is not present in KS-derived cell lines, suggesting that **autocrine** or paracrine factors may be involved in the formation of KS. There is some evidence that HHV-8 may also cause other tumours such as B-cell lymphomas ($\pm$ EBV as a 'helper').

Although many different virus infections may involve the liver, there are at least six viruses which seem specifically to infect and damage hepatocytes. No two of these belong to the same family! The identification of these viruses has been a long story:

- Hepatitis B Virus (HBV) (*Hepadnavirus*): 1960s
- Hepatitis A Virus (HAV) (*Picornavirus*): 1973

- Hepatitis Delta Virus (HDV) (*Deltavirus*, see Chapter 8): 1977
- Hepatitis C Virus (HCV) (*Flavivirus*): 1989
- Hepatitis E Virus (HEV): 1990
- GB viruses: In 1967, a surgeon (GB) in the USA was found to be suffering from hepatitis and a blood sample was taken and stored. In 1994, this blood sample was passaged through a tamarin monkey, which in turn developed hepatitis. Three viruses were isolated from the monkey: GBV-A, GBV-B and GBV-C. GBV-A and GBV-B have flavivirus-like genomes with 20–30% amino acid similarity to each other and to HCV but are believed to be simian viruses 'accidentally' isolated from the monkey. GBV-C, however, is associated with acute and chronic hepatitis in humans. What data is available suggests that GBV-C infection may give rise to a fulminant form of hepatitis, although this is still under investigation.

Reports continue to circulate about the existence of other hepatitis viruses (HFV, HGV). Some of the agents are reported to be sensitive to chloroform (i.e. enveloped) while others are not. This may suggest the existence of multiple viruses, as yet undescribed, although this is still uncertain. It is likely that further hepatitis viruses will be described in the future.

Summary

Viral pathogenesis is a complex, variable and relatively rare state. Like the course of a virus infection, pathogenesis is determined by the balance between host and viral factors. Not all of the pathogenic symptoms seen in virus infections are caused directly by the virus – the immune system also plays a part in causing cell and tissue damage. Viruses can transform cells so that they continue to grow indefinitely. In some but not all cases, this can lead to the formation of tumours. There are at least a few well-established cases where certain viruses provoke human tumours and possibly many others that we do not yet understand. The relationship between the virus and the formation of the tumour is not a simple one, but the prevention of infection undoubtedly reduces the risk of tumour formation. New pathogenic viruses are being discovered all the time and changes in human activities result in the emergence of new or previously unrecognized diseases.

Further Reading

Clerici, M. and Shearer, G. (1993). A T_H1–T_H2 switch is a critical step in the aetiology of HIV infection. *Immunol. Today* **14**: 107–111.

Fan, H.Y., Chen, I.S.Y., Rosenberg, N. and Sugden, W. (Eds) (1991). *Viruses That Affect the Immune System*. American Society for Microbiology, Washington, D.C.

Gubler, D.J. and Clark, G.G. (1995). Dengue hemorrhagic-fever – the emergence of a global health problem. *Emerg. Inf. Dis.* **1**: 55–57.

Ho, D. *et al.* (1995). Rapid turnover of plasma virions and CD4 lymphocytes in HIV-1 infection. *Nature* **373**: 123–6.

Johnson, H.M. *et al.* (1992). Superantigens in human disease. *Sci. Amer.* **266**: 20–6.

Kroemer, G. and Martinez-A, C. (Eds) (1995). Apoptosis. *Curr. Top. Microbiol. Immunol.*, Vol. **200**.

Kurane, I. *et al.* (1991). Antibody-dependent enhancement of dengue virus infection. *Rev. Med. Virol.* **1**: 211–21.

Levy, J.A. (1993). Pathogenesis of Human Immunodeficiency Virus infection. *Microbiol. Rev.* **57**: 183–289.

Mims, C.A., Dimmock, N., Nash, A. and Stephen, A. (1995). *Mims' Pathogenesis of Infectious Disease* (4th edition). Academic Press, London.

Morse, S.S. (Ed.). (1993). *Emerging Viruses*. Oxford University Press, Oxford.

Murphy, F.A. (1994). New, emerging and reemerging infectious-diseases. *Adv. Vir. Res.* **43**: 1–52.

Nowak, M.A. *et al.* (1991). Antigenic diversity thresholds and the development of AIDS. *Science* **254**: 963–9.

Roitt, I. (1994). *Essential Immunology.* (8th edition). Blackwell Scientific, Oxford.

Spector, S. *et al.* (Eds) (1989). *Virus-Induced Immunosuppression.* Plenum Press, New York.

Temin, H. (1988). Mechanisms of cell killing/cytopathic effects by non-human retroviruses. *Rev. Infect. Dis.* **10**: 399–405.

Tiollais, P. and Buendia, M. (1991). Hepatitis B virus. *Sci. Amer.*, April, 48–54.

Wei, X. *et al.* (1995). Viral dynamics in HIV-1 infection. *Nature* **373**: 117–22.

Self-Assessment Questions

(7.1) Are the following statements true or false?
- (a) Most disease states are multi-factorial.
- (b) Inflammation, fever, headaches and skin rashes are frequently not caused by virus replication, but the immune response to virus infection.
- (c) The vast majority of virus infections do not result in disease.
- (d) Eukaryotic cells must constantly synthesize a variety of macromolecules which require continual replacement.
- (e) Virus nucleic acids act as toxins and poison cells.

(7.2) The cytopathic effects of virus infection include (true or false?):
- (a) Altered shape.
- (b) Detachment from the substrate.
- (c) Lysis.
- (d) Membrane fusion and altered permeability.
- (e) Apoptosis.

(7.3) Are the following statements true or false?
- (a) Shutoff is the sudden and dramatic cessation of virus macromolecular synthesis.
- (b) Shutoff in poliovirus-infected cells occurs due to protease degradation of ribosomes.
- (c) Lysis of adenovirus-infected cells is caused by virus-encoded molecules with a toxin-like action.
- (d) Virus-induced membrane fusion causes cell transformation.
- (e) The HIV gag protein is responsible for syncytium formation.

(7.4) Are the following statements true or false?
- (a) All human herpes viruses are highly cytopathic.
- (b) Retroviruses cause a variety pathogenic conditions including paralysis, arthritis, anaemia in addition to cell transformation.
- (c) Lentiviruses are the group of retroviruses most frequently associated with cell transformation.
- (d) HIV and other lentiviruses escape immune recognition by infecting monocytes.
- (e) All isolates of HIV are highly cytopathic in $CD4^+$ cells.

(7.5) The following mechanisms may be involved in the pathogenesis of AIDS (true or false?):
- (a) New antigenic variants of HIV constantly arise during the course of infection.
- (b) T-cell anergy.
- (c) Apoptosis.
- (d) Superantigens.
- (e) Insertional activation of oncogenes.

(7.6) Are the following statements true or false?
- (a) Subacute sclerosing panencephalitis (SSPE) occurs in 10% of measles virus infections of adults.
- (b) Dengue haemorrhagic fever is largely due to immune-mediated damage of virus-infected cells.
- (c) Reye's syndrome can be cured by administration of aspirin during the initial illness.
- (d) Kawasaki disease is a rare post-infection complication of a number of different viruses, but most commonly influenza virus and VZV (chicken pox).
- (e) Guillain–Barré syndrome is a rare demyelinating disease.

(7.7) Transformed cells may display (true or false?):
- (a) Loss of anchorage dependence.
- (b) Loss of contact inhibition.
- (c) Colony formation in semi-solid media.
- (d) Decreased requirements for growth factors.
- (e) Increased expression of p53.

(7.8) Are the following statements true or false?
- (a) Carcinogenesis = oncogenesis = transformation.
- (b) All acutely transforming retroviruses are replication defective.
- (c) p53 causes the cell cycle to arrest at the G_1 phase.
- (d) Rb causes the cell cycle to arrest at the G_2 phase.
- (e) Adenovirus E1A, SV40 T antigen and HPV E7 protein all bind to p53.

(7.9) The normal functions of oncogenes may be (true or false?):
- (a) Cytoplasmic protein kinases.
- (b) Extracellular growth factors.
- (c) G proteins involved in signal transduction.
- (d) Transcription factors.
- (e) Tyrosine kinases.

(7.10) Are the following statements true or false?
- (a) At least six viruses are associated with the formation of human tumours.
- (b) EBV has a dual cell tropism for human B-lymphocytes (productive infection) and epithelial cells, in which a non-productive infection occurs.
- (c) EBV infection is associated with at least four human tumours: Burkitt's lymphoma, nasopharyngeal carcinoma, B-cell lymphomas and hepatocellular carcinoma.
- (d) Chronic HBV carriers are at 100–200 times the risk of non-carriers of developing hepatocellular carcinoma.
- (e) Chronic active hepatitis caused by hepatitis C virus (HCV) infection is associated with hepatocellular carcinoma.

Answers to Self-Assessment Questions are given in Appendix 3.

Chapter 8

Novel Infectious Agents: Genomes Without Viruses, Viruses Without Genomes

What is the minimum genome size necessary to sustain an infectious agent? Could a virus with a genome of 1700 nucleotides exist? Could an agent with a genome of 240 nucleotides exist? Could an infectious agent without any genome at all exist? Perhaps the first two alternatives might be possible, but the idea of an infectious agent without a genome seems bizarre and ridiculous. Strange as it may seem, such agents as these do exist and cause disease in animals (including man) and plants. By reading the earlier chapters of this book, readers will hopefully have grasped the idea of what constitutes a conventional virus in terms of structure, replication and disease-causing potential.

Satellites and Viroids

Satellites are small RNA molecules which are absolutely dependent on the presence of another virus for multiplication. Even viruses have their own parasites! Most satellites are associated with plant viruses, but a few are associated with bacteriophages or animal viruses, e.g. the dependoviruses, which are satellites of adenoviruses. Two classes of satellites can be distinguished, satellite viruses, which encode their own coat proteins, and satellite RNAs (or '**virusoids**'), which use the coat protein of the helper virus (Appendix 1). Typical properties of satellites are:

- A genome of approximately 500–2000 nucleotides of single-stranded RNA
- Unlike defective virus genomes, there is little or no nucleotide sequence similarity between the satellite and the helper virus genome

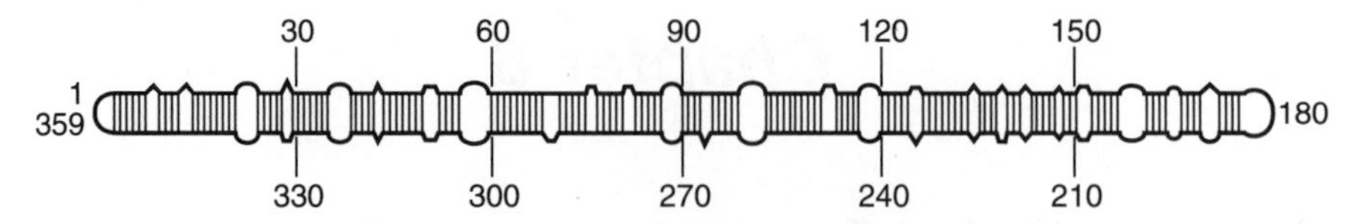

Figure 8.1 Structure of a viroid RNA.

Table 8.1 Satellites and viroids

Characteristic	Satellites	Viroids
Helper virus required for replication	Yes	No
Protein(s) encoded	Yes	No
Genome replicated by	Helper virus enzymes	Host cell RNA polymerase II
Site of replication	Same as helper virus (nucleus or cytoplasm)	Nucleus

- They cause distinct disease symptoms in plants which are not seen with the helper virus alone
- Replication of satellites usually interferes with the replication of the helper virus (unlike most defective virus genomes).

Viroids are very small (200–400 nt), rod-like RNA molecules with a high degree of secondary structure (Figure 8.1). They have no capsid or envelope and consist only of a single nucleic acid molecule. Viroids are associated with plant diseases and are distinct from satellites in a number of ways (Table 8.1). The first viroid to be discovered and the best studied is potato spindle tuber viroid (PSTVd – viroid names are abbreviated as 'Vd' to distinguish them from viruses). Viroids do not encode any proteins and are replicated by host cell RNA polymerase II. The details of replication are not understood, but it is likely to occur by a rolling circle mechanism followed by autocatalytic cleavage and self-ligation to produce the mature viroid.

There is considerable sequence variation between different viroids and this is used as an arbitrary classification to divide viroids into species. However, all viroids share a common feature, a conserved central region believed to be involved in their replication (Figure 8.2). One group of viroids is capable of forming a hammerhead structure, giving them the enzymatic properties of a ribozyme (an autocatalytic, self-cleaving RNA molecule). This activity is used to cleave the multimeric structures produced during the course of replication. Other viroids use unknown host nuclear enzymes to achieve this objective. Some viroids (e.g. cadang-cadang coconut viroid, CCCVd) cause severe and lethal disease in their host plants. Others range from no apparent pathogenic effects (e.g. hop latent viroid, HLVd) to mild disease symptoms (e.g. apple scar skin viroid, ASSVd). It is not clear how viroids cause pathogenic symptoms, but

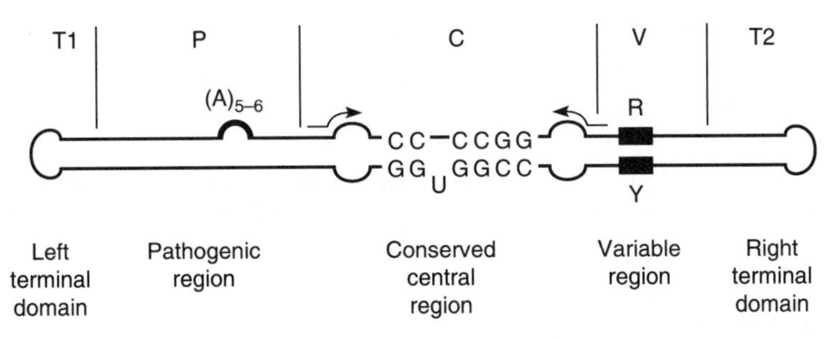

Figure 8.2 Functional regions of viroid RNA molecules.

obviously these must result from some perturbation of the normal host cell metabolism. They show some similarities with certain eukaryotic host cell sequences, in particular with an intron found between the 5.8S and 25S ribosomal RNAs and with the U3 snRNA which is involved in splicing. Therefore, it has been suggested that viroids may interfere with post-transcriptional RNA processing in infected cells. *In vitro* experiments with purified mammalian protein kinase PKR (Chapter 6) have shown that the kinase is strongly activated (phosphorylated) by viroid strains that cause severe symptoms, but far less by mild strains. Activation of a plant homologue of PKR could be the triggering event in viroid pathogenesis (see discussion of the hypersensitive response in Chapter 6).

Most viroids are transmitted by vegetative propagation, i.e. division of infected plants, although a few can be transmitted by insect vectors (non-propagative) or mechanically. Because viroids do not have the benefit of a protective capsid, viroid RNAs would be expected to be at extreme risk of degradation in the environment. However, their small size and high degree of secondary structure protects them to a large extent and they are able to persist in the environment for a sufficiently long period to be transferred to another host. The origins of viroids are obscure. One theory is that they may be the most primitive type of RNA genome – possibly leftovers from the 'RNA world' believed to have existed during the era of prebiotic evolution (see Chapter 3). Alternatively, they may have evolved at a much more recent time as the most extreme type of parasite. We may never know which of these alternatives is true, but viroids exist and cause disease in plants and in man.

Hepatitis delta virus (HDV, see Chapter 7) is a unique chimeric molecule with some of the properties of a satellite virus and some of a viroid (Table 8.2) which causes disease in humans. HDV requires hepatitis B virus (HBV) as a 'helper virus' for replication and is transmitted by the same means as HBV, benefiting from the presence of a protective coat composed of lipid plus HBV proteins. Virus preparations from HBV/HDV-infected animals contain heterologous particles distinct from those of HBV but with an irregular, ill-defined structure. These particles are composed of HBV antigens and contain the covalently closed circular HDV RNA molecule in a branched or rod-like configuration similar to that of other viroids (Figure 8.3). Unlike all other viroids, HDV encodes a protein, the δ antigen, which is a nuclear phosphoprotein. Post-transcriptional

Table 8.2 Properties of hepatitis δ virus (HDV)

Satellite-like properties	Viroid-like properties
Size and composition of genome, 1640 nt (about four times the size of plant viroids) Single-stranded circular RNA molecule Dependent on hepatitis B virus (HBV) for replication; HDV RNA is packaged into coats consisting of lipids plus HBV-encoded proteins Encodes a single polypeptide, the δ antigen.	Sequence homology to the conserved central region involved in viroid replication

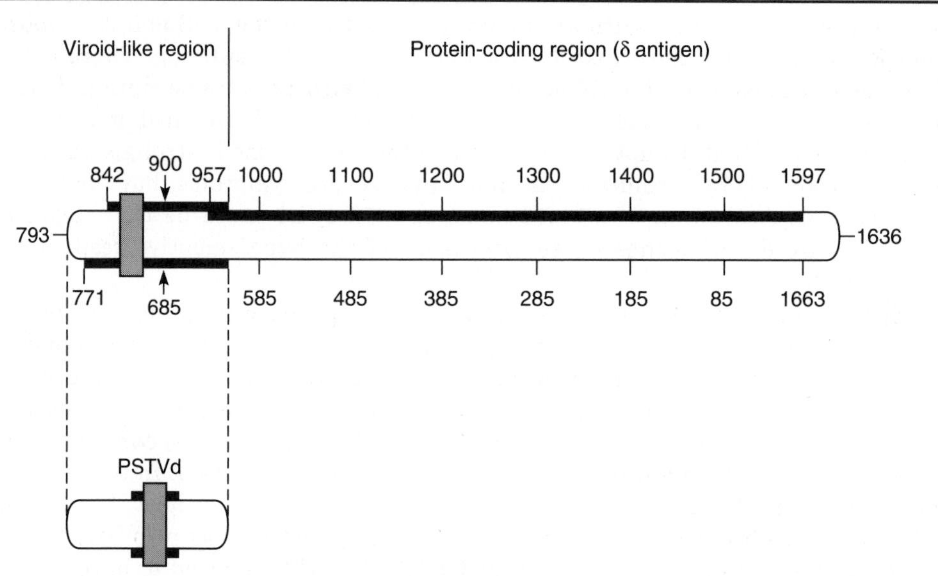

Figure 8.3 Structure of hepatitis δ virus RNA. A region at the left end of the genome strongly resembles the RNA of plant viroids such as potato spindle tuber viroid (PSTVd), which is shown for comparison.

RNA editing results in the production of two slightly different forms of the protein, δAg-S (195 amino acids) which is necessary for HDV replication and δAg-L (214 amino acids) which is necessary for the assembly and release of HDV-containing particles. HDV is found worldwide wherever HBV infection occurs. The interactions between HBV and HDV are difficult to study, but HDV seems to potentiate the pathogenic effects of HBV infection. Fulminant hepatitis (mortality rate ca. 80%) is 10 times more common in co-infections than with HBV infection alone. Because HDV requires HBV for replication, it is being controlled by HBV vaccination (Chapter 6).

Prions

There are a group of transmissible, chronic, progressive infections of the nervous system which show common pathological effects and are invariably fatal. Their pathology is reminiscent of amyloid diseases such as Alzheimer's syndrome and to distinguish them from such conditions, they are known as transmissible spongiform encephalopathies (TSE). The earliest record of any TSE dates from several centuries ago, when a disease called scrapie was first observed in sheep (see TSE in Animals). Long considered to be caused by viruses, the first doubts about the nature of the infectious agent involved in TSEs arose in the 1960s. In 1967, Tikvah Alper was the first to suggest that the agent of scrapie might replicate without nucleic acid and in 1982, Stanley Prusiner coined the term **prion** (proteinaceous infectious particle), which according to Prusiner is pronounced as 'pree-on'. The molecular nature of prions has not been unequivocally proved (see Molecular Biology of Prions), but the evidence that they represent a new phenomenon outside the framework of conventional scientific understanding is growing steadily stronger.

Pathology of prion diseases

All prion diseases share a similar underlying pathology, although there are significant differences between various conditions. A number of different diseases are characterized by the deposition of abnormal protein deposits in various organs, e.g. kidney, spleen, liver or brain. These 'amyloid' deposits consist of accumulations of various proteins in the form of plaques or fibrils depending on their origin, e.g. Alzheimer's disease is characterized by the deposition of plaques and 'tangles' composed of β-amyloid protein. None of the 'conventional' amyloidoses are infectious diseases and extensive research has shown that they cannot be transmitted to experimental animals; they result from endogenous errors in metabolism caused by a variety of largely unknown factors. Amyloid deposits appear to be inherently cytotoxic. Although the molecular mechanisms involved in cell death are unclear, it is this effect which gives the 'spongiform encephalopathies' their name owing to the characteristic holes in thin sections of affected brain tissue viewed under the microscope; these holes are caused by neuronal loss and gliosis. Thus deposition of amyloid is end stage which links conventional amyloidoses and TSEs and explains the tissue damage seen in both types of disease, but does not reveal anything about their underlying causes. Definitive diagnosis of TSE cannot be made on clinical grounds alone and requires demonstration of PrP deposition by immunohistochemical staining of post-mortem brain tissue, molecular genetic studies, or experimental transmission to animals, as discussed in the following sections.

TSE in animals

A number of TSEs have been observed and intensively investigated in animals. In particular, scrapie is the paradigm for understanding of human TSEs. Some of these diseases are naturally occurring and have been known about for centuries, whereas others have only been observed during the last decade and are almost certainly causally related to one another.

Scrapie

Scrapie is a naturally occurring disease of sheep found in many parts of the world, although not universally distributed, which was first described more than 200 years ago. Scrapie appears to have originated in Spain and subsequently spread throughout Western Europe. The export of sheep from Britain in the nineteenth century is thought to have helped scrapie spread around the world. Scrapie is primarily a disease of sheep although it can also affect goats. The scrapie agent has been intensively studied and has been experimentally transmitted to laboratory animals many times (see Molecular Biology of Prions). Infected sheep show severe and progressive neurological symptoms such as abnormal gait, often repeatedly scraping against fences or posts, from which behaviour the disease takes its name. The incidence of the disease increases with the age of the animals. Some countries, such as Australia and New Zealand, have eliminated scrapie by slaughtering infected sheep and the imposition of rigorous import controls. Work in Iceland has shown that the land on which infected sheep graze may retain the condition and infect sheep up to three years later.

The incidence of scrapie in a flock is related to the breed of sheep. Some breeds are relatively resistant to the disease while others are prone to it, indicating genetic control of susceptibility. In recent years the occurrence of TSE in sheep in the UK closely parallels the incidence of BSE in cattle (Figure 8.4). This is probably due to infection with the BSE agent (to which sheep are known to be sensitive) via infected feed (see below). The natural mode of transmission between sheep is unclear. Lambs of scrapie infected sheep are more likely to develop the disease, but the reason for this is unclear. Symptoms of scrapie are not seen in sheep of less than one and a half years old, which indicates that the incubation period of scrapie is at least this long. The first traces of infectivity can be detected in the tonsils, mesenteric lymph nodes and intestines of sheep 10–14 months old, which suggests an oral route of infection. The infective agent is present in the membranes of the embryo but it has not been demonstrated in colostrum or milk or in the tissues of the newborn lambs.

Transmissible mink encephalopathy (TME)

TME is a rare disease of farmed mink caused by exposure to a scrapie-like agent in feed. The disease was first identified in Wisconsin, USA, in 1947 and has also been recorded in Canada, Finland, Germany and Russia. Like other TSEs, TME is a slow progressive neurological disease. Early symptoms include changes in habits and cleanliness as well as difficulty in eating or swallowing. TME-infected mink become hyperexcitable and begin arching their tails over their backs, ultimately losing locomotor coordination. Natural TME has a minimum incubation period of 7–12 months and although exposure is generally through oral routes, mink to mink transmission cannot be ruled out. The origin of the transmissible agent in TME appears to be contaminated foodstuffs but this is discussed further below (see BSE).

Feline spongiform encephalopathy (FSE)

FSE was recognized in the UK in May 1990 as a scrapie-like syndrome in domestic cats resulting in ataxia (irregular and jerky movements) and other symptoms typical of spongiform encephalopathies. By May 1995, a total of 67 cases had been reported in the UK. In addition, FSE has been recorded in a domestic cat in Norway and in three species of captive wild cats (cheetahs, puma and ocelot). Inclusion of cattle offal in commercial pet foods was banned in the UK in 1990, so the incidence of this disease is expected to decline rapidly (see BSE).

Chronic wasting disease (CWD)

A disease similar to scrapie which affects deer and captive exotic ungulates (e.g. nyala, oryx, kudu, etc). CWD was first recognized in captive deer and elk in the Western United States in 1967 and appears to be endemic in origin.

Bovine spongiform encephalopathy (BSE)

BSE was first recognized in dairy cattle the UK in 1986 as a typical spongiform encephalopathy. Affected cattle showed altered behaviour and a staggering gait, giving the disease its name in the press of 'mad cow disease'. On microscopic examination, the brains of affected cattle showed extensive spongiform degeneration. It was concluded that BSE resulted from the use of contaminated foodstuffs. To obtain higher milk yields and growth rates, the nutritional value of feed for farmed animals was routinely boosted by the addition of protein derived from waste meat products and bonemeal (MBM) prepared from animal carcasses, including sheep and cows. This practice was not unique to the UK but was widely followed in most developed countries. By March 1996, a total of 161 663 cases of BSE had been reported in the UK and 412 cases elsewhere

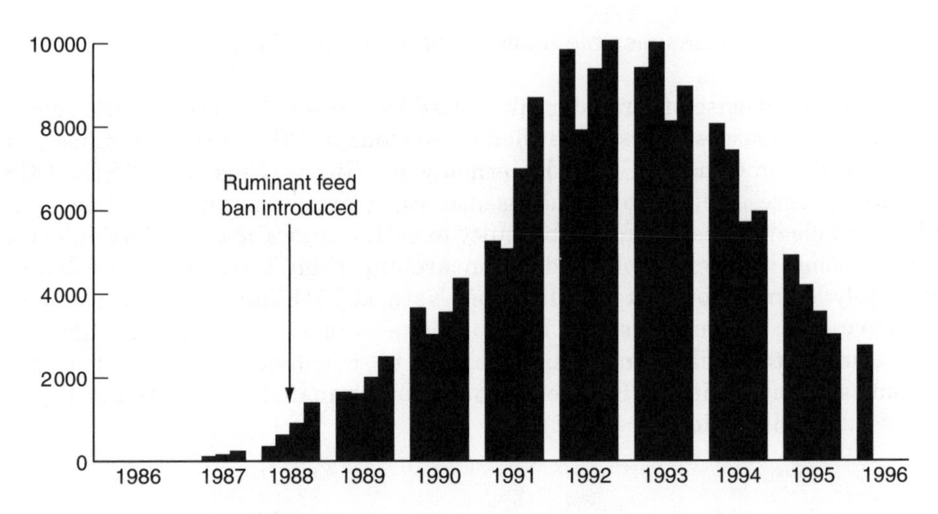

Figure 8.4 Reported incidence of BSE in the UK.

(including Switzerland, Ireland, Portugal, France, Germany, Italy, Oman, Canada, Denmark and the Falklands) (Figure 8.4).

The initial explanation for the emergence of BSE in the UK was as follows. Because scrapie is endemic in Britain, it was assumed that this was the source of the infectious agent in the feed. Traditionally, MBM was prepared by a rendering process involving steam treatment and hydrocarbon extraction, resulting in two products, a protein-rich fraction called 'greaves' containing about 1% fat from which MBM was produced and a fat-rich fraction called 'tallow' which was put to a variety of industrial uses. In the late 1970s, the price of tallow fell and the use of expensive hydrocarbons in the rendering process was discontinued, producing an MBM product containing about 14% fat in which the infectious material may not have been inactivated. As a result, a ban on the use of ruminant protein in cattle feed was introduced in July 1988 (Figure 8.4). In November 1989, human consumption of specified bovine offals thought most likely to transmit the infection (brain, spleen, thymus, tonsil and gut) was prohibited. A similar ban on consumption of offals from sheep, goats and deer was finally announced in July 1996 to counter concerns about transmission of BSE to sheep. The available evidence suggests that milk and dairy products do not contain detectable amounts of the infectious agent. The total number of BSE cases continued to rise, as would be expected from the long incubation period of the disease, and the peak incidence was reached in the last quarter of 1992, since when the number of new cases has started to fall. However, there are a number of false assumptions in the above reasoning:

- It is now known that none of the rendering processes used, before or after the 1980s, completely inactivates the infectivity of prions. Therefore cattle would have been exposed to scrapie prions in all countries worldwide where scrapie was present and MBM was used, not just in the UK in the 1980s. For example, the incidence of scrapie in the USA is hard to determine but in the 8 years after the level of compensation for slaughter of infected sheep was raised to $300 in 1977, the reported number of cases went up tenfold to a peak of about 50 affected flocks a year

- BSE is not scrapie. The biological properties of the scrapie and BSE agents are distinct, e.g. transmissibility to different animal species and pattern of lesions produced in infected animals (see Molecular Biology of Prions). There is no evidence to support the assumption that BSE is scrapie in cows. The only feasible interpretation based on present knowledge is that BSE originated as an endogenous bovine (cow) prion which was amplified by the feeding of cattle-derived protein in MBM back to cows. Thus the emergence of BSE in the UK appears to have been due to 'bad luck' compounded by poor husbandry practices, i.e. use of MBM in ruminant feed
- The reported international distribution of BSE is at odds with established facts. Approximately 40 000 tonnes of MBM were exported from the UK between 1985 and 1988; France alone imported at least 17 000 tonnes during this period and yet has only reported 20 cases of BSE compared with nearly 100 000 in the UK during the same period. During 1985–1990 the UK exported 57 900 cattle. These animals would have resulted in 1668 cases of BSE had they remained in Great Britain, but only a small fraction of these cases have been reported by the recipient countries. The United States Department of Agriculture (USDA) maintains that no cases of BSE have been confirmed in the USA, but TME in the USA has been found in mink fed on 'downer cows' and never fed on sheep which therefore could not have been exposed to scrapie (see McKenzie *et al.* in Further Reading).

Important unanswered questions remain concerning BSE. Many of these are raised by the large number of infected cattle (27 000) born after the 1988 feed ban. It is now generally acknowledged that the feed ban was initially improperly enforced and moreover, only applied to cattle feed. The same mills that were producing cattle feed were also producing sheep, pig and poultry food containing MBM, allowing many opportunities for contamination. As a result, in March 1996 the use of all mammalian MBM in animal feed was prohibited in the UK. It is now known that vertical transmission of BSE in herds can occur at a frequency of 1–10%. Similarly, there is a possibility of environmental transmission similar to that known to occur with scrapie (above). Apart from the economic damage done by BSE, the main concern at present is the possible risk to human health (discussed below).

Human TSEs

There are four recognized human TSEs (summarized in Table 8.3). Understanding of human TSEs is derived largely from studies of the animal TSEs already described. There are believed to be three sources from which human TSEs originate:

Table 8.3 Transmissible spongiform encephalopathies (TSEs) in humans

Disease	Description	Comments
Creutzfeldt–Jakob disease (CJD)	Spongiform encephalopathy in cerebral and/or cerebellar cortex and/ or subcortical grey matter, or encephalopathy with prion protein (PrP) immunoreactivity (plaque and/ or diffuse synaptic and/or patchy/ perivacuolar types)	Three forms: sporadic, iatrogenic (recognized risk, e.g. neurosurgery), familial (same disease in first degree relative)
Familial fatal insomnia (FFI)	Thalamic degeneration, variable spongiform change in cerebrum	Occurs in families with PrP_{178} asp–asn mutation
Gerstmann–Straussler–Scheinker disease (GSS)	Encephalo(myelo)pathy with multicentric PrP plaques	Occurs in families with dominantly inherited progressive ataxia and/or dementia
Kuru	Characterized by large amyloid plaques	Occurs in the Fore population of New Guinea due to ritual cannibalism (now eliminated)

- *Sporadic*: Creutzfeldt–Jakob disease (CJD) arises spontaneously at a frequency of about one in a million people per year with little variation worldwide. The average age at onset of disease is about 65 and the average duration of illness about 3 months. This category accounts for 90% of all human TSE, but only about 1% of sporadic CJD cases are transmissible to mice
- *Iatrogenic/acquired TSE*: This occurs due to recognized risks, e.g. neurosurgery, transplantation, etc. About 50 cases of TSE were caused in young people who received injections of human growth hormone or gonadotrophin derived from pooled cadaver pituitary gland extracts, a practice which has now been discontinued in favour of recombinant DNA-derived hormone.

Kuru was the first human spongiform encephalopathy to be investigated in detail and is possibly one of the most fascinating stories to have emerged from any epidemiological investigation. The disease occurred primarily in 169 villages occupied by the Fore tribes in the highlands of New Guinea. The first cases were recorded in the 1950s and involved progressive loss of voluntary neuronal control, followed by death less than 1 year after the onset of symptoms. The key to the origin of the disease was provided by the profile of its victims – it was never seen in young children, rarely in adult men and was most common in both male and female adolescents and in adult women. The Fore people practised ritual cannibalism as a rite of mourning for their dead. Women and children participated in these ceremonies but adult men did not take part, explaining the age/sex distribution of the disease. The incubation period for kuru can be in excess of 30 years, but in most cases is somewhat shorter. The practice of ritual cannibalism was discouraged in the late 1950s and the incidence of kuru has declined dramatically over the last few decades. Kuru is now very rare and only occurs in individuals who participated in cannibalistic feasts over 30 years ago.

- *Familial*: Approximately 10% of human TSEs are familial, i.e. inherited. There are a number of mutations in the human PrP gene which are known to give rise to TSE as an autosomal dominant trait acquired by hereditary mendelian transmission (Figure 8.5).

The above description covers the known picture of human TSEs which has been painstakingly built up over several decades. There is no evidence that any human TSE is traditionally acquired by an oral route, e.g. eating scrapie-infected sheep. There are good reasons why this should be (see Molecular Biology of Prions). However, in April 1996, a paper was published which described 10 cases of a new variant of CJD (nvCJD) in the UK (see Will *et al.* in Further Reading). Although relatively few in number, these cases share unusual features which distinguish them from other forms of CJD:

- An early age of onset or death (average 27, c.f. 65 for CJD)
- A prolonged duration of illness (average 13 months, c.f. 3 months for CJD)
- A predominantly psychiatric presentation including anxiety, depression, withdrawal and behavioural changes rather than neurological symptoms
- The subsequent development of a cerebellar syndrome with ataxia
- Forgetfulness and memory disturbance develop and progress to severe cognitive impairment
- Myoclonus (involuntary muscular contractions) develops in the majority of patients
- Typical EEG appearances of CJD are absent
- Neuropathologic spongiform change, neuronal loss and astrocytic gliosis most evident in the basal ganglia and thalamus. The most striking and consistent neuropathological feature is amyloid plaque formation reminiscent of that seen in kuru extensively distributed throughout the cerebrum and cerebellum.

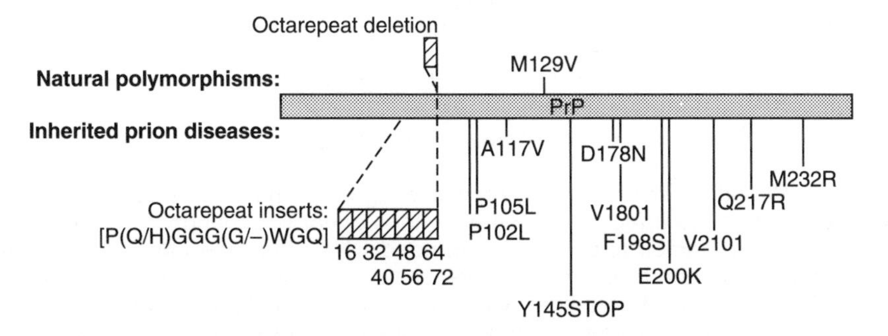

Figure 8.5 Mutations in the human PrP gene.

The official UK Spongiform Encephalopathy Advisory Committee concluded that vCJD is 'a previously unrecognised and consistent disease pattern' and that 'although there is no direct evidence of a link, on current data and in the absence of any credible alternative the most likely explanation at present is that these cases are linked to exposure to BSE before the introduction of the Specified Bovine Offal ban in 1989'.

Molecular biology of prions

The evidence that prions are not conventional viruses is based on the fact that nucleic acid is not necessary for infectivity since they show:

- Resistance to heat inactivation – infectivity is not inactivated by heating to 90°C for 30 min. Infectivity is severely reduced but not eliminated by high temperature autoclaving (135°C for 18 min). It has been reported that some infectious activity may even be retained after treatment for 1 h at 360°C

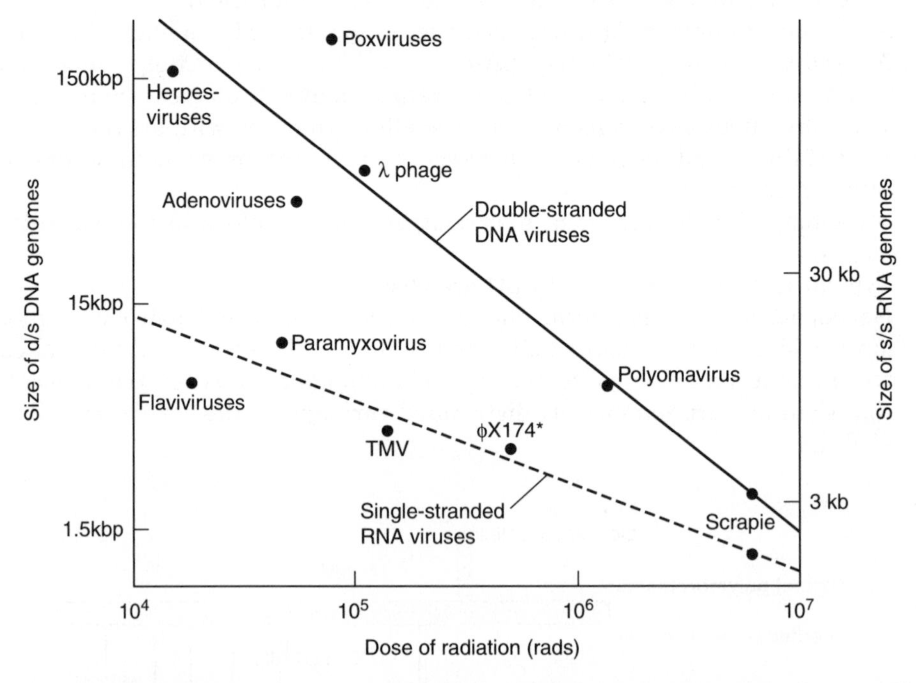

Figure 8.6 Radiation sensitivity of infectious agents. The dose of ionizing radiation required to destroy the infectivity of an infectious agent is dependent on the size of its genome. Larger genomes (e.g. double-stranded DNA viruses – upper line) present a larger 'target' and therefore are more sensitive than smaller genomes (e.g. single-stranded RNA viruses – lower line; N.B. ϕX174 has a single-stranded DNA genome). The scrapie agent is considerably more resistant to radiation than any known virus (note log scale on vertical axis).

- Resistance to radiation damage – infectivity was found to be resistant to short-wave ultraviolet radiation and to ionizing radiation. These treatments inactivate infectious organisms by causing damage to the genome. There is an inverse relationship between the size of target nucleic acid molecule and the dose of radioactivity or ultraviolet light needed to inactivate them, i.e. large molecules are sensitive to much lower doses than are smaller molecules (Figure 8.6). The scrapie agent was found to be highly resistant to both ultraviolet light and ionizing radiation, indicating that any nucleic acid present must be less than 80 nucleotides
- Resistance to DNAse and RNAse treatment, to psoralens and to Zn^{2+} catalysed hydrolysis – all of which treatments inactivate nucleic acids
- Sensitivity to urea, SDS, phenol and other protein-denaturing chemicals.

All of the above indicate an agent with the properties of a protein rather than a virus. A protein of 254 amino acids, PrP^{Sc}, is associated with scrapie infectivity. Biochemical purification of scrapie infectivity results in preparations highly enriched in PrP^{Sc} and purification of PrP^{Sc} results in enrichment of scrapie activity. In 1984, Prusiner determined the sequence of 15 amino acids at the end of purified PrP^{Sc}. This led to the discovery that all mammalian cells contain a gene (***Prnp***) which encodes a protein identical to PrP^{Sc}, termed PrP^{C}. No biochemical differences between PrP^{C} and PrP^{Sc} have been determined, although unlike PrP^{C}, PrP^{Sc} is partly resistant to protease digestion, resulting in the formation of a 141 amino acid protease-resistant fragment which accumulates as fibrils in infected cells (Figure 8.7). Only a proportion of the total PrP in diseased tissue is present as PrP^{Sc}, but this has been shown to be the infectious

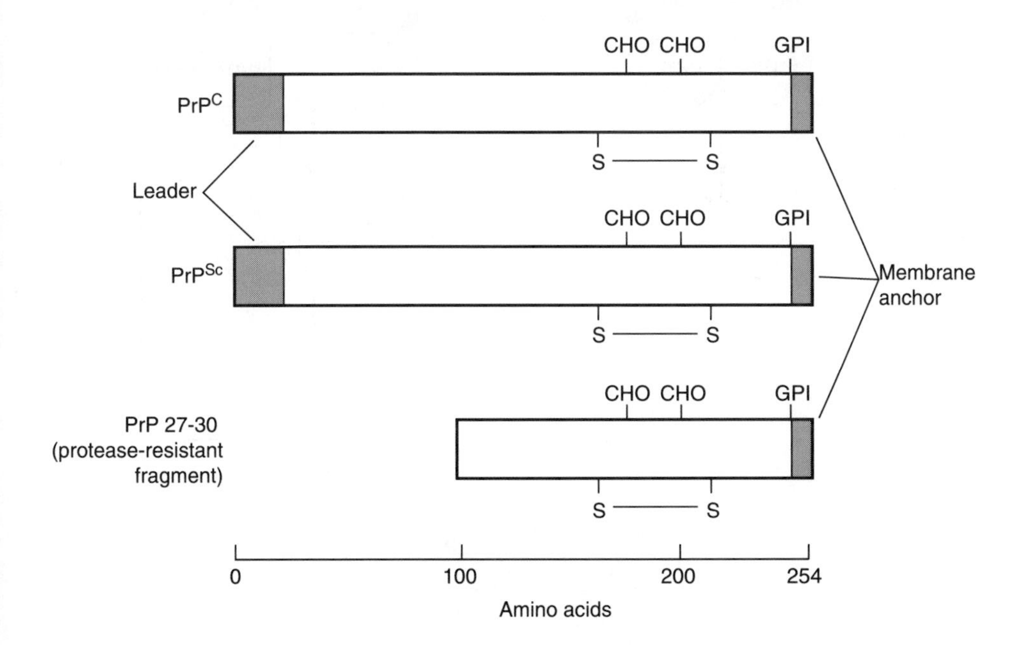

Figure 8.7 Structure of the prion protein, PrP.

form of the PrP protein, since highly purified PrP^{Sc} is infectious when used to inoculate experimental animals. Like other infectious agents, there is a dosage effect which gives a strong correlation between the amount of PrP^{Sc} in an inoculum and the incubation time until the development of disease.

Thus TSEs, which behave like infectious agents, appear to be caused by an endogenous gene/protein (Figure 8.8). Susceptibility of a host species to prion infection is co-determined by the prion inoculum and the PrP gene. Disease incubation times for individual prion isolates vary in different strains of inbred mice but for a given isolate in a particular strain are remarkably consistent. This variation depends on the mouse *Sinc* gene, which is very closely linked to or

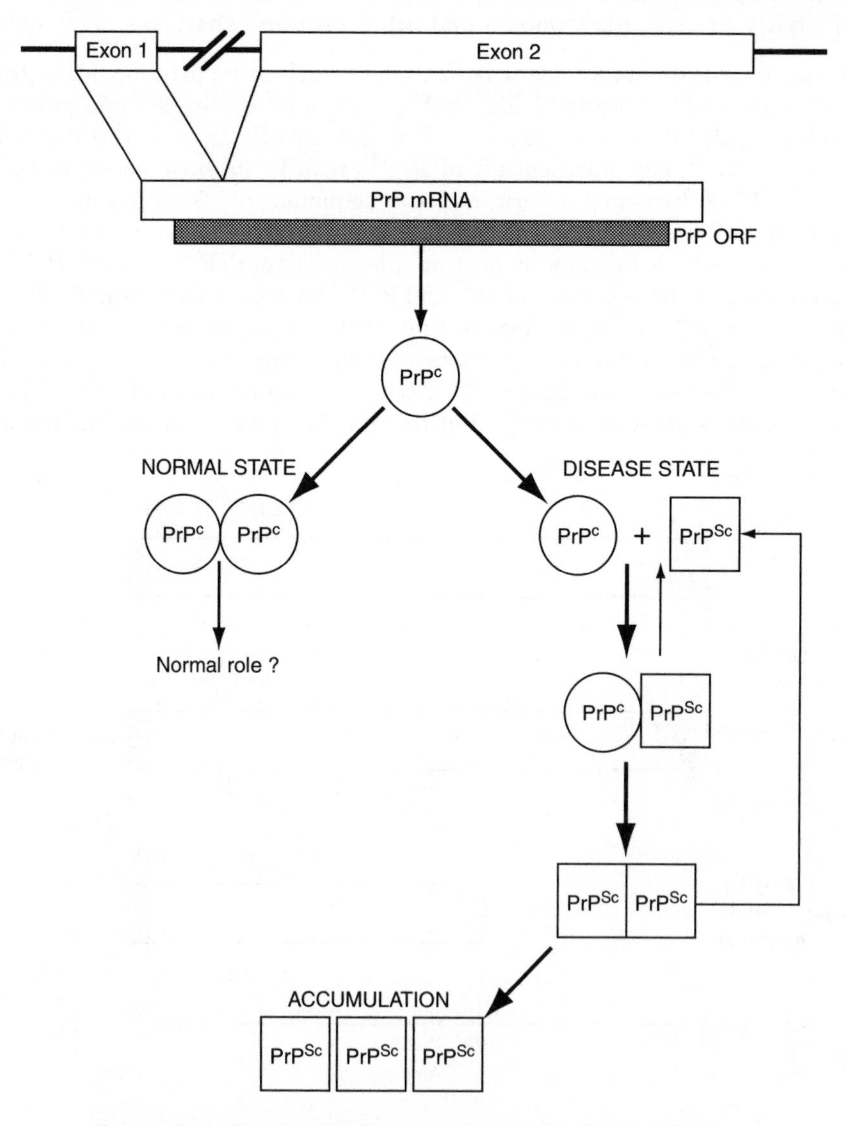

Figure 8.8 Schematic diagram of the role of PrP in TSEs.

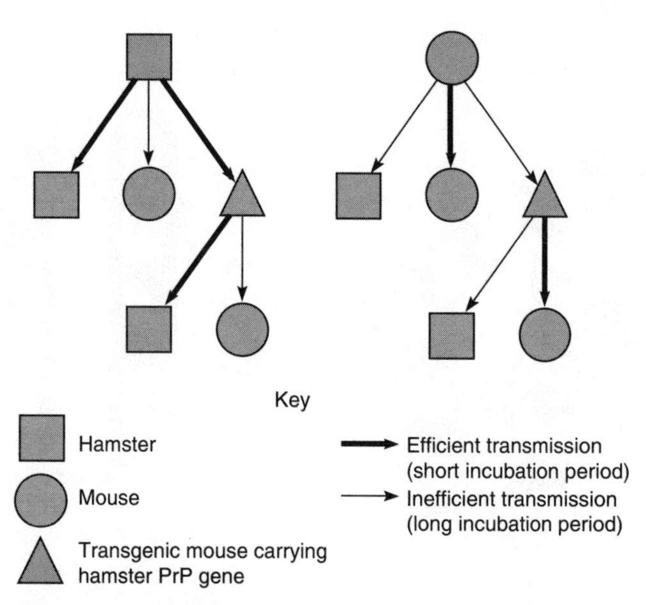

Figure 8.9 Experimental transmission of scrapie to animals demonstrates an apparent species barrier. Hamster-to-hamster and mouse-to-mouse transmission results in the onset of disease after a relatively short incubation period (75 days and 175 days, respectively). Transmission from one host species to another is much less efficient and disease occurs only after a much longer incubation period. Transmission of hamster-derived PrP to transgenic mice carrying several copies of the hamster PrP gene (the darker mice in this figure) is much more efficient, whereas transmission of mouse-derived PrP to the transgenic mice is less efficient. Subsequent transmission from the transgenic mice implies that some modification of the properties of the agent seems to have occurred.

coincident with the PrP gene itself, suggesting that some forms of PrP^{C} may be more easily converted to PrP^{Sc} than others. These observations have resulted in two important concepts:

(1) *Prion strain variation*: At least 15 different strains of PrP^{Sc} have been recognized. These can be determined from each other by the incubation time to the onset of disease and the type and distribution of lesions within the CNS in inbred strains of mice. Thus prions can be 'fingerprinted' and BSE can be distinguished from scrapie or CJD.

(2) *The species barrier*: When prions are initially transmitted from one species to another, disease develops only after a very long incubation period, if at all. On serial passage in the new species, the incubation time often decreases dramatically and then stabilizes. This species barrier can be overcome by introducing a PrP transgene from the prion donor, e.g. hamster, into the recipient mice (Figure 8.9).

However, PrP^{C} and PrP^{Sc} are not post-translationally modified and the genes which encode them are not mutated, i.e. this is distinct from mendelian inheritance of familial forms of CJD. How such apparently complex behaviour can be 'encoded' by a 254 amino acid protein has not been established, but there is evidence that the fundamental difference between the infectious, pathogenic

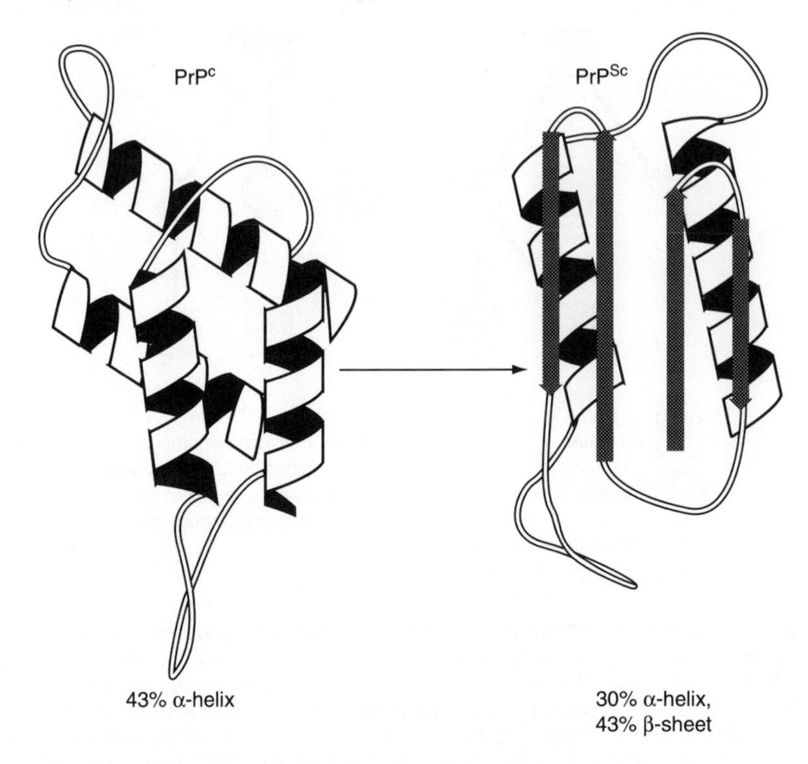

Figure 8.10 Conformational changes in PrP.

form (PrP^{Sc}) and the endogenous form (PrP^{C}) results from a change in the conformation of the folded protein (Figure 8.10).

The URE3 protein of the yeast *Saccharomyces cerevisiae* has properties very reminiscent of PrP. URE3 modifies a cellular protein Ure2p causing altered nitrogen metabolism. Cells 'infected' with URE3 can be 'cured' by treatment with protein-denaturing agents such as guanidium, which is believed to cause refolding of URE3 to the Ure2p conformation. The explanation for the inherited familial forms of prion disease is therefore presumably that inherited mutations enhance the rate of spontaneous conversion of PrP^{C} into PrP^{Sc}, permitting disease manifestation within the lifetime of an affected individual. This concept also suggests that the sporadic incidence of CJD can be accounted for by somatic mutation of the PrP gene and offers a possible explanation for the emergence of BSE, i.e. spontaneous mutation of a bovine PrP gene resulted in infectious prions which were then amplified through the food chain. In recent years, the construction of transgenic animals has cast further light on the above ideas:

- 'Knockout' $Prnp^{0/0}$ mice or sheep which do not possess an endogenous prion gene are completely immune to the effects of PrP^{Sc} and do not propagate infectivity to normal mice, indicating that production of endogenous PrP^{C} is an essential part of the disease process in TSEs and that the infectious inoculum of PrP^{Sc} does not replicate *per se*. Unfortunately, these experiments have given few clues to the normal role of PrP^{C}. $Prnp^{0/0}$ mice are developmentally

normal, although they may have altered sleep patterns. This indicates that loss of normal PrP^C function is not the cause of TSE and that the accumulation of PrP^{Sc} is responsible for disease symptoms
- Mice carrying a murine transgene with the $_P102_L$ mutation responsible for human GSS spontaneously develop a lethal scrapie-like disease which can be transmitted to normal mice
- HuPrP mice are not more susceptible to human TSEs than wild-type mice, i.e. the species barrier is intact. HuPrP differs from MoPrP at 28 of 254 amino acids. Mice carrying a chimeric mouse–human transgene which differs at only nine residues are susceptible to human TSEs. Mice which have been reconstituted with various alleles of the human Prnp gene and are currently being used as models to investigate human TSEs, in particular to investigate the transmissibility of the BSE agent to humans.

The prion hypothesis is revolutionary and has justifiably met with a somewhat sceptical reception. Although evidence in its favour is growing, there are a few inconsistencies which have not been adequately explained (Table 8.4). Certainly, the prion story is not complete; for example, there is sketchy evidence that PrP interacts with two presently unidentified cellular proteins known as X and Y which modify the 'phenotype' of affected animals. It is possible to construct numerous alternative theories of varying degrees of complexity (and plausibility!) to fit the experimental data. Science progresses by the construction of experimentally verifiable hypotheses. For many years, research into spongiform encephalopathies has been agonizingly slow because each individual experiment has taken at least one and in some cases many years to complete. With the advent of molecular biology, this has now become a fast-moving and dynamic

Table 8.4 Evidence for and against the prion hypothesis

For the prion hypothesis	Against the prion hypothesis
PrP^{Sc} co-purifies with infectivity	'Purified' PrP^{Sc} may still contain minimal amounts of nucleic acid
PrP^{Sc} accumulates in the brains of affected animals and correlates with pathology	The existence of multiple 'strains' of scrapie suggests modification by an additional agent, e.g. ubiquitous small nucleic acid molecule or widely-distributed microorganism
Prnp mutations are associated with familial transmissible spongiform enephalopathies (TSEs) in humans	Other amyloid diseases (e.g. Alzheimer's) are non-transmissible
Transgenic mice expressing foreign PrP^C have altered species tropism for PrP^{Sc}	Not all diseases with protease-resistant PrP are transmissible (e.g. familial fatal insomnia and some cases of Gerstmann–Straussler–Scheinker disease)
$Prnp^{0/0}$ mice are not susceptible to PrP^{Sc}	PrP^{Sc} is not always detectable in all cases of TSE

field. The next few years will undoubtedly reveal much more information about the cause of these diseases and will probably provide much food for thought about the interaction between infectious agents and the host in the pathogenesis of infectious diseases.

Summary

A variety of novel infectious agents cause disease in plants, in animals and in man. Several types of non-viral, sub-cellular pathogens have disease-causing potential. These include satellites, viroids and prions. Conventional strategies to combat virus infections, such as drugs and vaccines, have no effect on these unconventional agents. A full understanding of the biology of these novel infectious entities will be necessary before means of treatment for the diseases they cause will become available.

Further Reading

Diener, T.O. (Ed.) (1987). *The Viroids.* Plenum Press, New York.
McKenzie, D. *et al.* (1996). Transmissible mink encephalopathy. *Sem. Virol.* **7**: 201–6.
Prusiner, S.B. (1995). The prion diseases. *Scientific American* **272**: 48–57.
Prusiner, S.B. (Ed.) (1996). Prions, prions, prions. *Curr. Top. Microbiol. Immunol.* Vol **207**.
Will, R.G. *et al.* (1996). A new variant of Creutzfeldt–Jakob disease in the UK. *Lancet* **347**: 921–25.

Self-Assessment Questions

(8.1) Are the following statements true or false?
- (a) Satellites encode their own coat proteins.
- (b) Viroids use the coat protein of their helper virus.
- (c) The nucleotide sequence of satellites is related to the helper virus genome.
- (d) Viroids are 200–400 nucleotides long.
- (e) All viroids share a conserved region involved in their replication.

(8.2) Are the following statements true or false?
- (a) Viroids cause disease by interfering with cellular DNA replication.
- (b) Hepatitis delta virus (HDV) δAg-S is necessary for the assembly and release of HDV-containing particles.
- (c) HDV δAg-L is necessary for HDV replication.

(d) HDV has some of the properties of a satellite virus and some of a viroid.
(e) HDV infection without HBV coinfection is associated with fulminant hepatitis.

(8.3) The following diseases are caused by prions (true or false?):
(a) Alzheimers's syndrome.
(b) Coconut cadang-cadang disease.
(c) Chronic wasting disease (CWD).
(d) Creutzfeldt–Jakob disease (CJD).
(e) Fatal familial insomnia (FFI).

(8.4) Are the following statements true or false?
(a) Scrapie was first described in 1982.
(b) Scrapie is endemic in the UK and USA.
(c) Transmissible mink encephalopathy (TME) occurs only in the USA.
(d) More than 161 000 cases of BSE have been reported in the UK.
(e) BSE is caused by scrapie-infected feed.

(8.5) Are the following statements true or false?
(a) There are three sources of human TSEs: sporadic, iatrogenic/acquired and familial/inherited.
(b) Sporadic Creutzfeldt–Jakob disease (CJD) accounts for about 90% of human TSE.
(c) Kuru is an inherited form of TSE.
(d) The incidence of CJD is similar worldwide.
(e) Only about 1% of sporadic CJD cases are transmissible to mice.

(8.6) Prion infectivity is inactivated by the following treatments (true or false?):
(a) Heating to 90°C for 30 min.
(b) Heating to 135°C for 18 min.
(c) DNAse treatment.
(d) RNAse treatment.
(e) Phenol.

(8.7) PrP^{C} (true or false?):
(a) is a 254 amino acid protein.
(b) is encoded by a cellular gene.
(c) is converted by PrP^{Sc} to a protease-resistant form.
(d) consists of 30% α-helix and 43% β-sheet.
(e) is resistant to short-wave ultraviolet radiation and to ionizing radiation.

(8.8) Are the following statements true or false?
(a) Transgenic animals which do not possess an endogenous prion gene are immune to the effects of PrP^{Sc}.
(b) Transgenic animals which do not possess an endogenous prion gene develop a variety of tumours.
(c) Human PrP differs from mouse PrP at 28 of 254 amino acids.
(d) The BSE agent is indistinguishable from that of scrapie.

(e) The 'species barrier' prevents the efficient transfer of infectious prions from one species to another.

(8.9) Are the following statements true or false?

(a) The URE3 protein of yeast has the properties of a prion.
(b) The URE3 phenotype of yeast can be 'cured' by treatment with guanidium.
(c) PrP is believed to interact with at least two cellular proteins, X and Y.
(d) Prions cause disease in plants.
(e) The prion hypothesis is the only possible interpretation for TSEs.

Answers to Self-Assessment Questions are given in Appendix 3.

Appendix 1

Classification of Subcellular Infectious Agents

The importance of virus identification has been discussed in Chapter 4. Subcellular agents present a particular problem for taxonomists. They are too small to be seen without electron microscopes but very small changes in molecular structure may give rise to agents with radically different properties. The vast majority of viruses which are known have been studied because they have pathogenic potential for humans, animals, or plants. Therefore, the disease symptoms caused by infection are one criterion used to aid classification. The physical structure of a virus particle can be determined directly (by electron microscopy) or indirectly (by biochemical or serological investigation) and is also used in classification. However, the structure and sequence of the virus genome continues to increase in importance as molecular biological analysis provides a rapid and sensitive way to detect and differentiate many diverse viruses.

In 1966, the International Committee on Nomenclature of Viruses was established and produced the first unified scheme for virus classification. In 1973, this committee expanded its objectives and renamed itself the International Committee on Taxonomy of Viruses (ICTV), which meets every four years. Rules for virus taxonomy were established, some of which include:

- Latin binomial names (e.g. *Rhabdovirus carpio*) have now been abandoned, although existing names have been retained wherever possible. No person's name should be used in nomenclature. Names should have international meaning
- A virus name should be meaningful and should consist of as few words as possible. Serial numbers or letters are not acceptable as names
- A virus species is represented by a cluster of strains from a variety of sources or a population of strains from one particular source, all of which have in common a set pattern of stable properties that separates the cluster from other clusters of strains

- A genus is a group of virus species sharing common characters. Approval of a new genus is linked to the acceptance of a type species, i.e. a species which displays the typical characteristics on which the genus is based
- A family is a group of genera with common characters. Approval of a new family is linked to the acceptance of a type genus.

In general terms, groups of related viruses are divided into families whose names are capitalized, italicized, and end in the suffix *-viridae* (e.g. *Poxviridae*). In most cases, a higher level of classification than the family has not been established, although this is now changing with the consideration of evolutionary relationships between families and the arrival of the concept of 'superfamilies' equivalent to the 'orders' of formal nomenclature (see Chapter 3). In a few cases, very large families have been subdivided into subfamilies, written as above (e.g. *Chordopoxvirinae*). In most cases, families consist of a collection of genera whose names are capitalized, italicized and end in the suffix '*virus*' (e.g. *Orthopoxvirus*). In practice, this formal nomenclature is rarely used and vernacular usage such as 'the picornavirus family' or the 'enterovirus genus' is perfectly acceptable.

Many virus genera are sufficiently different from others to be recognized as a separate group in their own right and not as part of a larger family. Each genus contains a number of virus species, whose names are not capitalized or italicized (e.g. vaccinia virus). Some genera are monotypic, i.e. contain only one species. Subspecies, strains, isolates, variants, mutants, and artificially created laboratory recombinants are not officially recognized by the ICTV.

The sixth report of the ICTV was published in 1995 (F.A. Murphy, C.M. Fauquet, D.H.L. Bishop, S.A. Ghabrial, A.W. Jarvis, G.P. Martelli, M.A. Mayo, M.D. Summers (Eds.) Virus Taxonomy: Classification and Nomenclature of Viruses: Sixth Report of the International Committee on Taxonomy of Viruses. *Archives of Virology*, Supplement 10, Springer-Verlag, 1995). A total of 3465 virus species belonging to one order (Mononegvirales), 50 families, 9 subfamilies and 164 genera are recognized in this report. These well-characterized viruses are an unknown proportion of the total number of viruses which exist. By the time of the next ICTV meeting, undoubtedly many more viruses will have been added to this list.

Satellites and viroids are not officially classified by the ICTV in the same way as conventional viruses. However, criteria have been established which divide plant virus satellites into four groups:

- *Type A*: An RNA of more than 700 nt which encodes a structural capsid protein forming satellite-specific particles
- *Type B*: An RNA of more than 700 nt which encodes a non-structural protein
- *Type C*: A linear RNA of less than 700 nt which does not encode any proteins
- *Type D*: A circular RNA of less than 700 nt which does not encode any proteins.

Similarly, viroids are not officially recognized by the ICTV, but are grouped on the basis of nucleotide sequence conservation of the central region involved in replication. An arbitrary level of 90% sequence similarity is the criterion used to assign viroids to groups, each based on a type member (Fauquet, C.M., Martelli, G.P. Updated ICTV list of names and abbreviations of viruses, viroids and satellites infecting plants. *Archives of Virology* (1995), **140**: 393–413.).

Group I: dsDNA Viruses

Family	Genus	Type Species	Hosts
Adenoviridae	*Mastadenovirus*	Human adenovirus 2	Vertebrates
	Aviadenovirus	Fowl adenovirus 1	Vertebrates
	African swine fever-like viruses	African swine fever virus	Vertebrates
Baculoviridae	*Nucleopolyhedrovirus*	*Autographa californica* nucleopolyhedrovirus	Invertebrates
	Granulovirus	*Plodia interpunctella* granulovirus	Invertebrates
Corticoviridae	*Corticovirus*	*Alteromonas* phage PM2	Bacteria
Fuselloviridae	*Fusellovirus*	*Sulfolobus* virus 1	Bacteria
Herpesviridae	*Alphaherpesvirinae*		
	Simplexvirus	Human herpesvirus 1	Vertebrates
	Varicellovirus	Human herpesvirus 3	Vertebrates
	Betaherpesvirinae		
	Cytomegalovirus	Human herpesvirus 5	Vertebrates
	Muromegalovirus	Mouse cytomegalovirus 1	Vertebrates
	Roseolovirus	Human herpesvirus 6	Vertebrates
	Gammaherpesvirinae		
	Lymphocryptovirus	Human herpesvirus 4	Vertebrates
	Rhadinovirus	Ateline herpesvirus 2	Vertebrates
Iridoviridae	*Iridovirus*	Chilo iridescent virus	Invertebrates
	Chloriridovirus	Mosquito iridescent virus	Invertebrates
	Ranavirus	Frog virus 3	Vertebrates
	Lymphocystivirus	Flounder virus	Vertebrates
	Goldfish virus 1-like viruses	Goldfish virus 1	Vertebrates
Lipothrixviridae	*Lipothrixvirus*	*Thermoproteus* virus 1	Bacteria
Myoviridae	T4-like phages	Coliphage T4	Bacteria
Papovaviridae	*Polyomavirus*	Murine polyomavirus	Vertebrates
	Papillomavirus	Cottontail rabbit papillomavirus (Shope)	Vertebrates
Phycodnaviridae	*Phycodnavirus*	*Paramecium bursaria Chlorella* virus 1	Algae
Plasmaviridae	*Plasmavirus*	*Acholeplasma* phage L2	Mycoplasma
Podouiridae	T7-like phages	Coliphage T7	Bacteria

(continued)

Group I: dsDNA Viruses *(continued)*

Family	Genus	Type Species	Hosts
Polydnaviridae	*Ichnovirus*	*Campoletis sonorensis* virus	Invertebrates
	Bracovirus	*Cotesia melanoscela* virus	Invertebrates
Poxviridae	*Chordopoxvirinae*		
	Orthopoxvirus	Vaccinia virus	Vertebrates
	Parapoxvirus	Orf virus	Vertebrates
	Avipoxvirus	Fowlpox virus	Vertebrates
	Capripoxvirus	Sheeppox virus	Vertebrates
	Leporipoxvirus	Myxoma virus	Vertebrates
	Suipoxvirus	Swinepox virus	Vertebrates
	Molluscipoxvirus	*Molluscum contagiosum* virus	Vertebrates
	Yatapoxvirus	Yaba monkey tumour virus	Vertebrates
	Entomopoxvirinae		
	Entomopoxvirus A	*Melolontha melolontha* entomopoxvirus	Invertebrates
	Entomopoxvirus B	*Amsacta moorei* entomopoxvirus	Invertebrates
	Entomopoxvirus C	*Chironomus luridus* entomopoxvirus	Invertebrates
Siphoviridae	Lambda-like phages	Coliphage lambda	Bacteria
	Rhizidiovirus	*Rhizidiomyces* virus	Fungi
Tectiviridae	*Tectivirus*	Enterobacteria phage PRD1	Bacteria

Group II: The ssDNA Viruses

Family	Genus	Type Species	Hosts
Circoviridae	*Circovirus*	Chicken anaemia virus	Vertebrates
Geminiviridae	Subgroup I *Geminivirus*	Maize streak virus	Plants
	Subgroup II *Geminivirus*	Beet curly top virus	Plants
	Subgroup III *Geminivirus*	Bean golden mosaic virus	Plants
Inoviridae	*Inovirus*	Coliphage fd	Bacteria
	Plectrovirus	*Acholeplasma* phage L51	Bacteria
Microviridae	*Microvirus*	Coliphage ϕX174	Bacteria
	Spiromicrovirus	*Spiroplasma* phage 4	Spiroplasma
	Bdellomicrovirus	*Bdellovibrio* phage MAC1	Bacteria
	Chlamydiamicrovirus	*Chlamydia* phage 1	Bacteria

(continued)

Group II: The ssDNA Viruses *(continued)*

Family	Genus	Type Species	Hosts
Parvoviridae			
Parvovirinae	*Parvovirus*	Mice minute virus	Vertebrates
	Erythrovirus	B19 virus	Vertebrates
	Dependovirus	Adeno-associated virus 2	Vertebrates
Densovirinae	*Densovirus*	*Junonia coenia* densovirus	Invertebrates
	Iteravirus	*Bombyx mori* densovirus	Invertebrates
	Contravirus	*Aedes aegypti* densovirus	Invertebrates

Group III: dsRNA Viruses

Family	Genus	Type Species	Hosts
Birnaviridae	*Aquabirnavirus*	Infectious pancreatic necrosis virus	Vertebrates
	Avibirnavirus	Infectious bursal disease virus	Vertebrates
	Entomobirnavirus	*Drosophila* X virus	Invertebrates
Cystoviridae	*Cystovirus*	*Pseudomonas* phage ϕ6	Bacteria
Hypoviridae	*Hypovirus*	*Cryphonectria* hypovirus 1–EP713	Fungi
Partitiviridae	*Partitivirus*	*Gaeumarmomyces graminis* virus 019/6–A	Fungi
	Chrysovirus	*Penicillium chrysogenum* virus	Fungi
	Alphacryptovirus	White clover cryptic virus 1	Plants
	Betacryptovirus	White clover cryptic virus 2	Plants
Reoviridae	*Orthoreovirus*	Reovirus 3	Vertebrates
	Orbivirus	Bluetongue virus 1	Vertebrates
	Rotavirus	Simian rotavirus SA11	Vertebrates
	Coltivirus	Colorado tick fever virus	Vertebrates
	Aquareovirus	Golden shiner virus	Vertebrates
	Cypovirus	*Bombyx mori* cypovirus 1	Invertebrates
	Fijivirus	Fiji disease virus	Plants
	Phytoreovirus	Wound tumour virus	Plants
	Oryzavirus	Rice ragged stunt virus	Plants
Totiviridae	*Totivirus*	*Saccharomyces cerevisiae* virus L-A	Fungi
	Giardiavirus	*Giardia lamblia* virus	Protozoa
	Leishmaniavirus	*Leishmania* RNA virus 1-1	Protozoa

(continued)

Group IV: (+)sense RNA Viruses

Family	Genus	Type Species	Hosts
Astroviridae	*Astrovirus*	Human astrovirus 1	Vertebrates
	Arterivirus	Equine arteritis virus	Vertebrates
Barnaviridae	*Barnavirus*	Mushroom bacilliform virus	Fungi
	Marafivirus	Maize rayado fino virus	Plants
Bromoviridae	*Alfamovirus*	Alfalfa mosaic virus	Plants
	Ilarvirus	Tobacco streak virus	Plants
	Bromovirus	Brome mosaic virus	Plants
	Cucumovirus	Cucumber mosaic virus	Plants
Caliciviridae	*Calicivirus*	Vesicular exanthema of swine virus	Vertebrates
	Capillovirus	Apple stem grooving virus	Plants
	Carlavirus	Carnation latent virus	Plants
	Closterovirus	Beet yellows virus	Plants
Comoviridae	*Comovirus*	Cowpea mosaic virus	Plants
	Fabavirus	Broad bean wilt virus 1	Plants
	Nepovirus	Tobacco ringspot virus	Plants
Coronaviridae	*Coronavirus*	Avian infectious bronchitis virus	Vertebrates
	Torovirus	Berne virus	Vertebrates
	Dianthovirus	Carnation ringspot virus	Plants
	Enamovirus	Pea enation mosaic virus	Plants
Flaviviridae	*Flavivirus*	Yellow fever virus	Vertebrates
	Pestivirus	Bovine diarrhea virus	Vertebrates
	Hepatitis C-like viruses	Hepatitis C virus	Vertebrates
	Furovirus	Soil-borne wheat mosaic virus	Plants
	Hordeivirus	Barley stripe mosaic virus	Plants
	Idaeovirus	Raspberry bushy dwarf virus	Plants
Leviviridae	*Levivirus*	Enterobacteria phage MS2	Bacteria
	Allolevivirus	Enterobacteria phage Q-B	Bacteria
	Luteovirus	Barley yellow dwarf virus	Plants
	Machlomovirus	Maize chlorotic mottle virus	Plants

(continued)

Group IV: (+)sense RNA Viruses *(continued)*

Family	Genus	Type Species	Hosts
	Necrovirus	Tobacco necrosis virus	Plants
Nodaviridae	*Nodavirus*	Nodamura virus	Invertebrates
Picornaviridae	*Enterovirus*	Poliovirus 1	Vertebrates
	Rhinovirus	Human rhinovirus 1A	Vertebrates
	Hepatovirus	Hepatitis A virus	Vertebrates
	Cardiovirus	Encephalomyocarditis virus	Vertebrates
	Aphtovirus	Foot-and-mouth disease virus O	Vertebrates
	Potexvirus	Potato virus X	Plants
Potyviridae	*Potyvirus*	Potato virus Y	Plants
	Rymovirus	Ryegrass mosaic virus	Plants
	Bymovirus	Barley yellow mosaic virus	Plants
Sequiviridae	*Sequivirus*	Parsnip yellow fleck virus	Plants
	Waikavirus	Rice tungro spherical virus	Plants
	Sobemovirus	Southern bean mosaic virus	Plants
Tetraviridae	*Nudaurelia capensis* β-like viruses	*Nudaurelia capensis* β virus	Invertebrates
	Nudaurelia capensis ω-like viruses	*Nudaurelia capensis* ω-virus	Invertebrates
	Tobamovirus	Tobacco mosaic virus	Plants
	Tobravirus	Tobacco rattle virus	Plants
Tombusviridae	*Tombusvirus*	Tomato bushy stunt virus	Plants
	Carmovirus	Carnation mottle virus	Plants
Togaviridae	*Alphavirus*	Sindbis virus	Vertebrates
	Rubivirus	Rubella virus	Vertebrates
	Trichovirus	Apple chlorotic leaf spot virus	Plants
	Tymovirus	Turnip yellow mosaic virus	Plants
	Umbravirus	Carrot mottle virus	Plants

(continued)

Group V: (-) sense RNA Viruses

Family	Genus	Type Species	Hosts
	Order: Mononegavirales		
Filoviridae	*Filovirus*	Marburg virus	Vertebrates
Paramyxoviridae	*Paramyxovirinae*		
	Paramyxovirus	Human parainfluenza virus 1	Vertebrates
	Morbillivirus	Measles virus	Vertebrates
	Rubulavirus	Mumps virus	Vertebrates
	Pneumovirinae		
	Pneumovirus	Human respiratory syncytial virus	Vertebrates
Rhabdoviridae	*Vesiculovirus*	Vesicular stomatitis Indiana virus	Vertebrates
	Lyssavirus	Rabies virus	Vertebrates
	Ephemerovirus	Bovine ephemeral fever virus	Vertebrates
	Cytorhabdovirus	Lettuce necrotic yellows virus	Plants
	Nucleorhabdovirus	Potato yellow dwarf virus	Plants
Arenaviridae	*Arenavirus*	Lymphocytic choriomeningitis virus	Vertebrates
Bunyaviridae	*Bunyavirus*	Bunyamwera virus	Vertebrates
	Hantavirus	Hantaan virus	Vertebrates
	Nairovirus	Nairobi sheep disease virus	Vertebrates
	Phlebovirus	Sandfly fever Sicilian virus	Vertebrates
	Tospovirus	Tomato spotted wilt virus	Plants
Orthomyxoviridae	*Influenzavirus* A, B	Influenza A virus	Vertebrates
	Influenzavirus C	Influenza C virus	Vertebrates
	Thogoto-like viruses	Thogoto virus	Vertebrates
	Tenuivirus	Rice stripe virus	Plants

Group VI: RNA Reverse Transcribing Viruses

Family	Genus	Type Species	Hosts
Retroviridae	Mammalian type B retroviruses	Mouse mammary tumour virus	Vertebrates
	Mammalian type C retroviruses	Murine leukaemia virus	Vertebrates
	Avian type C retroviruses	Avian leukosis virus	Vertebrates
	Type D retroviruses	Mason–Pfizer monkey virus	Vertebrates
	BLV-HTLV retroviruses	Bovine leukaemia virus	Vertebrates
	Lentivirus	Human immunodeficiency virus 1	Vertebrates
	Spumavirus	Human spumavirus	Vertebrates

(continued)

Group VII: DNA Reverse Transcribing Viruses

Family	Genus	Type Species	Hosts
	Badnavirus	*Commelina* yellow mottle virus	Plants
	Caulimovirus	Cauliflower mosaic virus	Plants
Hepadnaviridae	*Orthohepadnavirus*	Hepatitis B virus	Vertebrates
	Avihepadnavirus	Duck hepatitis B virus	Vertebrates

Subviral Agents: Satellites, Viroids, Agents of Spongiform Encephalopathies (Prions)

Agent	Example	Hosts
Satellites	Tobacco necrosis virus satellite	Plants Vertebrates Invertebrates Fungi
Deltavirus	Hepatitis delta virus	Vertebrates
Viroids	Potato spindle tuber viroid	Plants
Prions	Scrapie agent	Vertebrates

Appendix 2

Glossary and Abbreviations

Terms shown in the text in **bold** print are defined in this glossary.

abortive infection The initiation of infection without completion of the infectious cycle and, therefore, without the production of infectious particles (c.f. **productive infection**).

adjuvant A substance included in a medication to improve the action of the other constituents; usually, a component of vaccines which boosts their immunogenicity, e.g. aluminium sulfate.

ambisense A single-stranded virus genome which contains genetic information encoded in both the positive (i.e. virus-sense) and negative (i.e. complementary) orientations, e.g. *Bunyaviridae* and *Arenaviridae* (see Chapter 3).

anergy An immunologically unresponsive state in which lymphocytes are present but not functionally active.

apoptosis The genetically programmed death of certain cells which occurs during various stages in the development of multicellular organisms and may also be involved in control of the immune response.

assembly A late phase of viral replication during which all the components necessary for the formation of a mature virion collect at a particular site in the cell and the basic structure of the virus particle is formed (see Chapter 4).

attachment The initial interaction between a virus particle and a cellular receptor molecule; the phase of viral replication during which this occurs (see Chapter 4).

attenuated A pathogenic agent which has been genetically altered and displays decreased virulence. Attenuated viruses are the basis of live virus vaccines (see Chapter 6).

autocrine The production by a cell of a growth factor which is required for its own growth. Such positive feedback mechanisms may result in cellular transformation (see Chapter 7).

avirulent An infectious agent which has *no* disease-causing potential. It is doubtful if such agents really exist – even the most innocuous organisms may cause disease in certain circumstances, e.g. in immunocompromised hosts.

bacteriophage A virus which replicates in a bacterial host cell.

bp Base pair – a single pair of nucleotide residues in a double-stranded nucleic acid molecule held together by Watson–Crick hydrogen bonds (see **kbp**).

budding A mechanism involving release of virus particle from an infected cell by extrusion through a membrane. The site of budding may be at the surface of the cell or may involve the cytoplasmic or nuclear membranes depending on the site of assembly. Virus **envelopes** are acquired during budding.

capsid The protective protein coat of a virus particle (see Chapter 2).

chromatin The ordered complex of DNA plus proteins (histones and non-histone chromosomal proteins) found in the nucleus of **eukaryotic** cells.

***cis*-acting** A genetic element which affects the activity of contiguous (i.e. on the same nucleic acid molecule) genetic regions; e.g. transcriptional promoters and enhancers are *cis*-acting sequences adjacent to the genes whose transcription they control.

complementation The interaction of virus gene products in infected cells that results in the yield of one or both of the parental mutants being enhanced while their genotypes remain unchanged.

conditional mutant A mutant phenotype which is replication competent under 'permissive' conditions but not under 'restrictive' or 'non-permissive' conditions; e.g. a virus with a temperature sensitive (t.s.) mutation may be able to replicate at the permissive temperature of 33°C but unable to replicate or is severely inhibited at the non-permissive temperature of 38°C.

conditional-lethal mutant A conditional mutation whose phenotype is (relatively) unaffected under permissive conditions but which is severely inhibitory under non-permissive conditions.

cytopathic effect (c.p.e.) Cellular injury caused by virus infection; the effects of virus infection on cultured cells, visible by microscopic or direct visual examination (see Chapter 7).

defective interfering (D.I.) particles Particles encoded by genetically deleted virus genomes which lack one or more essential functions for replication.

d/s Double-stranded (nucleic acid).

eclipse period An early phase of infection when virus particles have broken down after penetrating cells, releasing their genomes within the host cell as a prerequisite to replication. Often used to refer specifically to bacteriophages (see Chapter 4).

emergent virus A virus identified as the cause of an increasing incidence of disease, possibly as a result of changed environmental or social factors (see Chapter 7).

endemic A pattern of disease which recurs or is commonly found in a particular geographic area (c.f. **epidemic**).

enhancers *cis*-Acting genetic elements which potentiate the transcription of genes or translation of mRNAs.

envelope An outer (bounding) lipoprotein bilayer membrane possessed by many viruses. (N.B. Some viruses contain lipid as part of a complex outer layer, but these are not usually regarded as enveloped unless a bilayer unit membrane structure is clearly demonstrable).

epidemic A pattern of disease characterized by a rapid increase in the number of cases occurring and widespread geographical distribution (c.f. **endemic**). An epidemic which encompasses the entire world is known as a **pandemic**.

eukaryote An organism whose genetic material is separated from the cytoplasm by a nuclear membrane and divided into discrete chromosomes.

exon A region of a gene expressed as protein after the removal of **introns** by post-transcriptional splicing.

fusion protein A virus protein required and responsible for fusion of the virus **envelope** (or sometimes, the **capsid**) with a cellular membrane and consequently, for entry into the cell (see Chapter 4).

genome The nucleic acid comprising the entire genetic information of an organism.

helix A cylindrical solid formed by stacking repeated subunits in a constant relationship with respect to their amplitude and pitch (see Chapter 2).

haemagglutination The (specific) agglutination of red blood cells by a virus or other protein.

heterozygosis Aberrant packaging of multiple genomes may occasionally result in multiploid particles (i.e. containing more than a single genome) which are therefore heterozygous.

hnRNA 'Heterogeneous nuclear RNA' or 'heavy nuclear RNA' – the primary, unspliced transcripts found in the nucleus of eukaryotic cells.

hyperplasia Excessive cell division or the growth of abnormally large cells. In plants, results in the production of swollen or distorted areas due to the effects of plant viruses.

hypoplasia Localized retardation of cell growth. Numerous plant viruses cause this effect, frequently leading to **mosaicism** (the appearance of thinner, yellow areas on the leaves).

icosahedron A solid shape consisting of 20 triangular faces arranged around the surface of a sphere – the basic symmetry of many virus particles (see Chapter 2).

immortalized cell — A cell capable of indefinite growth (i.e. number of cell divisions) in culture. On rare occasions, immortalized cells arise spontaneously but are more commonly caused by mutagenesis as a result of virus **transformation** (see Chapter 7).

inclusion bodies — Subcellular structures formed as a result of virus infection. Often a site of virus assembly (see Chapter 4).

interference — Inhibition (partial or complete) of virus replication by another virus (see Chapter 6).

intron — A region of a gene removed after transcription by splicing and consequently not expressed as protein (c.f. **exon**).

IRES (Internal Ribosome Entry Site) — An RNA secondary structure found in the 5′ untranslated region (UTR) of (+)sense RNA viruses such as picornaviruses and flaviviruses, which functions as a 'ribosome landing pad', allowing internal initiation of translation on the vRNA.

isometric — A solid displaying cubic symmetry, of which the icosahedron is one form.

kb — 1000 nucleotide residues – a unit of measurement of single-stranded nucleic acid molecules. Sometimes (wrongly) used to mean **kbp** (below).

kbp — 1000 base pairs (see above) – a unit of measurement of double-stranded nucleic acid molecules.

latent period — The time after infection before the first new extracellular virus particles appear (see Chapter 4).

lysogeny — Persistent, latent infection of bacteria by **temperate bacteriophages** such as phage λ.

lytic virus — Any virus (or virus infection) which results in the death of infected cells and their physical breakdown.

maturation — A late phase of virus infection during which newly formed virus particles become infectious. Usually involves structural changes in the particle resulting from specific cleavages of capsid proteins to form the mature products, or conformational changes in proteins during assembly (see Chapter 4).

monocistronic A messenger RNA that consists of the transcript of a single gene and which therefore encodes a single polypeptide; a virus genome which produces such an mRNA (cf. **polycistronic**).

monolayer A flat, contiguous sheet of adherent cells attached to the solid surface of a culture vessel.

mosaicism The appearance of thinner, yellow areas on the leaves of plants caused by the cytopathic effects of plant viruses.

movement proteins Specialized proteins encoded by plant viruses which modify plasmodesmata (channels which pass through cell walls connecting the cytoplasm of adjacent cells) and cause virus nucleic acids to be transported from one cell to the next, permitting the spread of a virus infection.

mRNA Messenger RNA.

multiplicity of infection (m.o.i.) The (average) number of virus particles which infect each cell in an experiment.

necrosis Cell death, particularly that caused by an external influence (c.f. **apoptosis**).

negative-sense The nucleic acid strand with a base sequence complementary to the strand which contains the protein-coding sequence of nucleotide triplets *or* a virus whose genome consists of a negative-sense strand. (Also 'minus-sense' or '(−)sense.')

non-propagative transmission A term describing the transmission via secondary hosts (such as arthropods) of viruses which do not replicate in the vector organism, e.g. geminiviruses. Also known as 'non-circulative transmission', i.e. the virus does not circulate in the vector population.

nt A single nucleotide residue in a nucleic acid molecule.

nucleocapsid An ordered complex of proteins plus the nucleic acid genome of a virus.

oncogene A gene which encodes a protein capable of inducing cellular **transformation**.

ORF Open reading frame – a region of a gene or mRNA which encodes a polypeptide, bounded by an AUG translation start codon at the 5′ end and a termination codon at the 3′ end. Not to be confused with the poxvirus called orf.

packaging signal A region of a virus genome with a particular nucleotide sequence or structure which specifically interacts with a virus protein(s) resulting in the incorporation of the genome into a virus particle.

pandemic An **epidemic** which encompasses the entire world.

penetration The phase of virus replication at which the virus particle or genome enters the host cell (see Chapter 4).

phage See **bacteriophage**.

phenotypic mixing Individual progeny viruses from a mixed infection which contain structural proteins derived from both parental viruses.

plaque A localized region in a cell sheet or overlay in which cells have been destroyed or their growth retarded by virus infection.

plaque-forming units (p.f.u.) A measure of the amount of viable virus present in a virus preparation. Includes both free virus particles and infected cells containing infectious particles ('infectious centres').

plasmid An extrachromosomal genetic element capable of autonomous replication.

polycistronic A messenger RNA which encodes more than one polypeptide (c.f. **monocistronic**).

polyprotein A large protein which is post-transcriptionally cleaved by proteases to form a series of smaller proteins with differing functions.

positive-sense The nucleic acid strand with a base sequence which contains the protein-coding sequence of nucleotide triplets *or* a virus whose genome consists of a positive-sense strand. (Also 'plus-sense' or '(+)sense.')

primary cell A cultured cell explanted from an organism which is capable of only a limited number of divisions (c.f. **immortalized cell**).

prion A proteinaceous infectious particle, believed to be responsible for transmissible spongiform encephalopathies such as Creutzfeldt–Jakob disease (CJD) or bovine spongiform encephalopathy (BSE) (see Chapter 8).

productive infection A 'complete' virus infection in which further infectious particles are produced (c.f. **abortive infection**).

prokaryote An organism whose genetic material is not separated from the cytoplasm of the cell by a nuclear membrane.

promoter A ***cis*-acting** regulatory region upstream of the coding region of a gene which promotes transcription by facilitating the assembly of proteins in transcriptional complexes.

propagative transmission A term describing the transmission via secondary hosts (such as arthropods) of viruses which are able to replicate in both the primary host and the vector responsible for their transmission, e.g. plant reoviruses. Also known as 'circulative transmission', i.e. the virus circulates in the vector population.

prophage The lysogenic form of a temperate bacteriophage genome integrated into the genome of the host bacterium.

provirus The double-stranded DNA form of a retrovirus genome integrated into the **chromatin** of the host cell.

pseudoknot An RNA secondary structure which causes 'frame-shifting' during translation, produced a hybrid peptide containing information from an alternative reading frame.

pseudorevertant A virus with an apparently wild-type phenotype but which still contains a mutant genome – may be the result of genetic **suppression**.

pseudotyping Where the genome of one virus is completely enclosed within the capsid, or more usually the envelope, of another virus. An extreme form of **phenotypic mixing**.

quasi-equivalence A principle describing a means of forming a regular solid from irregularly shaped subunits in which subunits in *nearly* the same local environment form *nearly* equivalent bonds with their neighbours (see Chapter 2).

quasispecies A complex mixture of rapidly evolving and competing molecular variants of RNA virus genomes which occurs in most populations of RNA viruses.

receptor A specific molecule on the surface of a cell to which a virus attaches as a preliminary to entering the cell – may consist of proteins or the sugar residues present on glycoproteins or glycolipids in the cell membrane (see Chapter 4).

recombination The physical interaction of virus genomes in a **superinfected** cell resulting in progeny genomes which contain information in non-parental combinations.

release A late phase of virus infection during which newly formed virus particles leave the cell (see Chapter 4).

replicase An enzyme responsible for replication of RNA virus genomes (see **transcriptase**).

replicon A nucleic acid molecule containing the information necessary for its own replication; includes both **genomes** and other molecules such as **plasmids** and **satellites**.

retrotransposon A transposable genetic element closely resembling a retrovirus genome, bounded by long terminal repeats (see Chapter 3).

satellites Small RNA molecules (500–2000 nt) which are dependent on the presence of a helper virus for replication but unlike defective viruses, show no sequence homology to the helper virus genome. Larger satellite RNAs may encode a protein. (c.f. **viroids, virusoids**)

shutoff A sudden and dramatic cessation of most host cell macromolecular synthesis which occurs during some virus infections and results in cell damage and/or death (see Chapter 7).

splicing Post-transcriptional modification of primary RNA transcripts which occurs in the nucleus of **eukaryotic cells**

during which **introns** are removed and **exons** are joined together to produce cytoplasmic **mRNAs**.

superantigens Molecules which short-circuit the immune system, resulting in massive activation of T-cells by binding to both the variable region of the β-chain of the T-cell receptor (Vβ) and to MHC class II molecules, cross-linking them in a non-specific way (Figure 7.3).

superinfection Infection of a single cell by more than one virus particle, especially two viruses of distinct types, *or* deliberate infection of a cell designed to rescue a mutant virus.

suppression The inhibition of a mutant phenotype by a second suppressor mutation, which may be either in the virus genome or in that of the host cell (see Chapter 3).

syncytium A mass of cytoplasm containing several separate nuclei enclosed in a continuous membrane resulting from the fusion of individual cells.

systemic infection An infection involving multiple parts of a multicellular organism.

temperate bacteriophage A bacteriophage capable of establishing a **lysogenic** infection (c.f **virulent bacteriophage**).

terminal redundancy Repeated sequences present at the ends of a nucleic acid molecule.

titre A relative measure of the amount of virus present in any preparation.

***trans*-acting** A genetic element encoding a diffusible product which acts on regulatory sites whether or not these are contiguous with the site from which they are produced, e.g. proteins which bind to specific sequences present on any stretch of nucleic acid present in a cell, such as transcription factors (c.f. ***cis*-acting**).

transcriptase An enzyme, usually packaged into virus particles, responsible for the transcription of RNA virus genomes (see **replicase**).

transfection Infection of cells mediated by the introduction of nucleic acid rather than by virus particles.

transformation — Any change in the morphological, biochemical, or growth parameters of a cell.

transgenic — A genetically manipulated eukaryotic organism (animal or plant) which contains additional genetic information from another species. The additional genes may be carried and/or expressed only in the somatic cells of the transgenic organism, or in the cells of the germ line, in which case they may be inheritable by any offspring.

transposons — Specific DNA sequences which are able to move from one position in the genome of an organism to another (see Chapter 3).

triangulation number — A numerical factor which defines the symmetry of an icosahedral solid (see Chapter 2).

tropism — The types of tissues or host cells in which a virus is able to replicate.

uncoating — A general term for the events which occur after the penetration of a host cell by a virus particle, in which the virus capsid is completely or partially removed and the genome exposed, usually in the form of a nucleoprotein complex (see Chapter 4).

vaccine — A preparation containing an antigenic molecule or mixture of such molecules designed to elicit an immune response. Virus vaccines can be divided into three basic types: subunit, inactivated, and live vaccines (see Chapter 6).

vaccination — The administration of a **vaccine**.

variolation — The ancient practice of inoculating immunologically naive individuals with material obtained from smallpox patients – a primitive form of vaccination (see Chapter 1).

virino theory — An alternative to the **prion** theory which suggests that there is a small nucleic acid molecule associated with PrP (see Chapter 8).

virion — Morphologically complete (mature) infectious virus particle.

viroid — Autonomously replicating plant pathogens consisting solely of unencapsidated, single-stranded, circular (rod-like) RNAs of 200–400 nucleotides. Viroids do not encode any protein products. Some viroid RNAs have ribozyme activity (self-cleavage). (c.f. **satellites, virusoids**)

virulent bacteriophage A bacteriophage which is not capable of establishing a lysogenic infection and always kills the bacteria in which it replicates (c.f. **temperate bacteriophage**).

virusoids Small satellite RNAs with a circular, highly base-paired structures similar to viroids. Depend on a host virus for replication and encapsidation but do encode any proteins. All virusoid RNAs studied to date have ribozyme activity. (c.f. **satellites, viroids**)

virus-attachment protein A virus protein responsible for the interaction of a virus particle with a specific cellular receptor molecule.

zoonoses Infections transmitted from animals to humans.

Appendix 3

Answers to Self-Assessment Questions

Chapter 1

(1.1) Are the following statements true or false?

TRUE (a) Viruses are submicroscopic, obligate intracellular parasites. This simple definition excludes all other types of microorganism.

FALSE (b) Virus particles (virions) do NOT 'grow' or undergo division.

TRUE (c) Viruses lack the genetic information which encodes apparatus necessary for the generation of metabolic energy.

TRUE (d) Viruses lack the genetic information which encodes apparatus necessary for protein synthesis (ribosomes).

FALSE (e) Prions are infectious agents believed to consist of a single PROTEIN component.

(1.2) Are the following statements true or false?

FALSE (a) Virus particles cannot be seen using light microscopes.

TRUE (b) Edward Jenner first vaccinated a patient against smallpox on 14 May 1796.

TRUE (c) Louis Pasteur invented the term 'virus' in the 1890s.

TRUE (d) Dimitri Iwanowski showed that viruses could pass through filters fine enough to retain the smallest known bacteria.

FALSE (e) Viruses which infect bacteria are called BACTERIOPHAGES.

(1.3) These are Koch's Postulates:

TRUE (a) The agent must be present in every case of the disease.

TRUE (b) The agent must be isolated from the host and grown *in vitro*.

TRUE (c) The disease must be reproduced when a pure culture of the agent is inoculated into a healthy susceptible host.

TRUE (d) The same agent must be recovered once again from the experimentally infected host.

(1.4) The following are immunological methods of studying viruses (true or false?):

TRUE (a) Complement fixation.
FALSE (b) Northern blots use hybridization to analyse RNA.
FALSE (c) Southern blots use hybridization to analyse DNA.
TRUE (d) Western blot.
FALSE (e) PCR (polymerase chain reaction) is a method of amplifying small quantities of nucleic acid.

(1.5) Are the following statements true or false?

FALSE (a) Electron microscopes allow direct examination of viruses at magnifications of up to 100000 times.
TRUE (b) X-ray diffraction by crystalline forms of purified virus can be used to determine their atomic structure.
FALSE (c) The atomic structures of comparatively few viruses have been determined due to technical difficulties.
FALSE (d) Viruses DO NOT originate by spontaneous generation.
TRUE (e) More than 4000 different viruses have been identified.

Chapter 2

(2.1) Are the following statements true or false?

FALSE (a) Nucleic acids are SENSITIVE to physical damage such as shearing by mechanical forces and to chemical modification.
TRUE (b) The protein subunits in a virus capsid are multiply redundant, i.e. present in many copies per particle.
TRUE (c) The forces which drive the assembly of virus particles include hydrophobic and electrostatic interactions.
FALSE (d) The protein subunits in virus capsids are rarely held together by covalent bonds.
FALSE (e) In addition to protecting the genome from physical, chemical, or enzymatic damage, the outer shells of a virus particle initiate the process of infection by binding to receptors on host cells.

(2.2) Are the following statements true or false?

TRUE (a) A helix can be constructed by stacking repeated components with a constant relationship to one another.
TRUE (b) A helix can be defined by two parameters: amplitude and pitch.
FALSE (c) A helix can be defined mathematically by the equation: $P = \mu \times p$
FALSE (d) The 'amplitude' of a helix is its diameter.
FALSE (e) Bacteriophage M13 particles are about 900 nm long, but their precise length varies depending on their nucleic acid content.

(2.3) Are the following statements true or false?

FALSE (a) The particles of many plant viruses have helical symmetry, but many have icosahedral symmetry.
TRUE (b) There are no known non-enveloped helical viruses of animals.
TRUE (c) Rhabdovirus particles have a helical nucleocapsid.
TRUE (d) Bacteriophage M13 is a member of the 'Ff' coliphage group.
FALSE (e) Bacteriophage M13 requires the F pilus on the surface of *E. coli* for infection.

(2.4) Are the following statements true or false?
FALSE (a) An icosahedron is a solid shape consisting of 20 triangular faces arranged around the surface of a sphere.
FALSE (b) An icosahedron has an axis of TWOFOLD rotational symmetry through the centre of each edge.
FALSE (c) An icosahedron has an axis of THREEFOLD rotational symmetry through the centre of each face.
FALSE (d) An icosahedron has an axis of FIVEFOLD rotational symmetry through the centre of each corner.
FALSE (e) Some simple icosahedral virus capsids contain only 60 subunits (e.g. ϕX174), most contain more than this (e.g. $T = 3$ viruses have 180 subunits).

(2.5) All the following virus groups have icosahedral symmetry (true or false?):
FALSE (a) Orthomyxoviruses are pleiomorphic, enveloped viruses.
FALSE (b) Paramyxoviruses are pleiomorphic, enveloped viruses.
TRUE (c) Picornaviruses have icosahedral symmetry.
FALSE (d) Rhabdoviruses are helical, enveloped viruses.
FALSE (e) Tobamoviruses are helical, non-enveloped viruses.

(2.6) Are the following statements true or false?
FALSE (a) The lipid composition of a virus envelope is SIMILAR to that of the host cell.
TRUE (b) Virus envelopes are acquired during extrusion (budding) of the particle through the host cell membrane.
FALSE (c) Budding can occur from the host cell surface, cytoplasmic or even nuclear membranes.
FALSE (d) Some enveloped viruses have underlying icosahedral symmetry.
FALSE (e) Most envelope proteins are encoded by the virus, although some virus envelopes contain a proportion of host cell proteins.

(2.7) All the following virus groups possess lipid envelopes (true or false?):
TRUE (a) Orthomyxoviruses
TRUE (b) Paramyxoviruses
FALSE (c) Picornaviruses
TRUE (d) Rhabdoviruses
FALSE (e) Tobamoviruses

(2.8) Are the following statements true or false?
TRUE (a) Matrix proteins are internal virion proteins that link the internal nucleocapsid assembly to the envelope.

TRUE (b) Matrix proteins are often very abundant in the virus particle.
TRUE (c) External glycoproteins are usually the major antigens of enveloped viruses.
TRUE (d) External glycoproteins are usually the virus attachment proteins of enveloped viruses.
TRUE (e) The influenza virus M2 peptide is a transmembrane protein.

(2.9) Are the following statements true or false?
FALSE (a) Poxvirus particles have elements of helical symmetry, but overall are too complex to define in simple terms.
TRUE (b) Poxvirus particles contain more than 100 different proteins.
FALSE (c) The tailed phages of *E. coli* are released from the host cell by LYSIS.
TRUE (d) The tailed phages of *E. coli* have separate assembly pathways for the head and tail sections of the particle.
TRUE (e) Reovirus capsids are composed of a double protein shell.

(2.10) Are the following statements about genome packaging true or false?
FALSE (a) It is specific for virus nucleic acids, but viruses make mistakes!
TRUE (b) It requires interaction with a specific virus-encoded capsid or nucleocapsid protein.
TRUE (c) It requires specific nucleotide sequences in the virus genome, although the secondary structure of the genome is also important in some cases.
TRUE (d) It requires positively charged groups to overcome negative electrostatic charges on the virus genome.
TRUE (e) In TMV it is initiated by the OAS region of the genome.

Chapter 3

(3.1) Are the following statements true or false?
TRUE (a) Virus genomes vary in size from 3500 nucleotides to approximately 235 kilobase pairs.
FALSE (b) The genetic code of a virus must be SIMILAR to that of the host organism so that it can be decoded by the host cell biochemical machinery.
TRUE (c) Virus genomes consisting of (+)sense RNA are infectious when the purified vRNA is applied to cells in the absence of any virus proteins (except retroviruses).
TRUE (d) 'Reverse genetics' makes possible the manipulation of (−)sense RNA virus genomes.
TRUE (e) Infection of cells caused by nucleic acid alone is referred to as transfection.

(3.2) Are the following statements true or false?

FALSE (a) 'Superinfection' is infection of a single cell by more than one virus particle.
FALSE (b) The probability of recombination between two genetic markers is PROPORTIONAL to the distance between them.
TRUE (c) In viruses with segmented genomes, 'reassortment groups' are usually equivalent to the individual genome segments.
TRUE (d) Populations of virus genomes consisting of mixtures of molecular variants are known as quasispecies.
TRUE (e) Site-specific molecular biological methods are commonly used to mutagenize virus genomes.

(3.3) Are the following statements true or false?
FALSE (a) Mutation rates in herpesvirus genomes may be as LOW as 10^{-8} to 10^{-11} per nucleotide incorporated, similar to those seen in cellular genomes.
FALSE (b) Mutation rates in retrovirus genomes may be as HIGH as 10^{-3} to 10^{-4} per nucleotide incorporated, due to the low fidelity of reverse transcriptase.
TRUE (c) Every new HIV provirus formed contains, on average, at least one new mutation.
FALSE (d) Temperature-sensitive (t.s.) mutants tend to be unstable and frequently revert to the original phenotype since they are usually single base substitutions and frequently disadvantage viruses compared with the wild-type sequence.
TRUE (e) T.s. mutations usually result from mis-sense mutations in proteins.

(3.4) Are the following statements true or false?
TRUE (a) Deletion mutants are very useful for assigning structure–function relationships to virus genomes, since they are easily mapped by physical analysis.
TRUE (b) Deletion mutants can only revert to wild type by recombination, which only occurs at low frequencies.
TRUE (c) Complementation may be asymmetric or symmetrical.
FALSE (d) Phenotypic mixing is where individual progeny viruses from a mixed infection contain PROTEINS derived from both parental viruses but are genetically unaltered.
FALSE (e) Only retrovirus have completely diploid genomes, i.e. two alleles of each gene.

(3.5) Herpesvirus genomes (true or false?):
FALSE (a) Vary in size from 105 to 235 kbp.
TRUE (b) Consist of linear, double-stranded DNA.
TRUE (c) In herpes simplex virus (HSV), each gene is expressed from its own promoter.
TRUE (d) The HSV genome is approximately 152 kbp.
TRUE (e) The HSV genome contains about 80 genes, densely packed and with overlapping reading frames.

(3.6) Are the following statements true or false?

TRUE (a) Parvovirus genomes consist of linear, non-segmented, single-stranded DNA.

TRUE (b) The ends of parvovirus genomes are palindromic sequences of about 115 nt required for replication.

TRUE (c) Polyomaviruses are about 5 kbp in size and contain six genes.

TRUE (d) In the virus particle, polyomavirus DNA assumes a supercoiled form and is associated with cellular histones.

TRUE (e) The polyomavirus genome contains overlapping reading frames on both strands of the DNA.

(3.7) (+)sense RNA virus genomes (true or false?):

FALSE (a) . . . can be up to 30 kbp in length (Coronaviruses).

TRUE (b) . . . are usually modified at the 5′ end, by a methylated cap or a covalently-attached protein.

FALSE (c) . . . are sometimes polyadenylated at the 3′ end, not always.

FALSE (d) . . . are sometimes expressed as a polyprotein, not always.

FALSE (e) . . . do NOT require splicing for expression, since they are replicated in the cytoplasm.

(3.8) (−)sense RNA virus genomes (true or false?):

TRUE (a) . . . are sometimes segmented, but not in all cases.

FALSE (b) . . . may have an ambisense coding strategy, e.g. Bunyaviruses, Arenaviruses.

TRUE (c) . . . are never capped at the 5′ end.

TRUE (d) . . . are never polyadenylated at the 3′ end.

FALSE (e) . . . are NEVER circular.

(3.9) Retroviruses (true or false?):

FALSE (a) Reverse transcriptase is also encoded by Hepadnaviruses and Caulimoviruses.

TRUE (b) . . . are the only RNA viruses whose genome is produced by cellular transcriptional machinery.

TRUE (c) . . . are the only (+)sense RNA viruses whose genome does not serve directly as mRNA immediately after infection.

TRUE (d) . . . have high mutation rates, due to the low fidelity of reverse transcriptase.

TRUE (e) . . . have a high rate of recombination, due to reverse transcription.

(3.10) Are the following statements true or false?

TRUE (a) Segmentation of virus genomes reduces the probability of strand breakage due to shearing.

FALSE (b) Hepadnaviruses have SMALL genomes of about 3.5 kbp.

TRUE (c) The cauliflower mosaic virus (CaMV) genome consists of a gapped, circular double-stranded DNA molecule.

TRUE (d) The cauliflower mosaic virus (CaMV) genome is generated by reverse transcription.

FALSE (e) Viruses MAY have had a common ancestor, but the existence of virus superfamilies does not prove this – alternative explanations such as convergent evolution are possible.

Chapter 4

(4.1) Virus replication involves the following distinct phases (true or false?):

TRUE (a) Initiation of infection.
TRUE (b) Replication of the virus genome.
TRUE (c) Expression of the virus genome.
FALSE (d) Cleavage of the virus genome.
TRUE (e) Release of mature virions from the infected cell.

(4.2) Are the following statements true or false?

FALSE (a) The 'single burst' experiment was performed by Ellis and Delbruck in 1939, not 1989!
TRUE (b) The 'single burst' experiment demonstrates that new virus particles from the intracellular assembly of preformed components are released in a batch-like process.
TRUE (c) The 'single burst' experiment demonstrates the essential phases of virus replication.
FALSE (d) The 'single burst' experiment demonstrates the essential phases of virus replication.
TRUE (e) The 'Hershey–Chase' experiment demonstrates that nucleic acid is the genetic material of viruses, but this conclusion can be extended to all other organisms.

(4.3) Are the following statements true or false?

FALSE (a) Virus receptor molecules may be proteins or the sugar residues present on glycoproteins or glycolipids.
FALSE (b) Many viruses have more than one receptor molecule, e.g. different receptors on different cell types.
FALSE (c) No known plant virus uses a cellular receptor due to the surface structure of plant cells.
FALSE (d) Sialic acid is the influenza virus receptor. The human rhinovirus receptor is either ICAM-1 or the LDL receptor.
TRUE (e) The expression of receptors on the surface of cells is the major factor which determines the tropism of viruses.

(4.4) The following are mechanisms by which virus particles penetrate cells (true or false?):

TRUE (a) Translocation.
FALSE (b) Transfection is the infection of cells mediated by the introduction of nucleic acid rather than by virus particles.
FALSE (c) Exocytosis is one mechanism by which viruses are released from cells.

TRUE (d) Endocytosis.
TRUE (e) Fusion.

(4.5) Are the following statements true or false?
TRUE (a) Viral penetration of cells is an energy-dependent process.
TRUE (b) Influenza haemagglutinin is a virus fusion protein.
FALSE (c) Uncoating only occurs after penetration of the host cell.
FALSE (d) The product of uncoating is frequently (not always) a complex of nucleic acids and proteins.
FALSE (e) Nucleoproteins closely associated with virus genomes in the core of the particle are responsible for nuclear localization of virus genomes.

(4.6) Are the following statements true or false?
FALSE (a) There are SEVEN classes of virus genome structure.
FALSE (b) Class I viruses have double-stranded DNA genomes which may be linear or circular.
TRUE (c) Class II viruses have (−)sense single-stranded DNA genomes.
FALSE (d) Class VI viruses (retroviruses) have (+)sense single-stranded RNA genomes (with a DNA intermediate in the replication cycle).
FALSE (e) Class VII viruses have (+)sense single-stranded DNA genomes (with an RNA intermediate in the replication cycle).

(4.7) Are the following statements true or false?
TRUE (a) Adenoviruses encode their own DNA polymerase but are dependent on cellular factors for genome replication.
FALSE (b) Poxviruses encode their own DNA polymerase and are INDEPENDENT of cellular factors for genome replication (although some cellular factors are required for unknown steps in the maturation process).
FALSE (c) Reovirus genomes are transcribed by a viral enzyme present in the particle.
FALSE (d) Some RNA virus genomes replicate in the nucleus of the host cell, e.g. influenza virus.
TRUE (e) Retrovirus genomes in the form of a DNA provirus are transcribed by cellular RNA polymerase.

(4.8) The following viruses replicate their genomes in the cytoplasm of infected cells (true or false?):
TRUE (a) Picornaviruses, e.g. poliovirus.
TRUE (b) Poxviruses, e.g. vaccinia virus.
TRUE (c) Rhabdoviruses, e.g. rabies virus.
TRUE (d) Paramyxoviruses, e.g. respiratory syncytial virus.
FALSE (e) Hepadnaviruses, e.g. hepatitis B virus.

(4.9) The following viruses replicate their genomes in the nucleus of infected cells (true or false?):
TRUE (a) Herpesviruses, e.g. Epstein–Barr virus.

TRUE (b) Parvoviruses, e.g. B19.
TRUE (c) Orthomyxoviruses, e.g. influenza virus.
FALSE (d) Rotaviruses, e.g. group A rotaviruses.
FALSE (e) Coronaviruses, e.g. OC43.

(4.10) Are the following statements true or false?
TRUE (a) Maturation usually involves structural changes in the virus particle.
TRUE (b) Maturation frequently involves protease cleavage of virus proteins.
TRUE (c) Enveloped viruses acquire their lipid membranes by budding through cellular membranes.
TRUE (d) Virus envelope proteins are acquired during the process of budding.
TRUE (e) Virus envelope proteins are involved in release from as well as attachment to host cells.

Chapter 5

(5.1) Are the following statements true or false?
FALSE (a) The course of virus replication is determined by tight control of gene expression.
TRUE (b) Transcription promoters are *cis*-acting sequences located adjacent to the genes whose transcription they control.
TRUE (c) Transcription factors are *trans*-acting proteins which bind to specific sequences present anywhere in the cell.
TRUE (d) Transcription of bacterial operons typically produces polycistronic mRNAs.
FALSE (e) Gene expression in bacteria is regulated at the level of transcription and at subsequent (post-transcriptional) stages of gene expression.

(5.2) Viruses have counteracted their genetic limitations by evolving of a range of solutions, including (true or false?):
TRUE (a) Powerful positive *cis*- and *trans*-acting signals which promote gene expression.
TRUE (b) Powerful *cis*-acting negative signals which repress gene expression.
TRUE (c) Highly compressed genomes with overlapping reading frames common.
TRUE (d) Control signals which are frequently nested within other genes.
TRUE (e) Several strategies designed to create multiple polypeptides from a single mRNA.

(5.3) Bacteriophage λ (true or false?):

TRUE (a) Can alternate between the lysogenic and lytic growth cycles.
FALSE (b) The λ N protein acts as a POSITIVE transcription regulator (anti-terminator).
TRUE (c) In the absence of the N protein, RNA polymerase stops at the nut site.
FALSE (d) The cI repressor protein binds to BOTH O_R AND O_L.
TRUE (e) Lysogeny is maintained by autoregulation of *cI* transcription.

(5.4) Are the following statements true or false?
TRUE (a) Transcription of eukaryotic cellular genes results in monocistronic mRNAs, each transcribed from its own promoter.
TRUE (b) Post-transcriptional control of gene expression in eukaryotes is achieved by splicing of hnRNA and differential export of mRNA from the nucleus to the cytoplasm.
FALSE (c) Eukaryotic mRNAs are translated with variable efficiency mainly due to the efficiency with which sequences surrounding the AUG initiation codon are recognized by ribosomes.
TRUE (d) The stability of eukaryotic mRNAs depends on the speed at which they are degraded.
TRUE (e) Translation enhancers perform a function analogous to that of transcription enhancers.

(5.5) Are the following statements true or false?
TRUE (a) Papovaviruses are heavily dependent on cellular machinery for gene expression.
FALSE (b) Herpesviruses make three classes of mRNA: Immediate-Early (IE) mRNAs (*trans*-acting regulators of virus transcription); Early mRNAs (further non-structural regulatory proteins and some minor structural proteins); Late mRNAs (major structural proteins).
TRUE (c) Geminivirus genomes are replicated and expressed in the nucleus of infected cells because they are heavily reliant on host cell functions.
TRUE (d) Reovirus transcription occurs inside virus core particles.
TRUE (e) Segmented virus genomes are usually transcribed to produce monocistronic mRNAs.

(5.6) Viruses with (+)sense RNA genomes (true or false?):
FALSE (a) Are largely INDEPENDENT of cellular mechanisms for control of gene expression.
FALSE (b) SOME are translated to produce a long polyprotein, OTHERS are transcribed to produce subgenomic RNAs.
FALSE (c) SOME are translated to produce a long polyprotein, OTHERS are transcribed to produce subgenomic RNAs.
FALSE (d) SOME are non-segmented, others segmented.
FALSE (e) Can produce varying ratios of the polypeptides they encode by mechanisms such as alternative cleavages of polyproteins and variable translation efficiency of subgenomic RNAs.

(5.7) Are the following statements true or false?

TRUE (a) Viruses with non-segmented (−)sense genomes produce monocistronic mRNAs.

TRUE (b) Viruses with (−)sense genomes all contain a virus-specific transcriptase/replicase within the virus nucleocapsid.

FALSE (c) Viruses with non-segmented (−)sense genomes can produce variable ratios of the polypeptides they encode, e.g. paramyxoviruses (Figure 5.10).

TRUE (d) Retroviruses are the only RNA viruses whose genomes (in the form of the DNA provirus) are transcribed by cellular RNA polymerase.

TRUE (e) Hepadnavirus and caulimovirus genomes contain a number of overlapping reading frames.

(5.8) Are the following statements true or false?

TRUE (a) SV40 large T-antigen binds to the origin of replication in the virus genome.

TRUE (b) SV40 large T-antigen represses transcription of the virus early genes.

FALSE (c) SV40 large T-antigen does not promote export of virus mRNAs from the nucleus to the cytoplasm.

FALSE (d) The HTLV tax protein binds to cellular transcription factors, not virus mRNAs.

FALSE (e) The HIV tat protein binds to the TAR sequence in the provirus genome.

(5.9) Are the following statements true or false?

TRUE (a) With the exception of retroviruses, no RNA virus genome undergoes splicing.

TRUE (b) The HTLV rex protein promotes the export of virus structural protein mRNAs from the nucleus.

FALSE (c) The abbreviation 'IRES' stands for 'Internal Ribosomal Entry Site'.

FALSE (d) SOME (+)sense RNA viruses contain an IRES element, others do not.

FALSE (e) Very few eukaryotic cellular mRNAs contain an IRES element.

(5.10) Are the following statements true or false?

TRUE (a) Ribosomal frameshifting allows viruses to produce several different proteins from a polycistronic mRNA.

TRUE (b) Ribosomal frameshifting allow viruses to produce varying amounts of different proteins from a polycistronic mRNA.

TRUE (c) RNA pseudoknots are produced by *trans*-acting proteins such as the HIV tat protein.

TRUE (d) RNA pseudoknots increase the frequency of ribosomal frame shifting.

TRUE (e) Suppression of termination codons allows some viruses to produce multiple proteins from a polycistronic mRNA.

Chapter 6

(6.1) Plant viruses (true or false?):

FALSE (a) Are responsible for an estimated US$60 000 000 000 (!) worth of damage each year.

TRUE (b) Can be transmitted by infected seeds.

TRUE (c) May be transmitted to new hosts as wind-blown dust or as rain-splashed mud.

FALSE (d) Sometimes replicate in insect vectors (propagative transmission), e.g. plant rhabdoviruses.

FALSE (e) May or may not cause systemic infections involving the whole plant.

(6.2) Are the following statements true or false?

FALSE (a) Movement proteins allow plant viruses to be transmitted from infected cells to neighbouring cells.

TRUE (b) Damage to virus-infected plants occurs due to necrosis of cells, hypoplasia and hyperplasia.

TRUE (c) The hypersensitive response in plants is functionally analogous to the production of interferons in animals.

TRUE (d) The hypersensitive response in plants tends to limit the occurrence of systemic infection.

TRUE (e) Virus-resistant plants have been created by genetic manipulation resulting in the endogenous expression of virus genes.

(6.3) Are the following statements true or false?

TRUE (a) The major impact of the humoral immune response on virus infections is the eventual clearance of virus from the body.

FALSE (b) Virus particles ARE directly neutralized by antibodies and phagocytosed by white blood cells.

TRUE (c) Overall, cell-mediated immunity is more important than humoral immunity in controlling virus infections.

TRUE (d) Natural killer (NK) cells are the first cell-mediated defence mechanism active against virus infection.

FALSE (e) Cytotoxic T-lymphocytes (CTL) ARE MHC-restricted, i.e. recognize a specific antigen only in the context of MHC class II antigen plus the T-cell receptor/CD3 complex on the surface of the CTL.

(6.4) Are the following statements true or false?

FALSE (a) Interferon production was first observed from chick chorioallantoic membranes in 1957.

FALSE (b) Biochemical purification of interferons from natural sources was difficult because interferons are highly potent and therefore non-abundant molecules.

TRUE (c) There are three distinct types of interferon – α, β, and γ.

FALSE (d) Only IFN-β is synthesized predominantly by fibroblasts; IFN-α and IFN-γ are synthesized predominantly by lymphocytes.
FALSE (e) Double-stranded RNA is a potent inducer of interferon synthesis; all other nucleic acids do not induce IFN synthesis.

(6.5) Are the following statements true or false?
TRUE (a) All interferons have similar antiviral capacities, but act differently as cellular regulators.
TRUE (b) 2′,5′-oligo A activates RNAse L, which digests all RNAs.
TRUE (c) PKR phosphorylates eIF2α, which is required for the initiation of translation.
TRUE (d) Interferons are used to treat chronic viral hepatitis.
TRUE (e) Several viruses have evolved mechanisms which enable them to resist the effects of interferons.

(6.6) Are the following statements true or false?
TRUE (a) Mucosal membranes are favourable routes of access for viruses to the tissues of the body.
TRUE (b) 'Coughs and sneezes spread diseases.'
TRUE (c) The natural environment is a considerable barrier to virus infections.
FALSE (d) Primary LOCALIZED infection by viruses is frequently followed by SYSTEMIC secondary infection.
FALSE (e) Spread of virus to the nervous system usually PRECEDED by primary viraemia.

(6.7) Viruses have evolved many tricks to fool the immune system (true or false?):
TRUE (a) Antigenic variation.
FALSE (b) DOWNREGULATION of MHC class I gene expression.
FALSE (c) DOWNREGULATION of accessory molecules involved in immune recognition.
TRUE (d) Infection of immunocompromised sites within the body.
FALSE (e) Few viruses infect the cells of the skin and this does not enable them to escape the immune system.

(6.8) Are the following statements true or false?
TRUE (a) Abortive infection of cells can have significant pathological consequences e.g. transformation.
TRUE (b) In acute infections, much viral replication occurs before the onset of any symptoms.
TRUE (c) In persistent infections, the virus adjusts its replication and pathogenicity to avoid killing the host.
TRUE (d) Persistent infections may result from the production of defective-interfering (D.I.) particles.
TRUE (e) Latent infections are typified by strictly limited gene expression without ongoing virus replication.

(6.9) Are the following statements true or false?

FALSE (a) Subunit vaccines are the LEAST effective and MOST expensive virus vaccines.
FALSE (b) The hepatitis B virus vaccine currently in use is produced by genetic engineering.
TRUE (c) Inactivated virus vaccines are sometimes not as effective as 'live' virus vaccines because they fail to stimulate mucosal immunity.
TRUE (d) The majority of successful virus vaccines are based on attenuated viruses.
TRUE (e) Retroviruses have been successfully used as vectors for gene therapy.

(6.10) Are the following statements true or false?
FALSE (a) Prevention of virus infection by drug treatment is rarely effective compared with vaccination.
TRUE (b) Any of the stages of viral replication can be a potential target for antiviral chemotherapy.
TRUE (c) The majority of antiviral drugs in use are nucleoside/nucleotide analogues.
FALSE (d) Ribavirin is active against a wide spectrum of viruses.
TRUE (e) The frequency with which viruses become resistant to antiviral drugs depends on the drug and the virus concerned and varies considerably.

Chapter 7

(7.1) Are the following statements true or false?
TRUE (a) Most disease states are multi-factorial.
TRUE (b) Inflammation, fever, headaches and skin rashes are frequently not caused by virus replication, but the immune response to virus infection.
TRUE (c) The vast majority of virus infections do not result in disease.
TRUE (d) Eukaryotic cells must constantly synthesize a variety of macromolecules which require continual replacement.
FALSE (e) Virus nucleic acids do not act as toxins.

(7.2) The cytopathic effects of virus infection include (true or false?):
TRUE (a) Altered shape.
TRUE (b) Detachment from the substrate.
TRUE (c) Lysis.
TRUE (d) Membrane fusion and altered permeability.
TRUE (e) Apoptosis.

(7.3) Are the following statements true or false?
FALSE (a) Shutoff is the sudden and dramatic cessation of HOST CELL macromolecular synthesis.

FALSE (b) Shutoff in poliovirus-infected cells occurs owing to protease degradation of the p220 component of eIF–4F, a complex of proteins required for cap-dependent translation of messenger RNAs by ribosomes.
TRUE (c) Lysis of adenovirus-infected cells is caused by virus-encoded molecules with a toxin-like action.
FALSE (d) Virus-induced membrane fusion may cause cell death but not transformation.
FALSE (e) The HIV env protein is responsible for syncytium formation.

(7.4) Are the following statements true or false?
FALSE (a) Some human herpesviruses are highly cytopathic, e.g. HSV, but others are not, e.g. EBV.
TRUE (b) Retroviruses cause a variety pathogenic conditions including paralysis, arthritis, anaemia in addition to cell transformation.
FALSE (c) Lentiviruses are NOT associated with cell transformation.
TRUE (d) HIV and other lentiviruses escape immune recognition by infecting monocytes.
FALSE (e) Some laboratory-adapted isolates of HIV are highly cytopathic in $CD4^+$ cells but may primary isolates of the virus are not.

(7.5) The following mechanisms may be involved in the pathogenesis of AIDS (true or false?):
TRUE (a) New antigenic variants of HIV constantly arise during the course of infection.
TRUE (b) T-cell anergy.
TRUE (c) Apoptosis.
TRUE (d) Superantigens.
FALSE (e) Insertional activation of oncogenes causes cell transformation, not AIDS.

(7.6) Are the following statements true or false?
FALSE (a) Subacute sclerosing panencephalitis (SSPE) occurs in about 1 in 300000 cases of measles.
TRUE (b) Dengue haemorrhagic fever is largely due to immune-mediated damage of virus-infected cells.
FALSE (c) Reye's syndrome MAY be worsened by administration of aspirin during the initial illness.
FALSE (d) Reye's syndrome is a rare post-infection complication of a number of different viruses, but most commonly influenza virus and VZV (chicken pox).
TRUE (e) Guillain–Barré syndrome is a rare demyelinating disease. Kawasaki disease is of unknown origin.

(7.7) Transformed cells may display (true or false?):
TRUE (a) Loss of anchorage dependence.
TRUE (b) Loss of contact inhibition.
TRUE (c) Colony formation in semi-solid media.
TRUE (d) Decreased requirements for growth factors.

FALSE (e) Transformed cells display normal or decreased expression of p53, or express altered p53 proteins.

(7.8) Are the following statements true or false?

FALSE (a) Transformation is a change in the morphological, biochemical, or growth parameters of a cell. Carcinogenesis or oncogenesis multi-step process in which cellular transformation may be only the first, although essential, step along the way.

FALSE (b) Most acutely transforming retroviruses are replication defective, Rous sarcoma virus, the first to be discovered (1911), is replication competent.

TRUE (c) p53 causes the cell cycle to arrest at the G_1 phase.

FALSE (d) Rb causes the cell cycle to arrest at the G_1 phase.

FALSE (e) SV40 T antigen binds to p53 AND Rb. Adenovirus E1A and the HPV E7 protein bind Rb.

(7.9) The normal functions of oncogenes may be (true or false?):

TRUE (a) Cytoplasmic protein kinases.

TRUE (b) Extracellular growth factors.

TRUE (c) G proteins involved in signal transduction.

TRUE (d) Transcription factors.

TRUE (e) Tyrosine kinases.

(7.10) Are the following statements true or false?

TRUE (a) At least six viruses which are associated with the formation of human tumours (EBV, HBV, HCV, HHV-8, HPVs, HTLV).

FALSE (b) EBV has a dual cell tropism for human epithelial cells (PRODUCTIVE infection) and B-lymphocytes (NON-PRODUCTIVE infection).

FALSE (c) EBV infection is associated with at least four human tumours: Burkitt's lymphoma, nasopharyngeal carcinoma, B-cell lymphomas and some clonal forms of Hodgkin's disease, but not HCG.

TRUE (d) Chronic HBV carriers are at 100–200 times the risk of non-carriers of developing hepatocellular carcinoma.

TRUE (e) Chronic active hepatitis caused by hepatitis C virus (HCV) infection is associated with hepatocellular carcinoma.

Chapter 8

(8.1) Are the following statements true or false?

TRUE (a) Satellites encode their own coat proteins.

TRUE (b) Viroids use the coat protein of their helper virus.

FALSE (c) There is little or no nucleotide sequence similarity between satellites their helper virus genomes.

TRUE (d) Viroids are 200–400 nucleotides long.

TRUE (e) All viroids share a conserved region involved in their replication.

(8.2) Are the following statements true or false?

FALSE (a) The mechanisms by which viroids cause disease is not known, but it is not by interfering with cellular DNA replication.

FALSE (b) Hepatitis delta virus (HDV) δAg-S is necessary for HDV replication.

FALSE (c) HDV δAg-L is necessary for the assembly and release of HDV-containing particles.

TRUE (d) HDV has some of the properties of a satellite virus and some of a viroid.

FALSE (e) HDV infection cannot occur without HBV co-infection.

(8.3) The following diseases are caused by prions (true or false?):

FALSE (a) Alzheimers's syndrome is a NON-TRANSMISSIBLE AMYLOID DISEASE.

FALSE (b) Coconut cadang-cadang disease is cause by a VIROID.

TRUE (c) Chronic wasting disease (CWD).

TRUE (d) Creutzfeldt–Jakob disease (CJD).

TRUE (e) Fatal familial insomnia (FFI).

(8.4) Are the following statements true or false?

FALSE (a) Scrapie was first described more than two hundred years ago.

TRUE (b) Scrapie is endemic in the UK and USA.

FALSE (c) Transmissible mink encephalopathy (TME) has been reported in the USA, Canada, Finland, Germany and Russia.

TRUE (d) More than 161 000 cases of BSE have been reported in the UK.

FALSE (e) BSE is NOT caused by scrapie-infected feed since the BSE and scrapie agents are distinct

(8.5) Are the following statements true or false?

TRUE (a) There are three sources of human TSEs: sporadic, iatrogenic/acquired and familial/inherited.

TRUE (b) Sporadic Creutzfeldt–Jakob disease (CJD) accounts for about 90% of human TSE.

FALSE (c) Kuru is an acquired form of TSE, resulting from ritual cannibalism.

TRUE (d) The incidence of CJD is similar worldwide.

TRUE (e) Only about 1% of sporadic CJD cases are transmissible to mice.

(8.6) Prion infectivity is inactivated by the following treatments (true or false?):

FALSE (a) Heating to 90°C for 30 min.

FALSE (b) Heating to 135°C for 18 min.

FALSE (c) DNAse treatment.

FALSE (d) RNAse treatment.

TRUE (e) Phenol, which denatures proteins.

(8.7) PrP^C (true or false?):

TRUE (a) is a 254 amino acid protein.

TRUE (b) is encoded by a cellular gene.
TRUE (c) is converted by PrP^{Sc} to a protease-resistant form.
FALSE (d) PrP^{C} consists of 43% α-helix and no β-sheet. PrP^{Sc} consists of 30% α-helix and 43% β-sheet.
TRUE (e) is resistant to short-wave ultraviolet radiation and to ionizing radiation.

(8.8) Are the following statements true or false?:
TRUE (a) Transgenic animals which do not possess an endogenous prion gene are immune to the effects of PrP^{Sc}.
FALSE (b) Transgenic animals which do not possess an endogenous prion gene do not develop any obvious symptoms.
TRUE (c) Human PrP differs from mouse PrP at 28 of 254 amino acids.
FALSE (d) The BSE agent is distinct from that of scrapie and has a unique 'fingerprint' in inbred strains of mice.
TRUE (e) The 'species barrier' prevents the efficient transfer of infectious prions from one species to another.

(8.9) Are the following statements true or false?
TRUE (a) The URE3 protein of yeast has the properties of a prion.
TRUE (b) The URE3 phenotype of yeast can be 'cured' by treatment with guanidium.
TRUE (c) PrP is believed to interact with at least two cellular proteins, X and Y.
FALSE (d) There is no evidence that prions cause disease in plants.
FALSE (e) The prion hypothesis is the most likely interpretation for TSEs, but other interpretations are possible.

Index